ADVANCES IN COTTON SCIENCE

Botany, Production, and Crop Improvement

ADVANCES IN COTTON SCIENCE

Botany, Production, and Crop Improvement

Ratikanta Maiti
Ch. Aruna Kumari
Abul Kalam Samsul Huda
Debashis Mandal
Sameena Begum

Apple Academic Press Inc.
4164 Lakeshore Road
Burlington ON L7L 1A4
Canada

Apple Academic Press Inc.
1265 Goldenrod Circle NE
Palm Bay, Florida 32905
USA

First issued in paperback 2021

Exclusive worldwide distribution by CRC Press, a member of Taylor & Francis Group
No claim to original U.S. Government works

ISBN 13: 978-1-77463-507-0 (pbk)
ISBN 13: 978-1-77188-819-6 (hbk)

Library and Archives Canada Cataloguing in Publication

Title: Advances in cotton science : botany, production, and crop improvement / Ratikanta Maiti, Ch. Aruna Kumari, Abul Kalam Samsul Huda, Debashis Mandal, Sameena Begum.

Names: Maiti, R. K., 1938- author. | Aruna Kumari, C. H., 1972- author. | Huda, Abul Kalam Samsul, author. | Mandal, Debashis, author. | Begum, Sameena, author.

Description: Includes bibliographical references and index.

Identifiers: Canadiana (print) 20190222719 | Canadiana (ebook) 20190222751 | ISBN 9781771888196 (hardcover) | ISBN 9780429283987 (ebook)

Subjects: LCSH: Cotton. | LCSH: Cotton growing.

Classification: LCC SB249 .M35 2020 | DDC 633.5/1—dc23

Library of Congress Cataloging-in-Publication Data

Names: Maiti, R. K., 1938- author. | Aruna Kumari, C. H., 1972- author. | Huda, Abul Kalam Samsul, author. | Mandal, Debashis, author. | Begum, Sameena, author.

Title: Advances in cotton science : botany, production, and crop improvement / Ratikanta Maiti, Ch. Aruna Kumari, Abul Kalam Samsul Huda, Debashis Mandal, Sameena Begum.

Description: Palm Bay, Florida, USA : Apple Academic Press, [2020] | Includes bibliographical references and index. | Summary: "Cotton is one of the most important fiber and cash crops throughout the world, and it plays a dominant role in the industrial and agricultural economies of many countries. This volume, Advances in Cotton Science: Botany, Production, and Crop Improvement, is a rich resource of information on the cultivation and production of cotton. It provides an overview of its origin and evolution, and its physiological basis and characterization, and goes on to discuss methods of cultivation, biotic stresses, harvesting and postharvest technology, and new research on breeding and biotechnology. The authors take an interdisciplinary approach, providing a multi-pronged approach to information necessary to increase cotton productivity to meet the world's growing demands. The volume answers the need to understand the roles of cotton, the nature of the crop, and advancements in research for best cultivation methods, effective utilization of resources, and operations for achieving higher yields, thus achieving higher productivity. The volume will be immensely helpful for growers, students, academicians, teaching faculty, and other professionals in the field to gain knowledge and understanding of the crop"-- Provided by publisher.

Identifiers: LCCN 2019054172 (print) | LCCN 2019054173 (ebook) | ISBN 9781771888196 (hardcover) | ISBN 9780429283987 (ebook)

Subjects: LCSH: Cotton. | Cotton growing.

Classification: LCC SB249 .M24 2020 (print) | LCC SB249 (ebook) | DDC 633.5/1--dc23

LC record available at https://lccn.loc.gov/2019054172
LC ebook record available at https://lccn.loc.gov/2019054173

Apple Academic Press also publishes its books in a variety of electronic formats. Some content that appears in print may not be available in electronic format. For information about Apple Academic Press products, visit our website at **www.appleacademicpress.com** and the CRC Press website at **www.crcpress.com**

About the Authors

Ratikanta Maiti, PhD, DSc, was a world-renowned botanist and crop physiologist. He worked for nine years on jute and allied fibers at the former Jute Agricultural Research Institute (ICAR), India, and then he worked as a plant physiologist on sorghum and pearl millet at the International Crops Research Institute for the Semi-Arid Tropics for 10 years. After that he worked for more than 25 years as a professor and research scientist at three different universities in Mexico. He also worked for six years as a Research Adviser at Vibha Seeds, Hyderabad, India, and as Visiting Research Scientist for five years in the Forest Science Faculty, Autonomous University of Nuevo León, Mexico. He authored more than 40 books and about 500 research papers. He won several international awards, including an Ethno-Botanist Award (USA) sponsored by Friends University, Wichita, Kansas; the United Nations Development Programme; a senior research scientist award offered by Consejo Nacional de Ciencia y Tecnología (CONNACYT), Mexico; and a gold medal from India in 2008, offered by ABI. He is Chairman of the Ratikanta Maiti Foundation and Chief Editor of three international journals. Dr. Maiti died in June 2019.

Ch. Aruna Kumari, PhD, is an Assistant Professor in the Department of Crop Physiology at Agricultural College, Jagtial, Professor Jaya Shankar Telangana State Agricultural University (PJTSAU), India. She has seven years of teaching experience at PJTSAU and seven years of research experience at varied ICAR institutes and at Vibha Seeds. She has received a CSIR fellowship during her doctoral studies and was awarded a Young Scientist Award for best thesis presentation on at the "National Seminar on Plant Physiology." She teaches courses on plant physiology and environmental science for BSc (Ag.) students. She has taught seed physiology and growth, and yield and modeling courses to MSc (Ag.) students. She also acted as a minor advisor to several MSc (Ag) students and guided them in their research work. She is the author of book chapters in four books. She is also one of the editors of the book *Glossary in Plant Physiology* and an editor of six international books, including *Advances in Bio-Resource and Stress Management; Applied Biology of Woody Plants; An Evocative Memoire: Living with Mexican Culture, Spirituality and Religion*; and *Gospel of Forests*. She has published over 50 research articles in national and international journals. Her field of specialization is seed dormancy of rice and sunflower.

Abul Kalam Samsul Huda, PhD, is Associate Professor in the School of Science & Health, Western Sydney University, Australia. He did his doctoral degree in Agronomy at the University of Missouri, Columbia, USA. He was previously a research scientist in the South Australian Department of Primary Industries and Agro-climatologist and Modeller, International Crops Research Institute for the Semi-Arid Tropics (ICRISAT), Hyderabad, India. In addition, he acted as visiting professor and scientist in academic and research institutes, including University of California, Davis, USA; the University of Sydney, Australia; Texas A&M the University, Temple, Texas, USA. His research work spans the fields of agro-climatology, agronomy, modeling, and systems thinking. He is specialized in using climate information to improve agricultural productivity while reducing production risks and maximizing opportunities. He has published more than 286 research articles in highly reputed journals and handled 24 research projects while working for more than two decades in Western Sydney University and attracted research funding over $5 million. He is an internally renowned scientist in climate-smart agriculture and working in multiple countries like Australia, India, China, Qatar, Cambodia, Indonesia, and Zimbabwe for extensive agro-climatological research. For his enormous contribution in the field, he was conferred as Fellow, at the American Society of Agronomy.

Debashis Mandal, PhD, is an Assistant Professor in the Department of Horticulture, Aromatic and Medicinal Plants at Mizoram University, Aizawl, India. He is a young academician and research fellow working in sustainable hill farming for the past nine years. He was previously Assistant Professor at Sikkim University, India, and has published 35 research papers and book chapters in reputed journals and books. He has also published four books. He is currently chief editor for four volumes on fruits: production, processing, and marketing, and associate editor for four volumes on production, processing, and therapeutics of medicinal and aromatic plants. In addition, he is working as a member in the working group on Lychee and Other Sapindaceae Crops of the International Society for Horticultural Science, Belgium, and is also a member in the ISHS section on tropical-subtropical fruits and organic horticulture and the commission on quality and postharvest horticulture. Currently he is working as Editor-in-Lead (Horticulture) for the *International Journal of Bio Resources & Stress Management* (IJBSM). He is also the founding Managing Editor for a new international publication *Chronicle of Bioresource Management.* Dr. Mandal is an editorial advisor for Horticulture Science for Cambridge Scholar Publishing, UK, and a regular reviewer of many journals. He is also a consultant horticulturist to the Department of Horticulture & Agriculture (Research & Extension), Govt. of Mizoram, India, and Himadri Specialty Chemicals Ltd. He also is handling externally funded research projects. He was the Convener for the International Symposium on Sustainable Horticulture, 2016, India; and Co-Convener, International Conference of Bio-Resource and Stress Management, 2017, Jaipur, India. He was a session moderator and keynote speaker at the ISHS Symposium on Litchi, India, 2016; on Post Harvest Technology, Vietnam, 2014 and at South Korea, 2017; and AFSA Conference, 2018, Cambodia. He has visited many countries for professional meetings, seminars, and symposia. His thrust areas of research are organic horticulture, pomology, postharvest technology, plant nutrition, and micro irrigation.

He did his PhD from BCKV, India, and was postdoctoral project scientist in IIT, Kharagpur.

Sameena Begum, young researcher, has completed a BSc Agriculture with distinction in the year 2016 and and an MSc in Genetics and Plant Breeding with distinction in the year 2018 from the College of Agriculture, Professor Jayashankar Telangana State Agricultural University, Hyderabad, India. During her master's degree program, she conducted research on combining ability, gall midge resistance, yield, and quality traits in hybrid rice (*Oryza sativa* L.) and identified two highly resistant hybrids.

Contents

Abbreviations

ABA	abscisic acid
AF	alternative furrow
AFIS	analysis information system
APEP	alkaline peroxide extrusion pulping
AVG	amino ethoxy vinylglycine
BI	border irrigation
CAT	catalase
CEF	cyclic electron flow
CFML	cotton fiber middle lamellae
CLCuD	cotton leaf curl disease
CLCuKoV-Burcotton	leaf curl Kokhran virus strain Burewala
CLCuMB	cotton leaf curl Multan beta satellite
CLCV	cotton leaf curl virus
CT	conventional plough tillage
DEGs	differentially expressed genes
DPA	days postanthesis
GA	gibberellic acid
GCA	general combining ability
GFP	green fluorescent protein
GM	genetically modified
GMS	genetic male sterility
HNR	height to node ratio
HV	high volume instrumentation
IAA	indole-3-acetic acid
IPM	integrated pest management
LAI	leaf area index
LIR	light interception rate
MAPK	mitogen-activated protein kinase
MDA	malondialdehyde
MTA	leaf inclination angle
NAWB	nodes above white bloom
NCC	nanocrystalline cellulose
PB	permanent raised beds

POD	peroxidase
PPO	polyphenol oxidase
PVP	polyvinylpyrrolidone
QTL	quantitative trait loci
QTL	quantitative trait loci
RFLP	restriction fragment length polymorphism
ROS	reactive oxygen species
SCA	specific combining ability
SDI	surface drip irrigation
SEM	scanning electron microscopy
SFC	short fiber content
SLW	specific leaf weight
SOD	superoxide dismutase
SWC	soil water content
TrAP	transcriptional activator protein
TUE	thermal use efficiency
VPD	vapor pressure deficit
WAE	weeks after cotton emergence
WUE	water use efficiency

Preface

Cotton (*Gossypium* sp. L.) is one of the most important fiber and cash crops across the world and plays a dominant role in the industrial and agricultural economy of most countries. It provides the basic raw material (cotton fiber) to the cotton textile industry, but also plays a role in the feed and oil industries with its seed, rich in oil (18–24%) and protein (20–40%). Worldwide cotton provides a direct livelihood to several million farmers, and an estimated 350 million people are employed in cotton production either on-farm or in transportation, ginning, baling, and storage.

In terms of global production, cotton is the foremost fiber crop. Present world production is some 25.5 million tons of seed cotton from 34.8 million ha. China, the United States, and India are the world's major cotton-producing countries, accounting for nearly 60% of the world production. Cotton is grown in more than 100 countries, accounting for 40% of the world fiber market. Cotton is a major export revenue source for several developing and some developed countries. The cotton is grown in diverse climates such as tropical, sub-tropical, and temperate climates. Australia and Egypt generate the best quality cotton in the world. The world's lowest cost cotton producers are Australia, China, Brazil, and Pakistan.

The history of the domestication of cotton is very complicated and is not known accurately. Several isolated civilizations in both the Old and New World independently domesticated and converted cotton into fabric. *Gossypium barbadense*, known as "Pima" or "Egyptian" cotton, was domesticated in the Peruvian Andes between 4000 and 5000 years ago. "Upland" cotton, *Gossypium hirsutum*, makes up the bulk of the world's cotton crop and was domesticated at approximately the same time in the Yucatan Peninsula. During last two decades, tremendous progress and innovations have been attained in all fields of cotton science, including "development of high-yielding varieties and hybrids, lodging resistant, big boll size, excellent boll opening, easy picking, improved fiber quality, hybrids suit for high-density planting, Bt genes for worm control commonly known as boll guard technology, mechanical harvesting (synchronous flowering), herbicide resistance commonly known as Round-up Ready®, multiple disease resistance including Lygus, or plant bug resistance, Reniform nematode resistance and drought tolerance gene for arid regions, and all these attempting to create desirable traits and increase production of cotton to meet the world's demands. Based on the importance of the crop, farmers, students, and the scientific community (like researchers, scientists) need to understand about the role, nature of the crop and advancements in research for best cultivation methods, effective utilization of resource, and operations for getting higher yields, thus achieving higher productivity.

From the above point of view, the authors decided to provide a resource of complete information and research literature of several disciplines on cotton in the form of book as a guide for students, teachers, researchers, as well as scientists.

The authors have provided information on all aspects of several disciplines of cotton and recent literature together under one umbrella, namely *Advances in Cotton Science: Botany, Production, and Crop Improvement.* This book attempts to bring together recent advances in different disciplines of cotton science. This book covers almost the aspects of cotton starting from background, production, origin to domestication, ideotype, botany, physiology of crop growth and productivity, abiotic and biotic factors affecting crop productivity, methods of cultivation, postharvest management, fiber quality analysis, improvement of cotton crop, research advancements in breeding and biotechnology till 2018. Researchers need to be concerned that the productivity of cotton is affected by several biotic and abiotic stresses, which require a concerted interdisciplinary research. Every aspect of each chapter is described extensively and enriched with recent research literature.

This book was written in a lucid style and in a mode of presentation will help students graduates, academicians, and teaching faculty to gain knowledge and understanding about the crop. Especially this book guides researchers working on cotton and cotton scientists to understand the relation between several disciplines and implementation of new methods and technology covered in recent literature in cotton crop for crop improvement in order to get higher productivity. A multi-pronged approach needs to be used to increase cotton productivity to meet the world's demands.

The authors strongly believe that libraries of schools or colleges of undergraduate, graduate, postgraduate, research institutes of public and private sectors must have this book. It also should occupy a distinctive place in libraries for its versatile contents, extensive descriptions, and enriched research literature of recent advancements. This book effectively is helpful in the aspect of gaining knowledge and explaining subject matter. It will also be a quick reference for teaching staff, professors, the research community, and cotton scientists. It fulfills most of their requirements.

Some problems were faced during the course of writing this book. But the authors' dedication and determination played major driving force in overcoming the problems and successfully completing the book.

Ratikanta Maiti
Ch. Aruna Kumari
Abul Kalam Samsul Huda
Debashis Mandal
Sameena Begum

Acknowledgment

The authors sincerely thank Apple Academic Press for accepting the manuscript and publishing this book within the prescribed timeline. The authors also thank Deasaru Rajkumar for his courtesy in supplying original photographs of cotton plants, Miss R. C. Lalduhsangi for the compilation of research abstracts, and Ing. Jeff Cristopher González Diaz for organizing photographs and references. We heartily thank our chief author, Dr. Ratikanta Maiti, eminent dedicated scientist for his continuous efforts, motivation, and initiation in writing this book and making this book unique, and providing important information for future generations.

Most of the books published on cotton are on specific aspects; very few books have attempted to bring together all disciplines in a concrete form like the present book, and of those, some of them are old. We believe this volume will be a valuable resource.

The authors acknowledge several public and private institutes for playing an important role in society and for providing an excellent platform for their carriers and continuous support in the area of research and publications:

1. ICRISAT, Patancheru, Telangana State, India
2. Jute Agricultural Research Institute, Barrackpore (ICAR), Kolkata, West Bengal, India
3. Universidad Autónoma de Nuevo León, Forest Science Faculty, UANL, NL., Mexico
4. Research and Development Centre, Neo Seeds India Private Limited, Hyderabad, Telangana State, India
5. Vibha Agro tech Pvt. Ltd, Madhapur, India

—Ratikanta Maiti
Ch. Aruna Kumari
Abul Kalam Samsul Huda
Debashis Mandal
Sameena Begum

CHAPTER 1

Background and Importance

ABSTRACT

This chapter presents the importance of cotton (including its various industrial products) at global level and discusses a brief outline of research advances in various aspects of cotton crops. Cotton has high demand across the world for manufacture of comfortable dresses. Additionally, various products of high economic values are derived from cotton plants.

1.1 ECONOMIC IMPORTANCE

Cotton is a popular fiber crop grown widely across the globe. It is popularly referred to as "White gold" because of the silky white fibers that are produced. These silky white cotton fibers are popular as "Kapok" and since the olden times, they were put to use in the filling up of mattresses, cushions, and pillows. Accordingly, during World War II in Europe, cotton fibers were used as padding sources of jackets (which were life saving during the war), aiding in providing buoyancy. The cotton seeds are rich in oil and protein, both edible and are useful for soap and lighting. The remnants of cotton seeds are used as feed for livestock. Most of the products of cotton have different industrial uses. They are used in chemicals, food, and in textile production.

The fruits of cotton are often called capsules or bolls. These contain many seeds. Two types of fibers cover seeds, namely, those which are of short length fuzz and long length, lint (Fig. 1.2). Among these short and long fibers, only the lint (Fig. 1.3) have a major use in the textile industry in the production of clothes. In many areas since man began its cultivation, it is harvested manually, is ginned, and later processed.

After the completion of the ginning process, these cotton fibers are made flexible by padding with a wooden bow. Before the commencement of the spinning process, the fibers are carded by the use of a hand comb so that these get separated from one another. And finally during the spinning process, these individual fibers are twisted into yarn. This can be achieved by a hand spindle or on a spinning wheel.

Figure 1.1 shows a cotton plant ready for harvest.

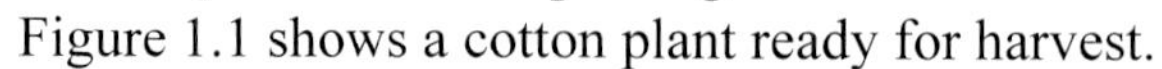

FIGURE 1.1 Cotton plant at boll bursting stage, ready for harvest.

Source: Photo courtesy of Dr. Sadasivan Manickam, Central Cotton Research Institute, ICAR, India.

Across the world, cotton is the only fiber being utilized to a large extent. This acts as a leading cash crop in the US, earning much of the economy to the country. Cotton at the farm level every year involves a large amount of purchase of produce of worth greater than $5.3 billion. Because of its various industrial uses, it acts in stimulation of business activities for most of the industries spread within the country. More business activity is rendered after its processing. More than $120 billion is earned as revenue in the United Sates from cotton and its products.

Monthly Economic Letter (2018) reveals that cotton is extensively used in our daily lives for its multiple uses. It has its main usage in clothing and in several household items. Cotton production reaches several thousands of bales, it is used widely in the production of many industrial products also.

FIGURE 1.2 Boll fully burst.

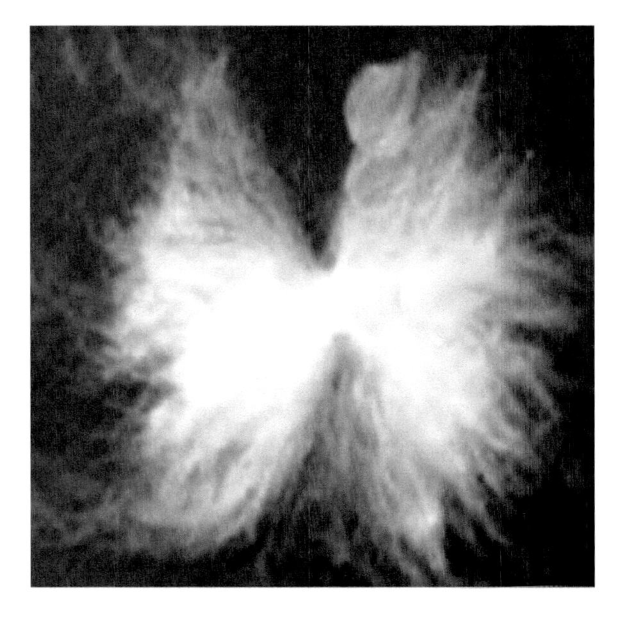

FIGURE 1.3 Cotton fiber.

As mentioned before, all parts of the cotton plant are used for one purpose or the other, often, most widely used part is the fiber or lint, which is commonly used for the manufacture of cotton cloth. The textiles prepared from cotton are comfortable to wear in all seasons because of their light texture and weight.

Similarly, the short fuzz fiber on the seed, the linters are rich in cellulose. These are therefore used for the manufacture of plastics, explosives, and other products. Fuzz fibers have their utility even in the production of high quality fabricated paper and its product, materials in padding mattresses, furniture and automobile cushions.

Three products are extracted from the crushed cotton seed, namely, oil, meal, and hulls. The seed oil of cotton has its high utility in salad dressing or sometimes used as cooking oil, while rest of the two products are used as poultry, livestock, fish feed, or as fertilizer source. After harvesting of cotton, left out plant debris are plowed under to increase the soil nutrients. Some of the baked food products also use cotton seed as a source of high protein concentrate.

Since the beginning of the Harappan civilization and much before that in the Indian subcontinent, the domestic and Asiatic cottons (*G. herbaceum* L. and *G. arboreum* L.) were under domestication or under commercial cultivation in this country, which is a traditional home of cotton and its textiles. All the four popular species of cotton, namely, *G. arboreum*, *G. herbaceum*, *G. hirsutum*, and *G. barbadense* are under commercial cultivation in this country.

Although the diploid cottons such as *G. arboretum* L. and *G. herbaceum* L. are mainly cultivated in dryland tracts, the Bengal *desi* is grown to a large extent mostly under the irrigated conditions in the northern states of West Bengal. *G. hirsutum* L., known as the American cotton, is most popular, with a number of varieties and hybrids. On the other hand, *G. barbadense* L. is popularly known as the Sea Island cotton.

Even though it is a widely cultivated fiber crop, its cultivation also faces unfavorable climatic conditions which are prevalent in most of the cotton growing regions. The cotton production in India is limited owing to the extreme variability in rainfall patterns and stream flows involved as the sources of water supply. There was a drastic reduction of 40% decrease in prices of world cotton during the period of 2001 and 2002 (Minot and Daniels, 2005). This has drawn the attention of Governments across the world to provide subsidies to all the cotton growing farmers who were upset with decreased world prices of cotton.

Van Esbroeck and Bowman (1998) studied about the germplasm diversity in cotton and its utility in the development of cultivars. In general, it is presumed in many crops that the parents which are genetically diverse have a great potential in serving as the parents for the creation or development of a superior progeny. In cotton, however, only a few existing studies have given information in establishing the relationship that existed between the parental genetic distance and the development of suitable cultivars which can perform successfully at different environmental conditions. One of the theories of genetic distance advocates that the matings that have been carried between the distantly related parents could generate more of transgressive segregates than that resulted from the parental lines which were related closely. In most crops, yield improvements were obtained in many cases from the matings that have been obtained from the closely related genotypes, rather than those which were distantly related. Van Esbroeck and Bowman (1998) undertook a study to establish a relationship between the parental genetic distance and the development of successful cotton (*Gossypium hirsutum* L.) cultivars. They observed the pedigrees of cultivars, these cultivars occupied greater than 1% of the total planting of US in 1987–1996. Then they estimated in final crosses, the genetic relatedness of the parents. It is expressed as coefficient of parentage. Sixty final crosses were found to be successful. These cultivars were obtained by two ways cross (60%) reselection products/germplasm lines (25%)/complex crosses (15%). In final cross, average coefficient of percentage is 0.29. It is more compared with random pairing of parents. They could successfully demonstrate the diversity and its level that was existing within these cotton cultivars which were more locally/regionally adapted.

Soyoung Kim et al. (2003) investigated Asian–American Consumers in Hawai', they studied their attitude and tendency toward ownership of apparels of ethnic nature. This research investigated the strength of ethnic identification, its influence, the attitudes of people toward apparel quality, etc. Approximately 167 of these consumers who had a frequent visit to apparel store were interviewed. They emphasized toward clothing and its features and also on display of apparels in the shop. The results highlighted that attitude about an apparel is much more important rather than understanding or attributes of display.

Cotton clothes with their majestic colors have earned glamor in Iranian and Indian cultures. Moraveji (2016) undertook a comparative study of graphic aspects of textiles in Indian Gurakani and Iranian Safavid eras. In these periods, valuable textiles played a significant role in expressing the individuals' social dignity provided with valuable fibers, rare colors, and particular designs. The authors made a comparative analysis of the Iranian textiles designs in Safavids and Indian ones in Gurkani dynasties during 9–11th centuries (AH), which enabled the assessment of similarities and differences in term of designs and color of the textiles in these two countries, and the level and reason for their effectiveness. Investigation into the political–cultural relationship between these two dynasties, techniques, the materials, and instruments used in textiles, itineraries and historical and research books and designs analysis on the basis of the available images and pictures were performed. The findings revealed that the level of effectiveness of technique and Iranian textile design were more than its vice versa state. Although in India and Iran colors of the textiles were often similar, more emphasis was placed on some Indian colors. There was a greater European influence on Indian designs compared with Iranian samples. However, the influence of religion and literature on Iranian artists and textile designs was greater, compared with those of Indian artists and textile designs.

Kern (2018) conducted research on the background, importance, and production volumes of fatty acids. The research observed that much importance of establishment of industry for fatty acid production was based on the production and economic gains that were realized during 1978 in the United States, wherein the production of fatty acid oils was 956 M lbs. The 1978 US production of various fatty acids was broken down into nine saturated categories and five unsaturated categories. These were (1) stearic and 127.2 M lbs. (13.3%); (2) hydrogenated animal and vegetable acids (2a) 97.3 M lbs. (10.2%), (2b) 158 M lbs. (16.5%), (2c) 32 M lbs. (3.4%); (3) high palmitic, 14.6 M lbs. (1.5%); (4) hydrogenated fish, 6.5 M lbs. (0.7%); (5) lauric acid types, 88.8 M lbs. (9.3%); (6) fractionated fatty acids, (6a) C_{10} or lower, 18.5 M lbs. (1.9%), (6b) C_{12} and C_{14}55% 17 M lbs. (1.9%); (7) oleic acid, 158.3 M lbs. (16.6%); (8) animal fatty acids other than oleic, 156.3 M lbs. (16.3%); (9) vegetable or marine fatty acids, 0.1 M lbs. (less than 1%); (10) unsaturated fatty acids, 57 M lbs. (6.0%); (11) unsaturated fatty acids IV over 130, 24.2 M lbs (2.5%). Reported 1977 fatty acid derivative production from fatty acids (not fats and oils) was 1980 M lbs. It was observed that the average price of fatty acids increased from 23¢/lb to 60¢/lb. within a short span in last 5 years.

In view of the great importance of cotton globally, enormous research inputs have been directed on various aspects of cotton such as world production, origin, evolution and domestication, ideotype, botany, physiology, abiotic stresses affecting cotton production, (drought, salinity, heat stress, cold tolerance); biotic tresses (insect pests, diseases), harvest and post-harvest management, research advances in cotton breeding and biotechnology; fiber quality and its management, etc., which are presented in subsequent chapters.

It is worth to note that Dr. Maiti, during his stay as Research adviser in Vibha Seeds, acquired an immense knowledge and conducted a series of research and developed techniques, which, although unpublished is intended to be included in this book. The various techniques developed are outlined below:

1. Screening for Abiotic stress tolerance
   a) Drought tolerance: Developed efficient technique for screening and selection of cotton crops (The selected lines shown good performance under drought prone areas).
   b) Salinity tolerance: Developed simple, efficient, and cheap technique (semi-hydroponic) for screening and selection of cotton. The selected cultivars have shown good performance in saline prone areas depicting the transfer of technique from lab to land.
   c) Heat stress tolerance: Developed techniques for evaluation and selection of cotton for heat stress tolerance, which were confirmed by breeders in actual field situation.
   d) Flooding stress tolerance: Developed techniques for evaluation and selection of cotton, maize, and tomato in the aspect of flooding stress tolerance.
2. Biotic stress tolerance: Studies on sucking pest tolerance in cotton for biotic stress tolerance. Morpho-anatomical selections for biotic stress tolerance in cotton, tomato, chilli elite lines, pipeline hybrids.
3. Root studies: Developed techniques for evaluation and selection of higher root growth for resistant to drought and salinity in Cotton.
4. Characterization of morphological and anatomical traits of several field and vegetable crops such as cotton, maize, rice, sunflower, castor, okra, tomato, and chilli, etc.

Many studies have been undertaken to demonstrate the utility of cotton plant and its byproducts for the benefit of mankind.

Zhijun and Xunzai (2014) reported the alkaline peroxide extrusion of pulping suitable for bast and stalks of cotton. Earlier, these were pulped together. This led to a decline in quality and homogeneity of the pulped material. Furthermore, it has complicated the processes, namely, bleaching involved in manufacture of paper. It led to its limitation for its use only for the preparation of paper of low quality grade. Alkaline peroxide extrusion pulping (APEP) was done to individual parts of bast and stalks of cotton. By usage of orthogonal analytical methods, they determined the parameters of pulping (optimal). There was an increase in yield of pulped material and better quality, much whitened pulp was obtained from the cotton stalk by this method of extraction. The amount of chemicals required for extrusion were less, it resulted in less energy requirement and less pollution. However, extrusion of bast by either by APEP or kraft pulp method did not exhibit any variation in the physical performance.

Mandhyan et al. (2015) studied wettability and wicking phenomenon in scoured cotton fiber. They noted that there were various factors affecting wetting and wicking of fibrous assemblies; micronaire value of cotton. Their study deliberated on the fiber properties and their effect on wetting and wicking of a fibrous assembly. An attempt was made to calculate the contact angle of the fiber, using the powder cell method, and compared it with the fineness of cotton fibers. Three categories of the cottons were prepared depending on their fineness which was further grouped into three, based on the coarse, fine, and super fine varieties. It was found that within the group the contact angle depends upon the micronaire value of the cottons. They extracted pulp from cotton and cotton stalk by acid hydrolysis. They were analyzed by Fourier-transform infrared spectroscopy. Pulp yields from cotton were more (77%) compared with that from cotton stalk (23%). Pulp extracted from raw cotton was of superior quality. Surface areas of aerogels from cotton nanocrystalline cellulose, C-NCE were 91.47 m^2/g, while that from cotton stalk was 93.89 m^2/g. There were no differences between these two products. Study revealed that preparation of nanocrystalline cellulose from cotton stalks is much cheaper and environmentally safe.

Egbuta et al. (2017) and McIntosh et al. (2017) reported about the biological importance of cotton by-products. The importance of these was discussed with reference to the chemical constituents. These were some chemical compounds, primary metabolites, some were terpenoids and terpenoid derivatives (51), fatty acids (four), and phenolics (six), isolated from cotton stalks, stems, leaves, and bolls. A few of these isolated chemical constituents were found to be associated with antimicrobial and anti-inflammatory properties. For example, β-bisabolol, extracted from flowers exhibits anti-inflammatory property. Apart from fiber, forming an important economic product, many biologically active ingredients were also found in large quantities, within the cotton plant, offering an immense scope for exploitation and utilization of these novel phytochemicals as potential value added products. These can be extracted through modern techniques at the time of fiber separation process only.

Nikolić et al. (2017) discussed that cotton fabric can effectively be used as a potential source of biofuel (renewable) for bioethanol extraction. There was an increase in glucose yield when cotton fabric was exposed to corona pretreatment of enzymatic hydrolysis. It also resulted in more ethanol yields. *Saccharomyces cerevisiae var. ellipsoideus* was the key microbe in fermentation process for bioethanol extraction. Ethanol productivity was higher and of superior grade from mercerized cotton fabric (0.900 g/L h) after 6 h of fermentation. Results indicated that mercidized cotton fabric scraps can act as potential alternatives sources for bioethanol production on large scale.

Manjunath et al. (2016) made a comparative economic analysis of Bt and non-Bt cotton. The results of the study indicated that Bt cotton farmers were relatively younger than non-Bt cotton farmers, who readily accepted the Bt cotton technology. It was found to be superior in terms of gross income and net income over non-Bt cotton. The partial budgeting analysis has suggested that farmers could benefit to the tune of Rs. 9882/ac by adopting Bt cotton technology. Furthermore, the study indicated that there has been rapid expansion of area under Bt cotton registering an impressive growth of 183% during the last decade (2002–2003 to 2009–2010).

Esteve-Turrillas et al. (2017) discussed that organic recover cotton from textile industry drastically reduces the usage of chemical compounds, reducing environmental pollution impact of these chemicals impact. However, during this process of recover cotton from textile industry, it is a prerequisite to see that ginning and dyeing steps are maintained with due optimal care. Recover cotton may effectively avoid the impacts on environment by bringing about a reduction in production of textile wasted. However, it may likely to result in some extra costs involved in the shredding and cutting processes. Recover cotton produced from high quality textiles enables in provision of valuable products and is safe from environmental point of view.

Qian et al. (2017) designed a method for purification of dyeing waste water obtained during cotton fiber processing. They initially identified that there were few crystallization spaces in natural cotton. These could be effectively reduced through pretreatment of triethanolamine, through the usage of cationic reagent: 3-chloro-2-hydroxypropylmethyldiallylammonium chloride (CMDA) and cationic monomer: dimethyldiallylammonium chloride (DMAC) on G cotton, they could effectively obtain polyatomic film coated cotton (PF-cotton), which was found to have wide universal applicability. It was found to be more efficient to act as filtering filler, and could adsorb to a large extent the dye solutions of different dyeing waste waters and also enabled in effective utilization of this dyeing waste waters through recyclization process.

Moorthy et al. (2017) developed a technique for effective production of ethanol from waste biomass of cotton. Many industries involved in textile production have a serious problem of management of solid wastes, which are obtained in large quantities during this textile production. These solid wastes were found to have an impact in causing health hazards, but also have serious environmental impacts. Cotton wastes are rich in lignin, hemicellulose, and cellulose, which can act as effective stocks for bioethanol production. They discussed and reviewed about various methods of pretreatment of cotton wastes and technologies involved for effective removal of lignin and solubilization of hemicellulose.

QunYing et al. (2017) quantified the economic impact of cultivated cotton in both rain fed and irrigated production systems in future. They evaluated the options of adaptation and also the negative impacts, opportunities available for effective production of these centered toward 2030. They assessed that returns obtained from various production systems would be variable. Furthermore, debt level of cotton production farms in Australia is likely to be reflected and carried in future years, which needs to be reduced for avoidance of fluctuations in the equity prices. They emphasized that in order to avoid risks in future CC, all cotton growers have to adapt effective management strategies even at all locations.

Vognan et al. (2017) carried out a comparative analysis of the profitability of the cultivation of organic, conventional, and transgenic cotton farming systems in Burkina Faso, each with a sample of 60 households. They observed high production costs in systems of farming of conventional and transgenic rather than organic system, with little variation in the profitability from among these three farming systems. Furthermore, they assessed that organic farming was more suitable and resilient financially to variable cotton prices or inputs, making it a strong viable option of cotton production to alleviate poverty and future climate changes.

A general practice adopted by many cotton growers is that after harvest of cotton, the left over debris of cotton stalks were burnt within the fields. This practice results not only in pollution of the environment, but also leads to soil fertility deterioration. In order to overcome these problems, Mageshwaran et al. (2017) reported that that the cotton growers can take up oyster mushroom cultivation or produce compost and turn out as successful entrepreneur by utilizing the viable options that are available to him.

1.2 CONCLUSION

Cotton is a crop of high commercial importance and great demand at global level. It is used in fabrication of cotton clothes and dresses which are comfortable. It also provides various products that are amply used for household and commercial and industrial uses on cotton industry. Cotton clothes have great glamor in Indian and Iranian cultures. Cotton, called as white gold, is a cash crop of high demand in the world for the manufacture of comfortable dresses and other important uses as surgical accessories, various commercial and industrial uses. Cotton is well known for wettability and wickening phenomenon. Cotton stalks are used for production of products like paper pulp, use of cotton stalk for mushroom, oyster culture. Fibers of cotton stalk are used for textile use. Bioethanol is produced from cotton waste fabric, etc. More research needs to be directed for utility of cotton plant and lint.

KEYWORDS

- **cotton fiber**
- **lint**
- **importance**
- **products**
- **cotton seed oil**
- **cotton stalk**

REFERENCES

Egbuta, M. A.; McIntosh, S.; Waters, D. L. E.; Vancov, T.; Liu, L. Biological Importance of Cotton By-Products Relative to Chemical Constituents of the Cotton Plant. *Molecules* **2017,** *22*, 93. (http://www.mdpi.com/1420-3049/22/1/93/htm).

Esteve-Turrillas, F. A.; Guardia, M. de la. Environmental Impact of Recover Cotton in Textile Industry. *Res. Recycling* **2017,** *116*, 107–115. (http://www.sciencedirect.com/science/article/pii/S0921344916302828).

Kern, J. C. *Background, Importance and Production Volumes*, Proceedings of the AOCS Short Course on Industrial Fatty Acids, 2018, p 6.

Mageshwaran, V.; Varsha, S.; Hamid, H.; Shukla, S. K.; Patil, P. G. Compost Production and Oyster Mushroom Cultivation: A Potential Entrepreneurship for Cotton Growing Farmers. *Int. J. Forestry Crop Improvement* **2017,** *8*, 149–156.

Mandhyan, P. K.; Banerjee, S.; Nagarkar, R. D.; Prabhudesai, R. S.; Pawar, B. R. Wettability and Wicking Phenomenon in Scoured Cotton Fibre. *J. Cotton Res. Dev*. **2015,** *29*, 315–322. (http://www.crdaindia.com/fileserve.php?FID=246).

Manjunath, K.; Kiran, K.; Patil, R.; Chinnappa. Comparative Economic Analysis of Bt and Non Bt Cotton B. *J. Cotton Res. Dev.* **2016,** *30*, 131–134. (http://www.crdaindia.com/fileserve.php?FID=125).

Minot; Daniels. Monthly Economic Letter, 2018.

Moraveji, E. Comparative study of Graphic aspects of textiles in Indian Gurakani and Iranian Safavid eras. *Bulletin de la Société Royale des Sciences de Liège* **2016,** *13*, 793–806.

Moorthy, R. K.; Rajarathinam, R. K.; Sankar, M. K.; Kumar, M. N.; Velayutham, T. An Effective Conversion of Cotton Waste Biomass to Ethanol: A Critical Review on Pretreatment Processes. *Waste and Biomass Valorization* **2017,** *8*, 57–68. (https://link.springer.com/article/10.1007/s12649-016-9563-8).

Nikolić, S.; Lazić, V.; Veljović, Đ.; Mojović, L. Production of Bioethanol From Pre-Treated Cotton Fabrics and Waste Cotton Materials. *Carbohydr. Polym*. **2017,** *164*, 136–144. (http://www.sciencedirect.com/science/article/pii/S0144861717301017).

Qian, J.; Chunli, S.; HongYan, L.; Yuanyuan, H.; LiNa, L.; YiKai, Y. Construction of Polycationic Film Coated Cotton and New Inductive Effect to Remove Water-Soluble Dyes in Water. *Mater. Des*. **2017,** *124*, 1–15. (http://www.sciencedirect.com/science/article/pii/S0264127517303039#!).

QunYing, L.; Behrendt, K.; Bange, M. Economics and Risk of Adaptation Options in the Australian Cotton Industry. *Agric. Syst.* **2017,** *150*, 46–53. (http://www.sciencedirect.com/science/article/pii/S0308521X16305601).

Soyoung, K.; Arthur, L. B. Asian-American Consumers in Hawaii: the Effects of Ethnic Identification on Attitudes Toward and Ownership of Ethnic Apparel, Importance of Product. *Clothing Textiles Res. J.* **2003,** *45*, 21–29.

Van Esbroeck, G.; Bowman, D. T. Cotton Germplasm Diversity and Its Importance to Cultivar Development. *J. Cotton Sci.* **1998,** *2*, 121–129.

Vognan, G.; Glin, L.; Bamba, I.; Ouattara, B. M.; Nicolay, G. Comparative Analysis of the Profitability of Organic, Conventional, and Transgenic Cotton Farming Systems in Burkina Faso. *Tropicultura* **2017,** *35*, 12–24. (http://www.tropicultura.org/text/v35n1/12.pdf).

Zhijun, H.; Xunzai, N. Alkaline Peroxide Extrusion Pulping of Cotton Bast and Cotton Stalk. *Bioresources* **2014,** *9*, 2856–2865. (http://ncsu.edu/bioresources/BioRes_09/BioRes_09_2_2856_Hu_N_Alk_Perox_Extrusion_Pulping_Cotton_Bast_Stalk_5365.pdf).

CHAPTER 2

World Cotton Production and Factors Affecting Production

ABSTRACT

This chapter presents country-wide production of cotton at global levels. Cotton production varies widely in different countries depending on prevalent agroclimatic conditions. It also discusses climatic, abiotic and biotic factors affecting crop production in different countries.

2.1 INTRODUCTION

Cotton is grown globally for its textile values and for other industrial products derived from cotton plant. The productivity of cotton crop varies largely depending on climatic conditions, biotic and abiotic factors prevailing in individual countries and agronomic practices followed for obtaining higher yields.

2.2 COUNTRY-WISE COTTON PRODUCTION

Cotton production by country worldwide in 2016/2017 (in 1000 metric tons) is given below and shown in Figure 2.1.

1. India: 5879
2. China: 4953
3. United States of America (USA): 3738
4. Pakistan: 1676
5. Brazil: 1574
6. Australia: 914

7. Uzbekistan: 811
8. Turkey: 697

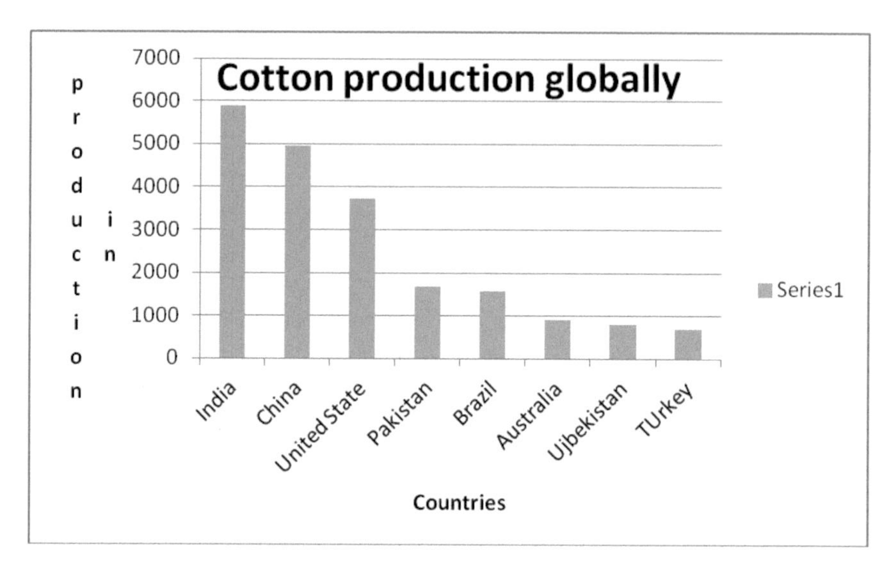

FIGURE 2.1 Cotton production globally.
Source: Monthly Economic Letter, May 2018.

Cotton is produced in huge quantities in many parts of the world. The top three countries in the world with good amount of cotton production are China, India, and the United States Table 2.1. The production yield in terms of thousand metric tons indicates that China had a production of 6532 thousand metric tons, while India and the United States had a production of 6423 and 3553 thousand metric tons of cotton, respectively, in the past years.

China: China is the world's leading producers of cotton. More than 1,000,000 farmers cultivate cotton in the country. It has about 7500 textile companies, which produce an annual income of US$73 billion.

India: India is the next largest producer of cotton after China, within this country the cotton usage dates back to the Indus valley civilization. Archeological excavations carried out at the Indus valley recovered some cotton threads. Every year, 6423 thousand metric tons of cotton is produced. Favorable congenial climatic conditions for cotton production are seen in the northern part of the country. Furthermore, the temperature of 25–35°C which is moderate for cotton is also prevalent in the country, attributing to the high production in the country. Based on the quality requirement parameters, the processing of cotton is done in huge quantities by the use of modern machinery.

TABLE 2.1 According to Cotton: World Markets and Trade—Foreign Agricultural Service: World Cotton Production/Export 480 lb Bales.

S. no.	Country	2013–14	2014–15	2015–16	2016–17	2017–18	2018–19
1	India	31.0	29.5	25.7	27.0	28.5	28.5
2	China	32.8	30.0	22.0	22.8	27.5	27.0
3	USA	12.9	16.3	12.9	17.2	20.9	19.5
4	Brazil	8.0	7.2	5.9	7.0	8.7	8.8
5	Turkey	2.3	3.2	2.7	3.2	4.0	4.3
6	Australia	4.1	2.3	2.9	4.1	4.8	4.0
7	Uzbekistan	4.1	3.9	3.8	3.7	3.9	3.8
8	Mexico	0.9	1.3	0.9	0.8	1.6	1.6
9	Burkina	1.3	1.4	1.1	1.3	1.3	1.4
10	Turkmenistan	1.6	1.5	1.5	1.3	1.4	1.4
11	Mali	0.9	1.0	1.0	1.2	1.4	1.4
12	Greece	1.4	1.3	1.0	1.0	1.2	1.3
13	Rest of world	9.8	9.7	7.7	8.4	9.2	9.5
14	African Franc zone	4.1	4.8	4.0	4.8	4.8	4.9
15	Eu-27	1.6	1.7	1.3	1.3	1.5	1.6
16	World	120.4	119.2	96.2	106.7	122.4	121.2

Source: Monthly Economic Letter, May 2018.

The United States: As it is the third largest producer of cotton, here, the harvesting of cotton is mostly through usage of modernized machinery. The major cotton producing states in the country are Texas, Mississippi, Arizona, Florida, and California. Cotton cultivation in these states is at a boom because of the presence of the favorable climatic conditions. The boll is harvested with machines without any damage to the existing plant, and usage of modern technology replaced the usage of spinning and power loom for processing of cotton to meet the required quality.

2.3 APPLICATIONS OF COTTON PRODUCTS IN MODERN TIMES

The fiber obtained from cotton is utilized to a large extent in the textile industry for the manufacture of clothing material of different qualities to suit varied uses. It is used in the manufacture of smaller textile of handkerchiefs, towels to T-shirts, jeans, trousers, shirts, etc.; most people prefer clothes manufactured from cotton as these textiles are lighter, soft-textured, and comfortable for wearing. The preference to the cotton dresses is mostly seen in countries such as Sri Lanka, India, and Pakistan where warmer climates are prevalent for most part of the year. Apart from the manufacture of textile cloth, cellulose which is obtained from the cotton fiber is mostly utilized in the manufacture of paper. Medical practitioners commonly use the puffs made of cotton for dressing of wounds. In some countries, it is also used in the manufacture of fishing nets. Some of the byproducts obtained from cotton are oil and cake. Cotton oil is also used in candle and soap preparations. Thus, it is clearly evident that this fiber crop has its utility in meeting the all round needs of all groups of people as a material of requirement and comfort in daily usage.

Yilmaz et al. (2004) in their analysis carried out in Turkey for the input costs incurred in cotton production and the energy usage observed that diesel energy of 49.73 GJha-1 was consumed of the cotton production accounting to 31.1%. The output–input energy ratio showed 0.74 kg of cotton MJ^{-1} while the energy productivity was only 0.06 kg of cotton MJ^{-1}. The cost analysis data has revealed that net returns earned from a kg of cotton seed were insufficient in meeting the costs of production that were incurred and there is a necessity to improve the energy management and its efficiency at the farm gross root level also.

Pray et al. (2002) in their survey carried out on Bt cultivation in China found that it was popular and in high demand among farmers because of its technology that plays a role not only in the reduction of the insecticide and pesticide applications but also gives the farmers ample time to spend for other tasks, as the scheduling of insecticide sprays and exposure time to them are much reduced with the cultivation of Bt cotton.

2.4 FACTORS AFFECTING COTTON PRODUCTION

Various factors affect cotton production in the world. Key factors among these are pests and diseases.

Precision agricultural technologies have a wide applicability in many areas of the world for management of many natural resources, analyzing the profitability of farm levels, zonal impacts, remote sensing and its usage, etc. The overview of this in the developmental aspects across the world and in China its present current status in the improvement of agriculture and its varied usages were discussed in brief by Zhang et al. (2002).

Mahanty et al. (2002) carried out a policy analysis and assessed the competitiveness that existed in cotton production in India. The researchers used the policy analysis matrix (PAM) approach to assess cotton production efficiency of five major cotton producing Indian states. Results of their research indicated that cotton production was not that efficient in a second largest cotton production state of the country. Furthermore, Samarendu et al. observed that there existed several inefficiencies in the cotton sector and that much of the policies were oriented toward the maintenance of the availability of cotton to the industrial sector at a cheaper price.

Oerke and Dehne (2004) discussed the importance of safeguarding cotton crop production by reducing production losses which arose from the infestation of insect pests, pathogens, animal pests, viruses, bacteria, and weeds. They observed that it was impossible to increase the productivity per unit area without implementation of the crop production and simultaneous intensification of the pest control methods.

Deguine et al. (2008) conducted a review on the importance of the sustainable pest management for cotton production. They observed that cotton lint production reached 20 million tons obtained from an area of 30–35 m ha from 69 countries. Furthermore, it was observed that there was an excessive usage of pesticides in cotton which was in one way responsible for an increase in the cost of production. They identified two methods of crop protection which were important in attaining sustainable cotton production. Those methods were (1) Total pest management, which involved pest eradication and (2) adoption of integrated pest management. The researchers observed that total pest management was feasible in the agricultural systems where pests lack alternate host crops in their vicinity. Furthermore, they observed that integrated pest management (IPM) was constrained both by intervention threshold techniques and specific nonchemical techniques which were frequently integrated in that method of control. Additionally, they noted that although IPM had certain constraints, it had an advantage as it takes the complete pest complex of a cropping system into consideration.

Beckert (2004) undertook an analysis on "emancipation and Empire: Reconstructing the Worldwide Web of Cotton Production in the Age of the American Civil War." He examined different aspects on cotton, merchants, famine, capitalism, occurrence of civil war, textile industry, and the history in the USA.

Cotton production is negatively affected by the incidences of pest and pathogens, which cause crop losses. These crop losses could be substantially reduced if suitable protective measures are adopted. Oerke (2006) recommended the adoption of IPM to achieve a substantial reduction in the usage of the pesticides and obtain economical yields with ecologically safe and acceptable levels of pesticide usage.

Production of any type of cotton crop involves water consumption. It has also an effect on water quality. In most of the countries these cotton product manufacturing units exhibit a series of chain impacts on resources of water. Chapagain et al. (2006) in their assessment on the water resources by the cotton products in several countries found that between 1997 and 2001 cotton products had a water consumption of 256 Gm^3 of water annually. Nearly 42% of the water consumed was from the blue waters, while 39% and 19% of water requirement was from the green water and dilution water, respectively. Their study revealed that 84% of water impact by cotton products was from the countries situated outside Europe, and the main impact arose from India and Uzbekistan. Furthermore, they mentioned that cotton product production brings about three types of impacts on the water resources, namely, an increased rate of evaporation from the infiltrated rain water used by a cotton crop for its growth, excessive withdrawal of ground and surface water sources for irrigating the crop and for the processing of the cotton products, and lastly the drastic increase in the pollution rates of water, pollutants being released during the processing of cotton products or those arising as surface run offs from the cotton fields, where excessive use of pesticides and insecticides were applied judiciously.

The productivity of cotton varies widely in different regions for various factors. A few examples of research for last 10 years are cited below.

Sabo et al. (2009) in their economic analysis of the cotton production in Nigeria had observed that in the cost of production of cotton, 50% of the variable cost arises from the labor costs only that were incurred in cotton production per hectare. Their regression analysis revealed that double log function was the best model. The coefficient of determination (R^2) was 0.61.

Furthermore, they observed that a negative correlation existed between the land rent cost and cost incurred toward ploughing and planting and poor pricing of cotton often challenged the cotton production in Adamawa state and fluctuations existed in the yields as well as cotton prices. They recommended the establishment of a cotton commodity board in the state for timely provision of inputs at subsidized prices to the farmers and rendered timely functional services, so that many more farmers show a bend or get stimulated toward taking up of cotton production.

Silvie et al. (2010) carried out an analysis of organic cotton production in Paraguay. They observed that arthropod and disease incidence and soil fertility aspects acted as constraint to the organic production of cotton. Analysis conducted in the form of interviews from the farm to the industrial level indicated that boll weevil *Anthonomus grandis* was a serious pest affecting at the cropping level. They recommended the necessity of conducting in-depth studies for the identification of the biological pathways and integration of the farmers practices both quantitative and qualitative with the impacts on cotton production over time and area.

Rao et al. (2012) discussed the strategies of soil fertility management involved in the maximization of the productivity levels in India. They noted that cotton that was grown in the red soils which were light textured had shown moderate to severe deficiency of magnesium. The researchers further noted that the magnesium deficiency led to a reduction in plant growth and boll number, and that a better nutrient management resulted in a BC ratio of 4 and significant economic benefits were realized from the nutrient foliar sprays. Furthermore, they noted that application of a small dose of nitrogen fertilizer after an intermittent drought spell proved to be beneficial by bringing about an increase in the foliar growth and the boll production and size of bolls. Similarly, application tank silt (40–60 t/ha) and FYM/compost in the light textured red soils brought about an increase in the tolerance levels in cotton against the drought spells.

MingYun et al. (2014) undertook an analysis of trends of cotton production in Binzhou, China. They suggested that in that province a decreased trend in total lint production occurred in the last recent 10 years and that decrease was around 51%. They observed that because of the policies adopted in importing cotton, disaster management, or reduced rate of competitions in the locally produced cotton, there was a substantial decrease in the cotton area and production in this region. They further predicted a drastic decline in cotton producing and planting areas in the next few years in Binzhou.

Kadam et al. (2014) assessed the importance of the input management in cotton production by carrying out a production function analysis in Vidarbha, India. There was a significant relationship between human labor, bullock labor, machine power, seed, and manure with productivity. The researchers observed that a production function analysis revealed a decreasing returns to scale (0.78) in cotton production with the value of R^2 (0.79), indicating that variables under study explained 79% contribution in cotton production.

Agarwal et al. (2015) examined the yield gap and constraints of cotton production in India by undertaking a survey during 2010–2011 and 2011–2012 to analyze the yield gap and constraints in cotton production in the country. Results revealed that the yield gap ranging from 6.66 to 18.4 q/ha was present in cotton production. Highest yield gap was found in Gujarat, followed by Madhya Pradesh and Maharashtra. Yield gap was comparatively low in Punjab, Haryana, Karnataka, and Tamil Nadu. Major constraints in north zone included a higher degree usage of inferior quality seed, incidence of the sucking pests, repeated occurrences of the dry spells, and infestation by cotton leaf curl virus. The researchers observed that weed infestations, lack of good seed, leaf reddening, delayed sowing led to a limitation in the cotton production in the central zone, while in the southern zone, apart from the above, the availability of human labor also acted as constraint to cotton production.

Fatima et al. (2016) evaluated the technical efficiency of cotton production in Pakistan by carrying out a comparative study on both non-BT and Bt cotton farms. Results of the inefficiency of their model revealed that factors such as education, experience, and distance from the main market were the most influential determinants of the technical efficiency in both non-Bt and Bt cotton farms. It was observed that extension of cotton production required the development of cottonseed that would be suitable for environmental conditions of Pakistan.

Ma Na (2016) conducted an analysis to understand the reasons for a decline in the cotton planting area in Henan province. They observed that the main reasons behind the decline in the cotton planting area were the presence of low mechanization technologies involved in cotton production, reduced prices for cotton, and requirement of high labor, and other production inputs coupled with the presence of unfocussed policies and support to the cotton growing farmers by the government. Besides, some factors, such as less attention to local government to cotton production, incomplete key technique application, and limited policy regulation. For solving these problems, some counter measures were suggested, including strengthening the support for cotton specialized cooperation organization, speeding up the pace of mechanization in the whole process of cotton production, increasing the subsidy for cotton production, and further improving the subsidy standards for the promotion of fine cotton varieties.

Karlı et al. (2017) conducted a cost and profit analysis in cotton production in Şanliurfa province, Turkey. The proper management of land, the appropriate seed variety, the use of adequate amounts of fertilizer, the correct application of the irrigation, and the effective control of plant diseases are important factors for sustainability.

Copur et al. (2017) observed that in Turkey, cotton cultivation is confined to three regions Southeastern Anatolia, the Mediterranean region, and Aegean regions. The researchers observed that cotton varieties that were grown in each of these regions were dissimilar owing to the prevalence of variable ecological environmental conditions. In their article, Copur et al. (2017) examined its production in Turkey and also the impact of Southeast Anatolia project (SAP) pertaining to suitable production technologies of farming, soil and environmental conditions suited for its production, etc.

Lu et al. (2017) undertook a review of the technology for high-yielding and efficient cotton cultivation in the northwest inland cotton-growing region of China. They suggested that there is a requisite for optimization of strategies involved in cotton cultivation in that region, if the heat and water potentials are explored for better utilization in designing of tolerant cultivars. Similarly, there could be an improvement in economic benefits and quality fiber production, if fertigation technologies, agronomic techniques are integrated and mechanized to a large extent. The researchers suggested that efficiencies can be improved, with reduced costs of production when much attention and care is oriented toward the processes involved in production of better quality seeds with good and improved seedling traits. Additionally, they suggested that comprehensive steps should be taken into consideration at all stages of production and improvement of cultivation systems so as to bring about the better quality fiber production in Xinjiang. Furthermore, the foreign subsidies policies must be enlightened to a larger degree to increase the cotton production in Xinjiang.

Feng et al. (2018) promoted cotton green production in Shandong through accelerating simplified cultivation technique. They noted that the fundamental reason was that the quality of cotton production development and the developments made were not suitable for the market demands. Furthermore, they suggested that cotton production and development must take the road of green development concept as the guidance, and improve the quality and benefits of cotton development. Finally, the researchers came up with the basic idea that technical approach and industrial mode of taking simplified cultivation technique should be adopted to support the promotion of green and efficient cotton production in Shandong Province.

2.5 CONCLUSIONS

Cotton (whose cultivation requires moderate rainfall) is cultivated in different countries for its high textile values of great demand of the world. China is reported to be the world's leading producer of cotton, followed by India. Favorable climates have been noted to lead to increased crop productivity. Water deficit affects crop production, and insects have been found to pose a great menace, reducing productivity drastically. Cotton requires heavy insecticides to protect it from insect devastation. Although the discovery of Bt cotton led to a boost in cotton production, its cultivation is restricted in some countries. Variation of cotton yields in different countries has been found to be due to climate change adoption of mitigation methods of green house gas emissions. Various factors such as cultural, technological, climatic, biotic factors affect cotton production. Innovative technologies have been developed to increase cotton productivity.

KEYWORDS

- **cotton**
- **yield**
- **global**
- **factors affecting yield**
- **abiotic factors**
- **insects**
- **diseases**

REFERENCES

Agarwal, I.; Reddy, A. R.; Singh, S.; Yelekar, S. M. Yield Gap and Constraints Analysis of Cotton in India. *J. Cotton Res.* **2015,** *29*, 333–338. (http://www.crdaindia.com/fileserve.php?FID=159).

Beckert, S. *Emancipation and Empire: Reconstructing the Worldwide Web of Cotton Production in the Age of the American Civil War. The American Historical Review*, 2004, 109, 1405–1438.

Chapagain, A.; Hoekstra, A. Y.; Savenije, H. H. G. The Water Footprint of Cotton Consumption: An Assessment of the Impact of Worldwide Consumption of Cotton Products on the Water Resources in the Cotton Producing Countries. *Ecol. Econ.* **2006,** *60*, 186–203.

CIRAD. What Impact Will Climate Change Have on Cotton Yields in Cameroon Agricultural Research for Development, 2013.

Copur, O.; Vila, S.; Antunovie, Z. Cotton production in Turkey; hrvatskii 12 medunarodni simpozij afronoma, Veljace 2017, Dubrovnik, Hrvatska, Zbornik radova, 2017.

Deguine, P. J.; Ferron, P.; Russell, D. Sustainable Pest Management for Cotton Production. A Review. *Agron. Sustainable Dev.* **2008,** *28*, 113–137.

Diarra, A.; Barbier, B.; Zongo, B.; Yacouba, H. Impact of Climate Change on Cotton Product in Burkina. CIRAD, UMRG.

Fatima, H.; Khan, M. A.; Zaid-Ullah, M.; Abdul-Jabbar; Saddozai, K. N. Technical Efficiency of Cotton Production in Pakistan: A Comparative Study on Non Bt and Bt-Cotton Farms. *Sarhad J. Agric.* **2016,** *32*, 267–274.

Feng, L.; Dai, J.; Tian, L.; Zhang, H.; Li, W.; Dong, H. Review of the Technology for High-Yielding and Efficient Cotton Cultivation in the Northwest Inland Cotton-Growing Region of China. *Field Crops Res.* **2017,** *208*, 18–26. (http://www.sciencedirect.com/science/journal/03784290).

Kadam, M. M.; Wagh, H. J.; Lamtule, J. A. Input Management in Cotton Production: A Production Function Analysis. *Int. Res. J. Agric. Econ. Stat.* **2014,** *5*, 216–219. (http://www.researchjournal.co.in/online/irjaes.htm).

Karlı, B.; Kart, F. M.; Gul, M.; Akpınar, M. G. Cost and Profit Analysis in Cotton Production in Şanliurfa Province, Turkey. Management. *Econ. Eng. Agric. Rural Dev.* **2017,** *17*, 207–220. (http://managementjournal.usamv.ro/pdf/vol.17_2/volume_17_2_2017.pdf).

Ma Na. *Reason Analysis and Countermeasures on Cotton Planting Area Decline in Henan Province*, Proceedings of the Annual Meeting of China Society of Cotton Sci-Tech in 2016, Jiangsu, China, 8–9 August, 2016, pp. 190–191.

Mahanty, S.; Fang, C.; Chaudhury, J. Assessing the Competitiveness of Indian Cotton Production: A Policy Analysis Matrix. *Res. Agric. Appl. Econ.* **2002,** *2*, 301.

MingYun, L.; QingNian, L.; XiaoHong, G.; Ren Na. *Analysis of Trends and Development Suggestion on Cotton Production in Binzhou of Shandong Province, China on Next Five Years*, Proceedings of the China Cotton Society in 2014, Inner Mongolia, China, 7–9 August, 2014, pp 268–272. Monthly Economic Letter, May 2019. Cotton Market Fundamentals & Price Outlook. https://lifestylemonitor.cottoninc.com/wp-content/uploads/2018/05/2018-05-Monthly-Economic-Letter.pdf

Oerke, E. C. Crop Losses to Pests. *J. Agric. Sci.* **2006,** *144*, 31–43.

Oerke, E. C.; Dehne, H. W. Safeguarding Production-Losses in Major Crops and the Role of Crop Protection. *Crop Protect.* **2004,** *23*, 275–285.

Pray, E. C.; Huang, J.; Hu, R.; Rozelle S. Five Years of Bt Cotton in China—The Benefits Continue. *Plant J.* **2002,** *31*, 423–430.

Rao, S. C.; Blaise, D.; Venugopalan, M. K.; Patel, K. P.; Biradar, D. P.; Aladakatti, Y. R.; Marimuthu, S.; Buttar, G. S.; Brar, M. S.; Kumari, S. R.; Reddy, V. C. Soil Fertility Management Strategies for Maximising Cotton Production of India. Fertiliser Association of India, New Delhi, India. *Ind. J. Fertilisers* **2012,** *8*, 80–95 (http://www.faidelhi.org).

Sabo, E.; Daniel, J. D.; Adeniji, O. T. Economic Analysis of Cotton Production in Adamawa State, Nigeria. *Afr. J. Agric. Res.* **2009,** *4*, 438–444. (http://www.academicjournals.org/ajar/PDF/pdf%202009/May/Sabo%20et%20al.pdf).

Silvie, P.; Martin, J.; Debru, J.; Vaissayre, M. Organic Cotton Production in Paraguay. 2. Agronomic Limitations for a Novel Industry; Les Presses Agronomiques de Gembloux, Gembloux, Belgium. Biotechnologie. *Biotechnol. Agron. Soc. Environ.* **2010,** *14*, 311–320. (http://www.bib.fsagx.ac.be/base).

Wang, G.; Wei, X.; Wang, Y.; Dong, W. Promoting Cotton Green Production in Shandong Through Accelerating Simplified Cultivation Technique. *Asian Agric. Res.* **2018,** *10*, 1–4.

Yilmaz, I.; Akcaoz, H.; Ozkan, B. An Analysis of Energy Use and Cotton Production in Turkey. Faculty of Agriculture Department of Agricultural Economics, University of Akdeniz, Turkey, 2004.

Zhang, N.; Wang, M.; Wang, N. Precision Agriculture—A Worldwide Overview. *Comput. Electron. Agric.* **2002,** *36*, 113–132.

CHAPTER 3

Origin, Evolution, and Domestication

ABSTRACT

Cotton, *Gossypium* spp., is a plant of ancient origin, which has evolved and been domesticated millions of years ago. This chapter deals with origin, evolutions, and domestication of cotton. It discusses research advances on the origin, evolution, and domestication of cotton based on phylogeny, cytology, and molecular levels. Sufficient research inputs have been directed on these important aspects.

3.1 INTRODUCTION

The origin, evolution, and domestication of crop plants date back to millions of years for different crops. Similarly, cotton, *Gossypium* species is a plant of ancient origin, which has evolved and been domesticated millions of years ago.

Many studies have been undertaken on the origin, evolution, and domestication of cotton in different countries.

This chapter presents the origin, evolution, and domestication of cotton and a few other crop plants.

3.2 ORIGIN

A few studies have been undertaken on the origin, genetic diversity, phylogeny, and cytological evidences of cotton. The origin of a crop species occurs in regions of maximum of diversity of the species.

Wendel (1989) reported that the present day new World tetraploid cottons (*Gossypium* spp.) originated from the ancestral diploid species, which had allopatric ranges in Asia–Africa (the *A* genome) and the New

World tropics and subtropics (the *D* genome). It is recognized that the new world tetraploid cotton originated through the process of hybridization of above ancestral diploid species. In spite of intensive studies conducted, parental identity of diploids, distant past polyploidization in cotton was not resolved completely. Wendel (1989) determined variations in the chloroplast genome which was maternally inherited in few species genomes, namely, parental genomes and tetraploids. An assessment was undertaken with 560 restriction sites in each of the accession, which revealed sequence information for about 3200 nucleotides. It was observed that the resulting maternal phylogeny did not have convergent restriction site mutations, which suggested that *A* genome diploid, had a chloroplast genome, was more identical to *Gossypium arboreum* and *Gossypium herbaceum*. The genome might be the cytoplasm donor for most of the tetraploid cotton species. Mutational differences between these two species were not detected, although a few mutations existed which enabled in distinguishing the chloroplast genomes of *A* genome diploids, from those of tetraploid taxa. On the contrary, they specified that even though there was an expected extensive diversity at the geographic, taxonomic, and genetic levels, very low/little of sequence divergence might have been accumulated, succeeding to polyploidization. It was observed that six apparent point mutations that occurred between one and six enabled in distinguishing the chloroplast genomes of each of these tetraploid species. Thus, through their research findings they suggested that the tetraploid cottons originated fairly quite recently, may be over the last 1–2 million years. They have undergone succeeding speedy evolutions and diversification throughout the areas of New World tropics.

DeJoode and Wendel (1992) undertook a study on the genetic diversity and also on origin of the Hawaiian Islands Cotton, *Gossypium tomentosum*, which has been known to be the only member endemic to the region. Though this species was distinct morphologically with other allopolyploid *Gossypium* species, its phylogenetic relationships with allopolyploid *Gossypium* species was not known. Variations in the chloroplast and ribosomal DNA restriction sites were taken into consideration in the determination of the phylogeny of the allopolyploids. *Gossypium mustelinum* is considered as sister to the leftover/residual allopolyploid species, including two species-pairs, *G. barbadense*-*G. darwinii* and *G hirsutum* (including *G. lanceolatum*)-*G. tomentosum*. The researchers concluded that *G. tomentosum* was sister to *G. hirsutum*. The hypothesis of the research was further supported by allozyme analysis. On the basis of biogeographic evidence and molecular data, DeJoode and Wendel proposed that *G. tomentosum* originated by transoceanic dispersal from a Mesoamerican progenitor. Among the allopolyploids they have detected were a few restriction site variants, suggesting that the there was a rapid divergence in current lineage following the process of polyploidization. Furthermore, the researchers undertook allozyme analysis of 30 *G. tomentosum* accessions that were obtained from seven islands. The results revealed the existence of relatively minor levels of genetic diversity among those 30 accessions. It was observed further that 11 of 50 loci were polymorphic, with a mean number of alleles per locus as 1.24, and a mean panmictic heterozygosity of 0.033. Allelic distributions exhibited very little of geographic patterning. In spite of the presence of ancient historic cultivation of *G. barbadense* and *G. hirsutum* in Hawaii, the evidence for their naturalized derivative presence or interspecific introgressions into *G. tomentosum* was observed. No allozyme analysis yet could give information of these interspecific introgressions.

Subsequently, Wendel et al. (1994) investigated the genetic diversity and phylogenetic relationships of the Brazilian endemic cotton, *G. mustelinum* (Malvaceae), and observed that *G. mustelinum* was one of five tetraploid species in the cotton genus whose distribution in NE Brazil was limited geographically to only a few states. They used allozyme analysis for assessing the intensities and types/patterns of genetic diversity in *G. mustelinum* and also their relationship with rest of the tetraploid species. Their study revealed that there existed low genetic variation. Furthermore, only 6 of 50 loci tested were found to be polymorphic, having a mean of 1.14 alleles per locus. There was a panmictic heterozygosity of 0.08. The researchers noted that although those estimates were low compared with other tetraploid cotton species, the estimates were typical to endemic species in this island. Uniformly high inter-populational genetic identities were observed. This supported the concept of presence of only one wild species of Brazilian cotton. The study revealed a correlation between the limited allelic diversity and geographical distribution, and that, in spite of the facts of limited variability, in the species, the populations confined to geographic marginal regions were found to be ordinary in nature, when detected electrophoretically. Furthermore, phylogenetic and phenetic analyses revealed that *G. mustelinum* was isolated and separated from other polyploid cotton species, which occupied one of the three basal clades, derived from an early initial radiation of polyploid taxa much later to the formation of polyploid species. The researchers suggested that *G. mustelinum* represented a paleoendemic species that exhibited widely scattered series of relictual populations. They observed that although there were many centuries of sympatric cultivation of *G. barbadense* and *G. hirsutum*, there was only a minor evidence of alleles of interspecific introgressions from cultivated cotton types into *G. mustelinum*.

3.3 EVOLUTION

Concerted research activities have been directed to determine evolution of cotton on the basis of archeological, cytological, and molecular levels. Some of the activities are presented in this section.

Ladizinsky (1985) who reported the founder effect in crop-plant evolution observed that seed-crop plants apparently have had their origin owing to the existence of sparse mutants. In these, there was a change in the mode of seed dispersal, which was quite different to that of populations which remained none domesticated. Furthermore, they noted that a single gene or a meager number of genes might control the seed non-shattering habit of the cultivated plants, and that the allopolyploid crop plants in the process were the derivatives of a limited number of interspecific hybridizations, these were also observed by chromosome doubling. The research observed a narrow genetic variability in the crop populations in comparison to its wild progenitor, as a result of this founder effect, and that during the process, various isolating mechanisms, and possible gene flow could inhibit the natural hybridization between the two, which seemed to be effective to a large extent, apparently in the direction from the cultivated to the wild populations. Thus, the research concluded that the founder effect in crop-plant evolution exhibited the value. It has also evidenced the presence of breeding potential that is seen in the form of the genetic variability that remained in others and its wild relatives.

Iqbal et al. (2001) used DNA fingerprinting to detect a genetic bottleneck in the evolution and underdomestication of upland cotton *G. hirsutum* L. They observed that it was essential to get accurate and reliable information of the evolutionary and genetic relationships of multiple various resources of germplasm to enable the establishment of key rational strategies, aiming for crop improvement at faster pace. They determined the genetic relationships of 43 cultivars of *G. hirsutum,* which represented the genomic composition of modern "Upland" cotton using AFLPs. They also involved some of the representatives of the related tetraploid species *G. barbadense*, and the diploid species *G. raimondii*, *G. incanum*, *G. herbaceum,* and *G. arboreum*. After testing 20 primer combinations, they obtained a total of 3178 fragments. Genetic similarities based on AFLPs which were also in agreement with the known taxonomic relationships at the species level and above were observed. The research revealed that similarity indices were in the range of 0.25–0.99. The representatives of germplasm resources of the *G. hirsutum* of North America, and also the secondary accessions, which were collected by breeders from Central America, namely, "Acala," "Tuxtla," "Kekchi" and those of the southwestern US ("Hopi Moencopi"), were grouped in a single cluster, which showed a very exceedingly limited genetic diversity. The pairwise similarity indices were many representing >0.96. It was concluded that the germplasm accessions were the derivatives of the same genetic pool. The researchers observed that the cultivars which were derived from early maturing or "latifolium" "Mexican Highlands" cultigens also had limited genetic diversity, probably owing the presence of a severe genetic bottleneck and result of the selection pressures of domestication. Above all, majority of *G. hirsutum* cluster support quite well the phylogenies of these accessions. Phylogenies were not defined well within the cluster, owing to limited diversity, lineage sorting, and reticulation. They discussed the importance of their findings to bring effective cotton improvement.

Later, Moulherat et al. (2002) observed the first evidence of cotton at Neolithic Mehrgarh, Pakistan based on analysis of mineralized fibers obtained from a Copper Bead from a Neolithic burial (6th millennium BC). They recovered several threads that were preserved by the process of mineralization, and noted that the mineralized fibers were characterized by a new procedure which involved the usage of both light reflected microscope and a scanning electron microscope. They identified that these fibers were that of cotton (*Gossypium* sp.). The Mehrgarh fibers represented the earliest known cotton of the Old World. It also has given the information about its first use date which is dated back to a millennium ago. The researchers observed that it was not possible that the fibers were derived from an already domesticated species, and ascertained that their evidences suggested an early origin, which might possibly in the Kachi Plain, representing one of the Old World cottons.

Grover et al. (2004) investigated the sizes of global and local genomes in their patterns evolutions in cotton, and observed significantly large variations in the genome sizes in cotton, which was attributed to active local and global forces combinations. These combinations involved some accumulations, removals or proliferations of some transposable elements. In order to unravel this mystery, the researchers compared to ~105 kb of contiguous sequence, which was found to surround the cellulose synthase gene *CesA1* with the two coresident genomes (A_T and D_T) of the allopolyploid cotton species, *G. hirsutum*. They observed that both of the genomes showed a two-fold difference in their size, which showed a divergence from a common ancestor that existed ~5–10 million years ago (Mya). The two genomes might have reunited in the same nucleus, at the time of formation of polyploids of ~1–2 Mya. Thereby, gene contents, its sequence, order, and spacing were highly conserved between these two genomes of this cotton species; however, a few transposable elements and a single cpDNA fragment can differentiate the two homologs. Thereby, they mentioned that there is high conservation of sequence, which might be possibly being present at regions of both genic and intergenic sequence regions. In this manner, they detected 14 conserved genes in both genomes, yielding with a density of 1 gene every 7.5 kb. Furthermore, researchers could not trace out any kind of disparity in size for the 105 kb region. In the DNA content, there was a difference of two folds. Consequently, they detected 555 indels that could differentiate the two homologous BACs, which were almost equally distributed between A_T and D_T genomes. These were equal in aggregate size and number during their distribution. The investigation revealed that the evolution of genome size occurring at this phylogenetic scale was not primarily derived by mechanisms that were operating uniform pattern all across the varied genomic regions and also the components. Also, the researchers observed that an overall two-fold difference that was observed in the content of DNA demonstrated the presence of the forces operating locally between these gene islands or across the large gene-free regions.

Wende et al. (2009) assessed the evolution and natural history of diversity of the Cotton Genus, *Gossypium*. Their overview helped to give a framework which facilitated insight about the fundamental aspects of plant biology and provided clues for the effective utilization of these genetic resources of cotton. Furthermore, it has paved guidance to explore the diverse genomes, based on diversities observed morphologically.

They mentioned that 50 species of *Gossypium* or more were found to be distributed in arid to semi-arid regions of the tropics and subtropics of the world, which involved the four species of *Gossypium* that were independently domesticated for the production of fiber. The researchers observed that among those, two each of the Africa–Asia and the Americas *Gossypium* species exhibited large variations in morphological traits which were trailing, and few were herbaceous perennial trees which could reach a height of ~15 m. Furthermore, there were also variations in features of vegetative and reproductive characteristics. The research revealed that a parallel level of cytogenetic and genomic diversity arose. This might have arisen during the global radiation of the genus, and led to the evolution of eight groups of diploid (n = 13) species (genome groups *A* through *G*, and *K*). The researchers, who observed an origin for *Gossypium* about 5–10 million years ago, followed by a rapid and quick diversification of all the major genome groups, opined that allopolyploid cottons appeared to have derived during the last 1–2 million years, as a result of the trans-oceanic dispersal of an *A*-genome taxon to the New World, and followed by hybridization with an indigenous *D*-genome diploid. They concluded that allopolyploids thus had radiated into three modern lineages, two of which belonged to the commercially important species *G. hirsutum* and *G. barbadense.*

Polyploidy plays a vital role in the evolution and contribution of fiber quality in spinnable cotton. Paterson and Wendel (2012) examined the occurrence of repeated polyploidization of *Gossypium* genomes and the evolution of cotton, and observed that polyploidy has often contributed to the important properties which led to the higher fiber productivity and of better quality of tetraploid cottons in comparison to the diploid cottons which were bred for planting in the same environments. The authors demonstrated that an abrupt five- to six-fold ploidy increase occurred approximately 60 million years (Myr) ago. This allopolyploidy combined divergent *Gossypium* genomes approximately 1–2 Myr ago and thereby could confer that about 30–36-fold duplication of ancestral angiosperm (flowering plant) genes were accumulated in elite cottons (*G. hirsutum* and *G. barbadense*). The researchers note that prior to occurrence of polyploidy, the evolution of nascent fiber was exhibited by comparing spinnable fiber of *G. herbaceum A* and non-spinnable *G. longicaly F* genomes to one another and the out group *D* genome of non-spinnable *G. raimondii*. They reported that the sequence of a *G. hirsutum* A_tD_t, a tetraplod cultivar, exhibited the occurrence of many non-reciprocal DNA exchanges between the subgenomes, which might have contributed largely to the phenotypic innovation and/to other emergent properties, thereby imparting ecological adaptation by polyploids. As a consequence, the novelty of most DNA-level novelty in *G. hirsutum* could recombine alleles from the D-genome progenitor which was thought to be native to its New World habitat and the spinnable fiber, might have its origin from the Old World *A*-genome progenitor. Furthermore, the researchers observed that the occurrence of coordinated expression changes in proximal groups of functionally distinct genes, involving a nuclear mitochondrial DNA block, could lead to clusters of cotton fiber quantitative trait loci which affect diverse traits.

Page et al. (2013) undertook a study on the evolution of cotton diploids and polyploids obtained from Whole-Genome Re-sequencing G3, and observed that it was complicated to understand the composition, evolution, and function of the *G. hirsutum* (cotton) genome owing to the occurrence in its nucleus the presence of two genomes (A_T and D_T genomes). Page et al. further noted that the two genomes were derivatives of progenitor *A*- and *D*-genome diploids which were involved in ancestral allopolyploidization. The authors undertook resequencing of the genomes of extant diploid relatives containing the A_1(*G. herbaceum*), A_2 (*G. arboreum*), or D_5 (*G. raimondii*) genomes, carried out a comparative analysis by using deep re-sequencing of multiple accessions of each diploid species, and identified 24M SNPs between the *A*- and *D*-diploid genomes. Their analyses helped them construct a robust index of conserved SNPs specially between the *A*- and *D*-genomes at all detected polymorphic loci. The researchers asserted that the index had its wide applicability in the undertaking of the mapping efforts of other diploid and allopolyploid *Gossypium* accessions. Furthermore, their analysis demonstrated the locations of putative duplications and deletions in the *A*-genome relative to the *D*-genome reference sequence. They also observed that the ~25,400 deleted regions included >50% deletion of 978 genes, many of which were thought to have a role in starch synthesis. They also detected in the polyploid genome, between homologous chromosomes, the conversion events (1472). They also included in the polyploid genome the events that have overlapped at least 113 genes. The researchers suggested that continued characterization of the *Gossypium* genomes would further increase their capacity for the manipulation of fiber and improving agronomic production of cotton.

3.4 DOMESTICATION

Many studies have been undertaken on the process of domestication of cotton based on archeology, cytology, and molecular levels.

Curt et al. (1994) undertook a study on the re-evaluation of the origin of domesticated Cotton (*G. hirsutum*; Malvaceae). They used nuclear Restriction Fragment Length Polymorphisms (RFLPs) for this evolution study. They noted that the origin of domestication of Mesoamerican *G. hirsutum* populations was highly obscured, and was attributed to many factors, namely, lack of properly identified wild progenitor, complexity of genetic structure of the population, and also the human intervention for several centuries for dispersal and gene flow of this species. In order to unravel the mystery, the authors undertook phenetic and the phylogenetic analyses of allelic variation at 205 restriction fragment length polymorphism (RFLP). The RFLP data obtained in their investigation, in conjunction with that of the previously published information on molecular, morphological, and anthropological data, suggested that coastal Yucatan populations were considered as really wild. They were the re-established feral derivatives. The researchers observed that the geographical origin of these wild coastal populations to agronomically primitive forms of *G. hirsutum* revealed that the Yucatan peninsula was considered as the primary site for the earliest initial stages of domestication of *G. hirsutum*, and presumed that cultivars which were advanced agronomically have been developed in southern Mexico and Guatemala. These were thought to be derived through some introductions of populations in the Yucatan peninsular forms. This has turned out to be seen as secondary center of diversity. Traditionally, this was also considered as the origin from geographical point of view, of domesticated *G. hirsutum*. Therefore, the researchers suggested that the gene pool of the modern, improved (Upland) cultivars had been derived from the populations of Mexican highlands, which, in turn, had their traces of origins to the southern Mexico regions and Guatemala. *G. hirsutum* is the first perennial tetraploid tested for the RFLP variations. The research revealed that the levels of RFLP variation in *G. hirsutum* (HT = 0.048, A = 1.24, and P = 22%) were low, in comparison to those of other plant taxa. The allozyme variation levels were greater than the levels of RFLP variations. It was also revealed that in spite of assaying 205 loci, only six of the 23 Upland cultivars were found to have unique multilocus genotypes.

Diamond (2002) observed that domestication was the most momentous change in human history of Holocene era. He questioned its operation in only a few wild species, its restriction to few geographical areas and the reasons of its adoption by human beings and its spread, and observed that answers of those questions revealed that modern cultivars were remade, due to spread of farmers, hunter–gatherers, and others.

Westengen et al. (2005) used amplified fragment length polymorphism fingerprinting to determine the genetic diversity of primitive South American *G. barbadense* cotton. They tried to explore a link that would give an indication of expansion to pre-Columbian era. They also collected new germplasm along coastal Peru and over an Andean transect in areas where most of the archaeological evidence relating to cotton domestication were found, and added to gene bank three diploid (*G. raimondii*, *G. arboreum*, and *G. herbaceum*) and four allotetraploid cotton species (*G. hirsutum*, *G. mustelinum*, *G. tomentosum,* and additional *G. barbadense*) for inter- and intra-specific comparison. The researchers observed that among the 131 accessions that were evaluated, 340 polymorphic bands could be produced from the eight primer combinations, and that the resulting neighbor and unweighted pair-group method with arithmetic means tallied with the known cytogenetics of the tetraploid cottons and their diploid genome donors. The study revealed that the four tetraploid species were distinct by taxonomic classification, and that the genetic diversity within *G. barbadense* showed geographic patterns. Furthermore, it was observed that the locally/regionally maintained cottons from regions of coastal Peru displayed a distinct genetic diversity. It reflected also the primitive agro-morphological traits, and accession from the northernmost coast of Peru and from southwestern (SW) Ecuador cluster basal to the east-of-Andes accessions and the remaining accessions from Bolivia, Brazil, Columbia, Venezuela, and the Caribbean and Pacific islands cluster with the east-of-Andes accessions. Northwestern Peru/SW Ecuador (the area flanking the Guayaquil gulf) appears to be the center of the primitive domesticated *G. barbadense* cotton from where it spread to other regions.

Hovav et al. (2008) assessed domestication, convergent evolution, and duplicated gene recruitment in allopolyploid cotton, and observed that the advantage of allopolyploidy gave the possibility of differential selection of duplicated (homologous) genes which had their origin from two different progenitor genomes. The researchers explored this hypothesis by the use of a SNP-specific microarray technology. It was applied to seed trichomes (cotton). These were harvested from three developmental time points in wild and modern accessions of two independently domesticated cotton species, *G. hirsutum* and *G. barbadense*, and showed that homeologpod expression ratios were dynamic both in developmentally and over the 1000-year period, which was encompassed by domestication and crop improvement. The domestication could have increased the modulation of homeologous gene expression. Furthermore, they detected that I-in both species, *D*-genome was preferentially expressed and its increase was found to be more with human selection pressure. Their data suggested that human selection might have operated on a number of varied genetic program involved in fiber development in *G. hirsutum* and *G. barbadense*, which in turn led to convergent alterations, rather than the parallel alterations in genetics and plant morphology.

3.5 CONCLUSION

Many research activities have been directed on origin, evolution, and domestication of cotton in different countries on the basis of archeological, phylogeny, cytological, and molecular levels. Different theories are put forth by different authors on these aspects. Different studies discussed the origin of cotton, the process of evolution, polyploidization, hybridization giving rise to origin of different species and gradual process of domestication, thereby giving rise to new world cotton. The evolution of modern polyploid upland cotton was derived from the result of hybridization and polyploidization of ancestral diploid cotton.

KEYWORDS

- **cotton**
- **origin**
- **evolution**
- **phylogeny**
- **cytology**
- **polyploidy**
- **molecular biology**

REFERENCES

Curt, L.; Wendel, J. F. Reevaluating the Origin of Domesticated Cotton (*Gossypium hirsutum*; Malvaceae) Using Nuclear Restriction Fragment Length Polymorphisms (RFLP). *Am. J. Bot.* **1994,** *81*, 1203–1208.

DeJoode, D. R.; Wendel, J. F. Genetic Diversity and Origin of the Hawaian *Gossypium tomentosum*. *Am. J. Bot.* **1992,** *79*.

Diamond, J. Evolution, Consequences and Future of Plant and Animal Domestication. *Nature* **2002,** *8*, 700–710.

Grover, C. E.; Kim, H. R.; Wing, R. A.; Paterson A. H.; Wendel, J. F. Incongruent Patterns of Local and Global Genome Size Evolution in Cotton. *Genome Res.* **2004,** *14*, 14.

Hovav, R.; Chaudhary, B.; Wendel, J. F. Parallel Domestication, Convergent Evolution and Duplicated Gene Recruitment in Allopolyploid Cotton. *Genetics* **2008,** *179*, 1725–1733.

Iqbal, M. J.; Reddy, O. U. K.; El Zik, K. M.; Pepper, A. E. A Genetic Bottleneck in the 'Evolution Under Domestication' of Upland Cotton *Gossypium hirsutum* L. Examined Using DNA Fingerprinting. *Theor. Appl. Genet.* **2001,** *103*, 547–554.

Ladizinsky, G. Founder Effect in Crop- Plant Evolution. *Econ. Bot.* **1985,** *39*, 191–199.

Moulherat, C.; Tengber, M.; Haquet, J-. F.; Mille, B. First Evidence of Cotton at Neolithic Mehrgarh, Pakistan: Analysis of Mineralized Fibres from a Copper Bead. *J. Archaeol. Sci.* **2002,** *29*, 1393–1401.

Page, J. T.; Huynh, M. D.; Liechty, Z. S.; Grupp, K.; Stelly, D. M.; Hulse, A. M.; Ashrafi, H.; Van Deynze, A.; Wendel, J. F.; Udall J. A. Insights into the Evolution of Cotton Diploids and Polyploids from Whole-Genome Re-sequencing. *G3: Genes Genomes Genetics* **2013,** *3* (10), 1809–1818.

Paterson, A. H.; Wendel, J. F. Repeated Polyploidization of *Gossypium* Genomes and the Evolution of Spinnable Cotton Fibres. *Nature* **2012,** *492*, 423–427.

Wendel, J. F. New World Tetraploid Cottons Contain Old World Cytoplasm. *Proc. Natl. Acad. Sci. USA* **1989,** *86*, 4132–4136.

Wendel, J. F.; Rowley, R.; Stewart, D. J. Genetic Diversity in and Phylogenetic Relationships of the Brazilian Endemic Cotton, *Gossypium mustelinum* (Malvaceae). *Plant Systematics Evol.* **1994,** *192*, 49–59.

Westengen, O. T.; Huamán, Z.; Heun, M. Genetic Diversity and Geographic Pattern in Early South American Cotton Domestication. *Theor. Appl. Genet.* **2005,** *110*, 392–402.

CHAPTER 4

Cotton Ideotype

ABSTRACT

Cotton is an important fiber yielding crop of high economic returns. An ideal ideotype characteristic feature in the crop varieties helps in the capture and interception of the solar radiation, more number of fine quality boll production with good quality fiber and yield. This chapter briefly highlights some of the research activities carried out in the development of ideotype concepts and characters in cotton suitable for different environmental conditions and planting densities.

4.1 INTRODUCTION

The productivity of a crop is dependent upon many plant characteristics, which to a large extent contribute to high yields. Among the various traits, the traits of leaf characteristics, absorbance of solar energy, stomatal conductance of carbon dioxide, assimilation and partitioning ability to economic parts, mechanism of stress tolerance, etc. influence the productivities that are realized in crops in various environments. Apart from this, it is important that the crop plants do have certain traits such as presence of trichomes, hairs, waxy coating, deep root system, etc., to enable them withstand certain stress conditions that are encountered during the growth and development of the crop and be able to produce the products effectively. Different concepts put forward by different authors about ideotypes of a crop plant including cotton are presented in this chapter.

4.2 PLANT IDEOTYPE CHARACTERS

An extensive survey of cotton germplasm accessions revealed a large variability in plant and reproductive characters as well as in boll and cotton fiber characteristics which are given in brief (Maiti and Vidyasagar, 2009).

4.2.1 PLANT TYPE CHARACTERS

Plant characters show great variations offering a great opportunity for the breeders for the selection of ideal plant/ideotype. Some of the characters are stated below:

1. Compact medium plant
2. Adaptive agronomic plant types
3. High sympodial branching
4. Early maturity
5. Plant canopy
6. Less vegetative growth
7. Strong erect stem
8. Short branched and cluster fruiting branches for gaining earliness and production advantage
9. Glabrous leaf against whitefly and boll worm incidence or whitefly induced leaf curl
10. Less gossypol glands types
11. Higher leaf area index
12. Long pedicel
13. Synchronous flowering habit
14. Faster flowering
15. Self-incompatibility
16. Insect pollination promoting mechanism like dark yellow corolla and dark yellow anthers in both parents and hybrids in cross pollination
17. Short duration
18. High partitioning efficiency of photosynthetates to bolls
19. High biomass potential
20. High harvest index
21. Resistance to insect diseases and drought and salinity
22. Wilt and rust resistance
23. Stable yield

4.2.2 BOLL CHARACTERS

1. High boll production potential
2. Increase in the boll number per plant
3. Evenness in boll development
4. High boll retention ability
5. Retention of early formed bolls
6. Full development of bolls
7. Moderate to big boll size
8. Increased boll weight
9. Low shedding, long stable
10. Nectarless bolls
11. Late shedding of square
12. An increase in the locule number
13. Non-shedding ability
14. Delayed morphogenesis of gossypol glands

4.2.3 SEED CHARACTERS

1. High seed per boll
2. Reduction in seed coat fragments
3. Presence of oil quality tracks
4. Increased seed oil and protein content
5. Glandless seed: Seed glandlessness is useful in introducing delayed morphogenesis of gossypol glands to get advantages of protection from bollworm and gossypol free oil protein
6. Naked seed types
7. Reduced motes

4.2.4 FIBER CHARACTERS

Fiber characters include the following:

1. High coarse absorbent types
2. Increased fiber length
3. Increased fiber intensity
4. High fuzzy types

5. High fiber strength
6. High fiber fineness
7. Long strong fiber
8. High fiber extensibility
9. Lower strength of attachment of fibers on seed
10. Higher uniformity of fiber length with lowest short fiber content
11. Medium staple length
12. Uniformity in fiber maturity
13. Color fastness of lint
14. Naturally colored cotton
15. Fineness in quality
16. High ginning ability

Generally, in a survey of cotton germplasm, an ideal ideotype suitable to high input situations, favorably should possesses an open leaf canopy, stout strong stem, medium internodal length, big sized bolls, more number of boll production, sympodial branches of medium sized in length deep rooted tap root. For adaptation to semiarid situations, the plants should have short stature with thick leaves, high density of trichomes, or high degree of glossiness. The presence of trichomes and glossiness reduce transpiration loss for adaptation under drought-prone areas.

An extensive survey conducted in evaluation of characteristics of cotton accessions and germplasm lines also supported the above concepts and it was found that huge variations in morphological, physiological, and biochemical traits exist among the germplasm lines, wild relatives, accessions, varieties, hybrids, etc. Most of the Egyptian and Greece accessions of the genotypes have a crowded and profusely overlapped leaves. They also bear bolls which are small in size. In contrast, genotypes with nonoverlapping leaves and with open leaf canopies exhibit greater penetration and interception of light deep into the canopy and produce large sized bolls. Most of the high yielding cultivars possess these characteristics. Cotton genotypes also show great variation in trichome morphology and density and leaf glossiness, which could be related to adaption under drought situations and tolerance to insect pests. The high yielding cultivars should have high number of short internodes to bear more bolls and also should have strong stem/mechanical tissue to support heavy load, otherwise it will lodge causing yield loss (generally observed). These cultivars should have deep root system for adaptation under drought situations. Very little attention has been directed in this direction. Therefore, cotton breeders should verify the feasibility of these concepts for breeding high yielding cultivars (Maiti hypothesis).

Various studies have been undertaken to put forward ideotype concepts of cotton.

NC (North Carolina) State assessed the importance of the development of a variety of cotton having an ideal ideotype of least structure, which helps in the enhancement of the increased boll production, its retention, yield, rot resistance, and penetrance to insecticidal sprays and other cotton characteristics. They mentioned that in cotton plants four types of leaf shapes are generally seen, namely, normal, sub-okra, okra, and super-okra. The type of leaf shape that is expressed on a cotton plant depicts its characteristics and tolerance and resistance levels to biotic and abiotic stresses.

They have identified the key genes contributing to leaf shape in cotton and have devised a method for manipulation of these genes to create a cotton leaf ideotype. This ideotype has an ability to produce normal leaves at the bottom of the canopy. The leaves help in the rapid closure of canopy, maximum light absorption, increased nutrient supply, and boll retention. The ideotype is then transformed to okra leaves on the top canopy. This in turn induces earlier flowering, enhanced increased boll production, increased chemical spray penetration, reduced incidence of boll rot, and improved yields.

The advantage of these innovations are: (1) Temporally controlled gene expression permit for variation in leaf shape formation; (2) Normal leaf shape on lower canopy helps for rapid canopy closure, weed suppression, maximum light capture, and improved nutrients; and (3) Okra leaf shape on top induce earlier flowering, reduced boll rot, and increased boll production and improved efficiency in chemical spray applications.

Loison (2017) designed cotton ideotypes suitable for low input rainfed situations/conditions in northern Cameroon so as to reduce the risks of future crop failures. He conducted several simulations through the usage of CSM-CROPGRO-Cotton model and concluded that the currently grown cultivars in the regions failed to cope with the future climatic conditions and no ideotype could maximize both the yield potential and yield resilience when planted under suboptimal planting conditions. Loison further suggested that the best ideotype should be an early flowering date with concomitantly longer reproductive phase, and that these should also have higher potential for net photosynthetic rates.

Various studies have been undertaken on breeding of cotton ideotypes. Some of such studies are presented in the paragraphs that follow.

Chunsong (1993) examined the ideotype cultivation which characteristically has a blooming stage for production of maximum fine quality bolls. For an ideal cultivation of this ideotype, it is essential to select a proper variety, the planting dates, and systems. Additionally, Tan observed that the presence of sympodia, number of nodes per plant, ideal plant population of the ideotype was helpful in assessing the production of a good number of good fiber quality bolls at the blooming stage so that more number of squares in the inner layer of the sympodia are produced to bloom within fine quality bolls blooming stage. Furthermore, the researchers noted that when the required total sympodia are reached, underfed vegetative growth should be minimized by some chemical control measures, so that it helps in the establishment of the ideotype in accordance with the plant population and bring about the production of the maximum fine quality bolls. The researcher opined that there is a requisite for a combination of optimal boll number production and boll weight in the ideotype so as to obtain top quality higher yields.

Sekloka et al. (2008) assessed the importance of breeding new cotton varieties, suitable to diverse cropping conditions in Africa, and observed that plant architecture feature, early and effective flowering in a cotton was late planted and were very important parameters for consideration. In their study carried out in 10 cotton (*Gossypium hirsutum*) varieties at three stand densities and two planting dates could identify two ideotypes suitable to different planting conditions and densities. The studies were carried out in four trials in two cotton growing areas in Benin in 2002 and 2003. The researchers used various parameters, such as first flower opening date (FF), effective flowering time (EFT), plant height at the time of harvest (HH), height to node ratio (HNR), fruiting branch length (LFB), number of vegetative branches (NVB), and average boll retention at the first position of the fruiting branches (RP1). They noted that one of the better ideotype was Mar 88-214, which performed well under late planting–high stand density conditions. They further observed that the ideotype had a less vegetative growth. It has come to flowering at an early date, with a short flowering period and retention of bolls of the first position fruiting branches was also low. H 279-1 which gave a good performance at low stand density and even at early planting produced more vegetative growth and has come to flowering late with a long period of effective flowering, and also had very high boll retention on the first formed fruiting branches. They further proposed a breeding strategy for both of these ideotypes with high heritability, taking into consideration the FF, plant height at harvest, more HNR; and for medium heritability the characteristics of number of vegetative branches, effective flowering time, and retention of bolls on first fruiting branch positions are more important.

An attempt was made to relate physiological traits to better cotton ideotypes in upland cotton by Bhardwaj (1990) under north Indian agroclimatic conditions. He noted that most of the upland cotton cultivars grown in arid and semi-arid regions of the northern India have small leaves, small bolls with low yield potential. In their study, the segregating lines originating from intervarietal crosses were subjected to selection and genotypes combining small leaves (a desirable characteristic) with big bolls have been identified. They observed a negative correlation between boll weight and boll number. Boll number was found to be dependent on the biomass production and LAI. Consequently, the researchers suggested that it is most reliable to improve the yields if the LAI was increased. Studies on the contributions of individual leaf area, specific leaf weight, and stomatal conductance to productivity revealed that increasing the specific leaf weight could serve as an alternative for increasing LAI for increasing boll numbers. A new ideotype combining small thick leaves with high boll number and boll weight should be identified.

4.3 CONCLUSION

Ideotype of a crop should have desirable plant characteristics such as leaf morphology, its orientation, branching patterns, canopy for efficient capture of sunlight leading to high production of the crops. Various ideotype concepts and ideotype characteristics were put forward by different authors for different types of cotton crop varieties that enabled in contributing to high yield. In our survey of breeding lines of cotton, we have observed that high yielding cotton cultivars have small to medium sized leaves with open canopy, which have capacity in capture of more solar energy for efficient photosynthesis. Genotypes with overlapping leaves are poor yielders.

KEYWORDS

- **cotton**
- **ideotype**
- **planting density**
- **plant type**
- **characteristics**

REFERENCES

Bhardwaj, S. N. Basis for Higher Productivity in Upland Cotton Under North Indian Agroclimatic Conditions, 1990.

Chunsong, T. On Ideotype Cultivation of Cotton. *Scientia Agricultura Sinica* 1993.

Loison, R.; Hoogen, G.; Oumarou, P.; Gérardeaux, E. Designing Cotton Ideotypes for the Future: Reducing Risk of Crop Failure for Low Input Rainfed Conditions in Northern Cameroon. *Eur. J. Agron.* **2017,** *90*, 162–173.

Maiti, R.; Parchuri, V. *Research Advances in Cotton (Gossypium* spp.*)*. New Delhi Publishers, 2009.

NC STATE. Development of a Cotton Variety with an Ideal Leaf Structure (Ideotype) to Improve Boll Production, Yield, Boll Rot Resistance, Chemical Spray Penetration, Boll Retention and Other Cotton Characteristics. Office of Technology Commercialization and New Ventures.

Sekloka, E.; Goze, J. E.; Hau, B. Breeding New Cotton Varieties to the Diversity of Cropping Conditions in Africa: Effect of Plant Architecture. Earliness and Effective Flowering Time on Late-Planted Cotton. *Exp. Agric*. **2008,** *44*, 197–200.

Wells, A. T.; Hearn, A. B. Developed OZCOT: a Cotton Crop Simulation Model. 1992.

CHAPTER 5

Cotton Botany and Characterization

ABSTRACT

This chapter presents the general characteristics of taxonomy, morphology, and anatomy of cottons, and also reviews the extant literature.

5.1 INTRODUCTION

Botany in general describes the taxonomy, morphological, anatomical characteristics of a plant species. Efficient management of a crop requires a good knowledge of the botany of the crop. The botany of a cotton plant is described in the next sub-section.

5.2 TAXONOMY

Cotton belongs to the family: Malvaceae and genus *Gossypium* spp.

According to Wright (2017), cotton is a tropical deciduous tree (previously known as *Ceibapentandra*). It blooms in winter, produces seed pods and cotton fibers on seed surface. It is also known as the giant kapok, silk-cotton tree, or lupuna.

Botanically, cotton is known as *Ceibapentandra*. Other botanical names are *Bombax pentandrum*, *Ceiba caribeae*, and *Ceiba casearia*. Traditionally, it is a member of the kapok family: *Bombacaceae*, but later modern taxonomist designated it as *Hibiscus* family, Malvaceae.

The cotton tree is considered as native to both tropical America and western Africa. In both continents, it is tropical or subtropical tree. The native tree extends from southern Mexico to the Amazon Basin across Bolivia, Peru, and Brazil. This tree is widely grown across the tropics, in Africa it was grown as a crop tree.

Cotton (*G. hirsutum* L.) is an important fiber, food, feed, and industrial crop of India in 72–80 other countries. It is one of the vital crops that is of commercial importance, is also known as king of fibers, and remains a pride in the Indian economy. It has a key role in the agricultural and industrial economy of a country. Production and processing of cotton, as well as trade in cotton goods provide employment to around 60 million people in the country. Furthermore, cotton is considered the backbone of agro-based industries in the country.

Cotton is also popularly referred to as the "White Gold." India has distinction of cultivating all the four cultivated species of *Gossypium* and has the largest acreage covered by *Gossypium hirsutum* sp., and is the center of origin of *arboreum* and *herbaceum* cottons. It is also the first in developing and cultivating commercial hybrid cottons.

In India, mainly four species of cotton are cultivated on a commercial scale. These are: *G. arboreum*, *G. herbaceum*, *G. hirsutum*, and *G. barbadense*. Cotton is grown and adapted to diverse soil and agroclimatic conditions. Crop production technologies vary widely according to varieties and hybrids, and also vary in nutrient and pest management practices. The geometry of the crop varies with the plant type, its architecture, soil fertility, and soil moisture.

5.3 SYSTEMATIC POSITION

Kingdom: Plant kingdom
Division: Phenerogams
Sub-division: Angiosperms
Class: Dicotyledons
Sub-class: Polypetalae
Series: Thalamiflorae
Order: Malvales
Family: Malvaceae
Genus: Gossipium
Species: herbacium

Seelanan et al. (1997) explored the evolutionary history of the Gossypieae and *Gossypium*. They used the biparentally and maternally inherited characters for phylogenetic analysis. They analyzed these data sets either individually or in combinations. Later they evaluated and quantified the incongruence that existed within these datasets. At the tribal level, nuclear ribosomal ITS sequences of phylogenetic analyses yielded tree. These were found to have a high congruence with the species which had their derivation from the plastid gene ndhF. This congruence was not as expected on those species which had a reticulate evolutionary history. It was also not seen in the results revealed in the clades supported by only a few characters. Later, they collected the problematic data sets taxa for inference of phylogeny. Results of their research revealed that (1) the Gossypieae was monophyletic, with one branch from the first split being represented by modern Cienfuegosia; (2) *Thespesia* was not monophyletic, and (3) *Gossypium* was monophyletic and sister to an unexpected clade consisting of the Hawaiian genus Kokia and the east African/Madagascan genus Gossypioides. On the basis of divergence in the magnitude of ndhF sequence, they suggested that Kokia and Gossypioides have diverged from one another in the Pliocene, much before their loss of chromosome pair, via chromosome fusion. cpDNA restriction site variation and ITS sequence data, for expression of phylogenetic relationship within the species, and genus and groups of *Gossypium* supported the monophyly that existed within each group of genome. The taxa which were known to have or suspected to have reticulate histories were pruned from the trees. They proposed that variations in cpDNA- and ITS-based resolutions of the genome groups in *Gossypium* reflected the presence of temporal closely spaced divergence events. These might have taken place in the early periods of this genus diversification. Thus, they specified that an identical and common cause of incongruence in the phylogeny may be because of the exhibition of the presence of "short internodes."

5.4 GENERAL MORPHOLOGY

The general morphology of cotton plant is characterized by:

Habit: Plant is herb.

Habitat: Mesophyte.

Root system: The plant possesses a tap root with inclined deep lateral roots;

Stem: It is erect, branched, and contains stellate hairs.

Leaves: Are simple, heart shaped, 3–5 lobed (okra type), margin entire, apex acute, stipulate. It is petiolate (short or long), palmately reticulate venation, alternate and pubescent, possesses stellate trichomes, and extra floral nectarines.

Inflorescence: Is racemose, axillary, and solitary.

Flower: Is bracteate, pedicellate, presence of epicalyx, complete, bisexual, pentamerous and hypogynous, and actinomorphic. Flowers are called squares.

Calyx: Sepals are 5, gamosepalous (united at the base), bowl shaped, trichiferous nectar is present at the base of calyx and volvate.

Corolla: Petals are 5, polypetalous, large, attractive, and twisted.

Androecium: Stamens are numerous, Monadelphous (united), Anthers are monothecous, kidney shaped, extrose, pollen grains are spherical in shape and protandry.

Gynoecium: It is multi-carpellary syncarpus, superior ovary, axile placentation, terminal style, stigma is lobed and number of stigma equal to number of locules.

Fruit: It is a capsule (pericarp breaks vertically along the middle of each locule), rounded or oblong in shape. Fruits are called bolls.

Seed: Seed is round in shape, grey-ash color, rough surface, and seed size 2.5–3 mm epidermal outgrowth are cellulosic fibers.

5.5 GROWTH HABIT

Cotton plant is in general a bushy, profusely branched with open or closed canopies (Fig. 5.1). It bears flowers of varying colors. Two types of growth habit are present, viz., determinate and indeterminate type.

The growth habit of cotton varies based on the amount of light interception. Three different types of canopies are found based on amount of light that has been intercepted into the crop, viz., compact canopy (with low light interception due to closeness and overlapping of leaves), semi-compact canopy (having medium light interception with medium overlapping leaves), and open canopy (maximum light interception in all the leaves). The plant with open canopy has maximum capacity to translocate the photosynthates to the developing boll. It is generally observed that a cotton plant that has a close canopy and more of high overlapping leaves is less efficient in the resource utilization (solar energy). The height of the plant is determined by the length and number of internodes present on the stem. Response to nitrogen is more in short and compact types rather than tall and medium types. Dwarf compact varieties require closer spacing with higher plant population.

Branching pattern: Cotton crops vary in branching pattern, monopodial or pseudo-monopodial. Branching may be compact, semi-compact, or open. It is pyramidal in shape.

5.6 MORPHOLOGY

The cotton plant shows a large variability in plant types and branching patterns (Fig. 5.2). It could be compact, semi-compact, and open type. The characteristics of cotton are shown below.

FIGURE 5.1 Cotton plant morphology.

FIGURE 5.2 Branching pattern.

The branching may be alternate, or opposite, denser space, with close or open canopy for absorption of solar energy which could be related to productivity of cotton. The branching may be monopodial or pseudopodial.

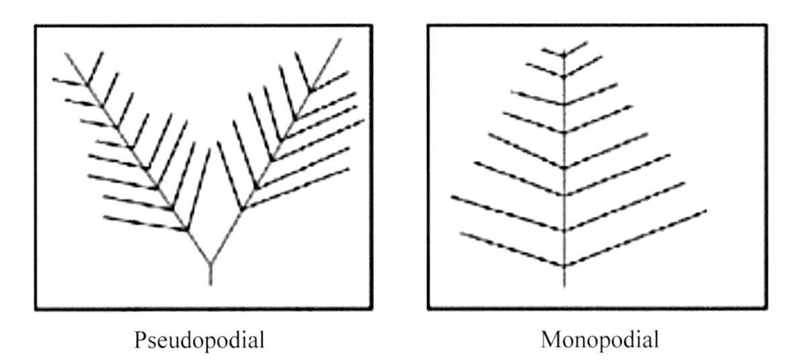

Pseudopodial Monopodial

Cotton shows large variability in branching habit, plant type, and leaf morphology.

5.6.1 STEM

The cotton plant contains an erect main stem and many lateral branches. Different varieties of cotton exhibit variations in coloration of stem, pigmentation, and pubescences. The stem has a growing point at its apex, with an apical bud. The cotton plant exhibits apical dominance, where the growth of lateral buds is restricted by the presence of an apical bud. The lateral buds remain dormant and until the apical bud is removed, they do not exhibit sprouting. A characteristic feature commonly observed in cotton is that its branches and leaves are only found on the main stem, without any flowers. The final height that a cotton plant is likely to reach can be estimated by the length and number of internodes which were present on the stem. In cotton, moisture supply determines the length of internodes, while the nitrogen supply determines the number of internodes. Most of the cotton plants possessing short internodes are early maturing types.

5.6.2 LEAF

The leaf is a simple petiolate cordately three to nine lobed. The venation type is palmate. Cotton plants vary in leaf size, shape, color, and pubescence. Petiole length varies with the size of the leaves. Long and stout petioles are related to higher light interception and higher yield. The angle between the stem and the leaf petiole may be acute or right angled, and the petiole inclination with the stem varies with the plant type. The leaf with long petiole has greater capacity for interception of light than the one with shorter petiole. Stomatal index of upper epidermis is 31 and lower epidermis is 36.

Leaves are generally hairy, but in some varieties and hybrids glabrous and velvety (densely hairy) type of leaves are also present. Though the presence of hairy leaves poses hindrance to the mechanical harvesting, its presence is more advantageous in those conditions where there is prevalence of high jassid populations. Under such conditions, presence of hairy leaves imparts tolerance to the crop from the jassid attack. Though the presence of hairy leaves is advantageous in imparting tolerance to jassid incidences, they are also disadvantageous, as they provide shelter for the white flies to reside within the hairs of the leaves.

The leaf varies in color and size from light green to dark green and from small to large. The shape of the leaves varies from normal to okra and super-okra types. Lobing pattern of shallow to deep in the leaves varies accordingly with the shape of the leaf. Leaf and petiole pigmentation variations are also present in cotton. Glands are present on the petiole, stem, leaves, bracts, and cotyledons. The concentration and number of glands vary with the cultivar. Nectar glands are also present on leaf, calyx, and bracts. Thick and smaller leaves with pubescence contribute to drought tolerance.

5.6.3 FLOWER

The flowers of cotton are called *squares*, which are solitary in nature with varied colors. They are borne singly in the axils of leaf. The flower is white or creamy at the time of opening, but toward the end of the day it changes its color to pink and then finally turns to red. Floral bud is enclosed and protected by three triangular bracts called square.

The flowers are bisexual and hermaphrodite. Each flower has five small calyx which are fused at the base. There are five petals, which are, large in shape, and vary in pigmentation. The color of the petal varies from light yellow, cream to deep yellow with or without of pigmentation. The flower contains compact to semi-compact stamens projected from staminal tube, and anthers are cream to yellow in color. The stigma is either embedded or protruded, and the ovary has two to six carpels with locules equal to the number of carpels. Each locule contains 8–12 ovules on axile placentation.

5.6.4 ROOT

Cotton contains a long taproot, which may even reach up to a depth of 20–25 cm much prior to the emergence of the seedling from soil. Lateral roots formation commences after emergence and unfolding of cotyledonary leaves. The taproot grows deep with several inclined lateral roots. Variation in the root growth is seen with the type and variety of cotton. The root usually reaches a soil depth of around 180–200 cm. However, in dry growing conditions, these roots may reach a deeper depth of even 3.0–4.0 m, while under adequate soil moisture conditions, the lateral roots are found in the top soil layer confined upto a depth of 30–35 cm and these may even show a lateral extension of 100 cm and more.

5.7 ANATOMY OF STEM

A transverse section of stem shows three regions, namely, epidermis, cortex, and vascular tissues. The epidermis is a uniseriate, compactly arranged outermost region surrounding the stem. It consists of a thick cuticle, a stellate and tuft trichomes. Multilayer cortex is present between the epidermis and stele. There are 3–5 layers of collenchyma below the epidermis, which stem its mechanical strength. Below the collenchyma are 5–8 layers of parenchymatous cortex, which contains mucilaginous cavities. The inner layer of the cortex is referred to as the endodermis.

The xylem and phloem are the vascular tissues, which are organized into vascular bundles. Between 8 and 10 vascular bundles are arranged in one ring called eustele. Each vascular bundle is wedge-shaped, conjoint, collateral, open, and endarch.

The central part of the stem is occupied by parenchymatous pith or medulla. The cells are round or oval with intercellular spaces. The medulla stores food materials and helps in lateral conduction.

Figure 5.3 is a transverse section of mature stem, showing the secondary xylem (secondary xylem vessels are uniseriate arranged below endodermis). The transverse section of the bark regions contain pyramid-shaped fiber cells.

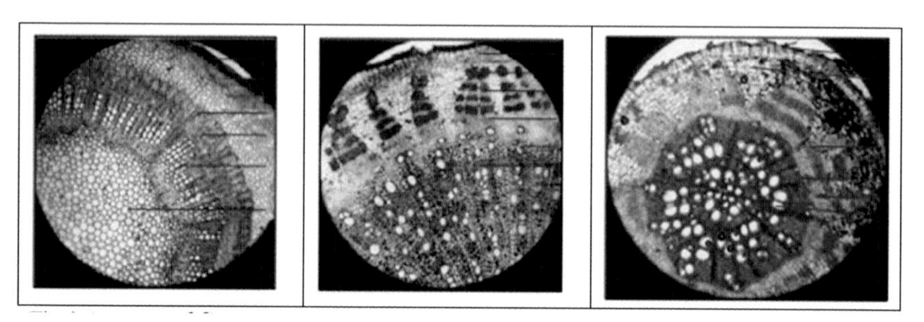

FIGURE 5.3 Anatomy of cotton stem.

5.8 ANATOMY OF THE LEAF

The cotton leaf is dorsiventral nature, that is, the leaf blade consists of distinct dorsal and ventral surfaces. The epidermis is single layered, having compactly arranged tabular cells that cover both sides of the leaf. A thin cuticle layer covers the epidermis and checks the transpiration. Both epidermal layers contain anisocytic type of stomata. Stomatal index of the upper epidermis is 18–20, and the lower epidermis is 26–31 under a 40 × magnification. The epidermis consists of gossypol glands and stellate trichomes. The functions of the high density trichomes and a thick cuticle are to reduce the loss of water due to transpiration. In some genotypes, high density of trichomes is also helpful in imparting sucking pest resistance. The mesophyll is divided into an upper palisade tissue and lower spongy tissue. The palisade tissue, which may be compact or spacious, is situated below the upper epidermis. The thick cuticle with compact palisade is related to drought resistance. The upper surface of the leaf is dark green in color due to the palisade tissue. The lower part of the mesophyll located toward the lower epidermis is the spongy tissue. Cells of the spongy tissue are thin-walled, irregular in shape, arranged loosely with large continuous intercellular spaces. The vascular bundles are closed and collateral. The xylem is located toward the upper epidermis and the phloem toward the lower epidermis. In the midrib, several layers of collenchyma cells are present both in the upper and lower regions offering mechanical resistance to the leaf.

5.8.1 EPIDERMAL IMPRESSION

The epidermal system of the cotton leaf shows the stomatal complex, which includes two kidney-shaped guard cells covering the aperture (stoma), and three unequal subsidiary cells covering the two guard cells. This type of stomata is called anisocytic type. Large variations occur among cotton cultivars in epidermal characteristics.

Both surfaces of the leaf contain different types of trichomes. They include stellate, bi-furcated and unicellular simple type. The lower epidermis contains the gossypol glands.

The cotton germplasm shows variability in stomatal index, size and density of gossypol glands, and trichomes. The gossypol glands, appearing as a dark spot on the lower surface of the leaf, are related to tolerance to boll worms and protection against fungal attack.

5.8.2 TYPES OF TRICHOMES

The different types of trichomes found on the cotton leaf are shown in Figure 5.4.

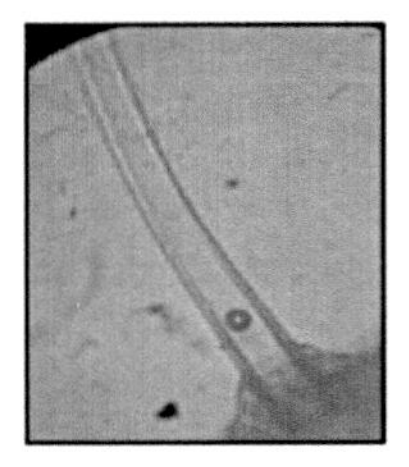
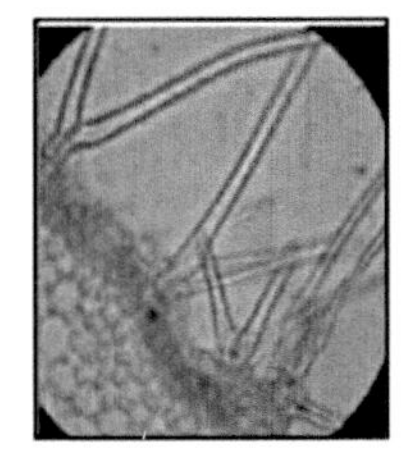
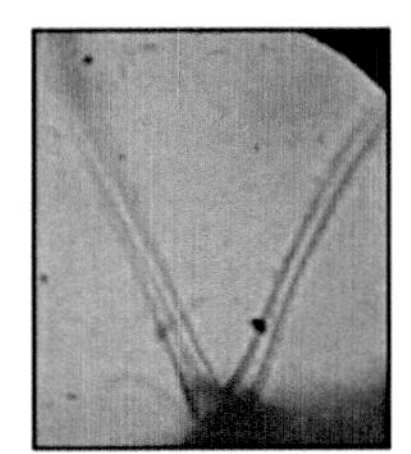

FIGURE 5.4 Different types of trichomes.

5.8.3 PETIOLE

The petiole is a part of the plant which connects the leaf with the shoot system (branch/stem). The vascular system of the shoot enters in the form of vascular bundles in to leaf through the petiole. The efficiency of translocation of mineral nutrients and water to the leaf is much dependent on the size and number of active vascular bundles located within the petiole.

The leaf of cotton plants consists of petiole. Morphology of the petiole shows variations in case of petiole length and thickness of the petiole, pigmentation, orientation, and trichome density.

5.8.3.1 TRANSVERSE SECTION OF THE PETIOLE

The tissue of petiole may easily be compared with the primary tissues of the stems. There is a close similarity between petiole and stem with regard to the structure of epidermis. The ground parenchyma of petiole is like the stem cortex in arrangement of cells and in number of chloroplast. The supporting tissue is collenchymas in relation to the arrangement of vascular tissues in the stem, the vascular bundles of the petiole are collateral.

A transverse section of petiole shows four regions namely: (1) Epidermis, (2) Hypodermis, (3) Ground tissue, and (4) Vascular bundles.

1. **Epidermis:** It is uniseriate and has a single layer of compactly arranged barrel-shaped living cells. The outer walls of the epidermal cells are cutinized. The cuticle reduces transpiration, multicellular hairs are found on the epidermis.
2. **Hypodermis:** Found immediately beneath the epidermis are multi-layered (2–6) hypodermis of collenchymas cells, which are living mechanical tissues. The hypodermis gives considerable strength, flexibility, and elasticity to young petioles. Having chloroplasts it may carry on photosynthesis. Due to the presence of thick layer of collenchyma, petiole is strong, stiff and has a good strength. The type of collenchyma is angular collenchyma.
3. **Ground tissue:** This is found just beneath the hypodermis. It consists of thin-walled parenchymatous cells having well defined intercellular spaces among them. Vascular bundles are arranged in half ring scattered in ground tissue.
4. **Vascular bundles:** The vascular bundles are of various sizes in the same petiole. Each vascular bundle is wedge-shaped. In the petiole, the xylem is often located toward upper side whereas phloem toward lower side (as in the leaf) (Fig. 5.5).

Features of special interest

- Usually a groove is present toward upper side.
- Mostly the vascular bundles are arranged in a semicircle in ground tissue.
- Mostly the central bundle is biggest and remains encircled by endodermal sheath.

- Stomatal structure

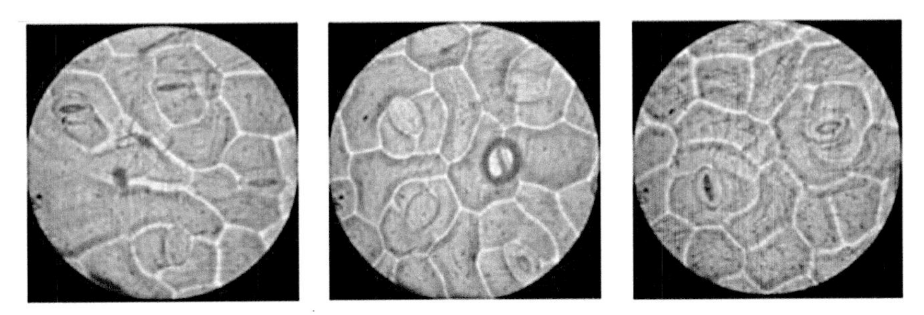

FIGURE 5.5 Vascular bundles.

The frequency of stomata is low and mostly sunken showing drought resistant trait (Fig. 5.6).

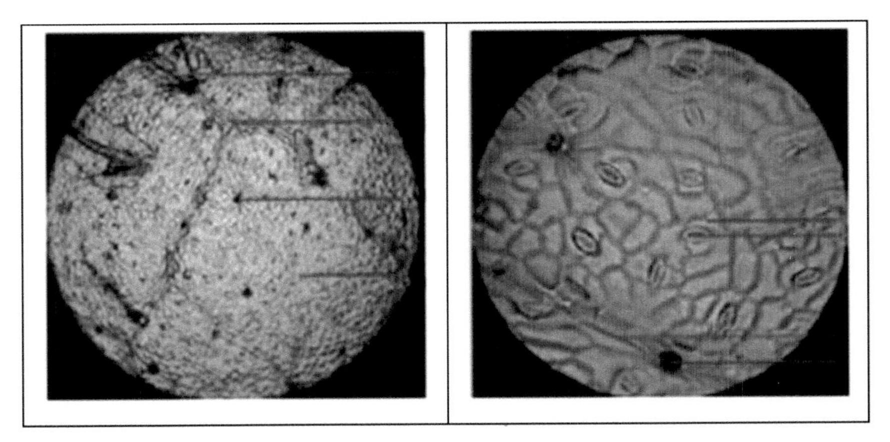

FIGURE 5.6 Frequency of stomata.

In a transverse section of midrib with leaf lamina (Fig. 5.7) indicates that midrib contains thick cuticle, one layer of ovoidal epidermal cells followed by one to two layers of collenchyma, followed by a massive cortical cells and then central vascular bundle. The upper epidermis contains thick cuticle, one layer of epidermal cells followed by compact palisade cells which is related to drought resistance.

FIGURE 5.7 Anatomy and venation pattern of the leaf.

5.9 ROOT ANATOMY

The root possesses uniseriate compactly arranged thin walled epidermis. It consists of root hairs which help in absorption of water and minerals. Below the epidermis is the cortex, which is multilayered and parenchymatous. The endodermis is the inner layer of cortex which consists of compactly arranged suberin thickened cells. Below the endodermis is the stele (vascular strands), consisting of the xylem and phloem. The vascular strands in the root are separate and radial. The xylem is exarch and triarch with protoxylem toward the pericycle and metaxylem toward the centre. The xylem is normally located toward the upper side and the phloem toward lower side.

The figure shows the anatomy of the root. The root epidermis consists of one layer of epidermal cells, followed by several layers of parenchyma. In the center is the vascular bundle, surrounded by the endodermis. In the center are four patches of xylem alternated by phloem.

Figure 5.8 shows a transverse section of stems showing patches of secondary phloic fibers pyramidal fiber patches arranged, surrounding secondary wood in the center. Wood contains vessel pores of different sizes, irregular in distribution. Wood vessels are round to ovoidal, mostly isolate, few in multiples. Wood parenchyma paratracheal. Primary xylem vessels are arranged radially (shown in Fig. 5.9).

A few research studies on the anatomy of structures of cotton and their physiological functions are presented in the next section.

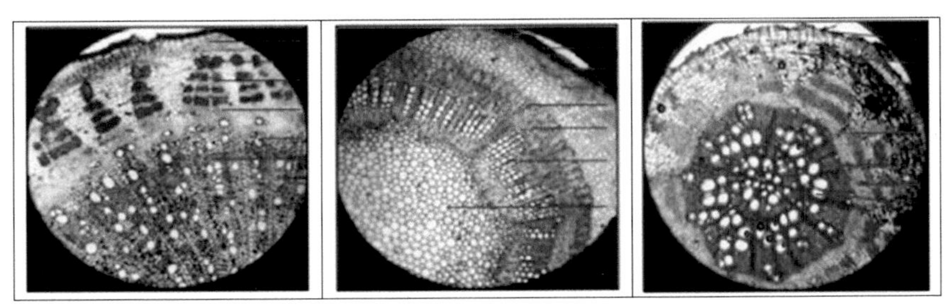

FIGURE 5.8 Transverse section of stem.

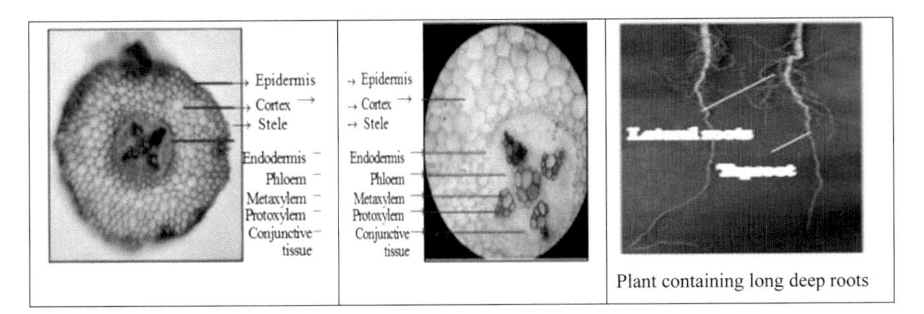

FIGURE 5.9 Cotton root anatomy.

Jensen (1965) in his study of synergids histochemistry and the ultrastructure in cotton observed that the cells were surrounded by a cellulosic partial wall, it also had Helianthella and pectins. It consisted of an unusual fiberfill arrangement. The fiberfill apparatus was surrounded by large amounts of JR and has many plastics and Michaelis. There was a large nucleus with a nucleolus and one or many micronuclei. Membrane bound vesicles were found in the nuclear membrane, where a few were found extending into the cytoplasm. The ER was found to be parallelly arranged to the long axis of the cell. Its concentration was low toward the Chaco end of the cell. Numerous dictyosomes were found to be closely associated with the ER. Vacuoles occupied larger portion of Chaco end of the cell. These were found to be rich in an inorganic compound leaving a residue of ash. There was more number of Sparganium-like bodies. They mentioned that synergieds are concerned with absorption, storage, and transport of various compounds from the Nucella, provides material to egg, endosperm and also to the developing embryo. Apart from this they might have a key role in pollen tube growth to reach into the embryo sac.

Longstreth et al. (1979) examined the effect of salinity on leaf anatomy and its consequences on photosynthesis. They observed that increasing salinity had substantially contributed to higher ratios of mesophyll surface area to leaf area (A^{mes}/A) in *G. hirsutum.* Furthermore, they noted that an increase in internal surface for CO_2 absorption did not bring about an increase in CO_2 uptake rates. This was attributed mostly to the CO_2 resistance, which was expressed on the basis of mesophyll cell wall area (r_{cell}). The CO_2 resistance increased with an increase in salinity. Thus, the differences that were observed among species, with reference to the sensitivity toward the photosynthesis, owe to the variable responses in variation of different Ames/A and r_{cell} responses against salinity.

Pettigrew et al. (1993) studied the differences in gas exchange and anatomy among the cotton leaf-type isolines. They suggested that line with higher photosynthetic rates and those with efficient partitioning of photosynthates between reproductive and vegetative growth may be identified keeping in view the photosynthesis rate as a selection criterion for plant breeders. They undertook the investigation on photosynthesis and leaf anatomy in a genotype of cotton (*Gossypium hirsutum*) "MD 65-11" of the youngest fully expanded leaf of the super-okra, okra, and normal leaf type isolines of the genotype. The genotype "MD 65−11," differed in leaf size and degree of lobing. These traits were compared with the normal leaf genotype "DPL 50," super-okra, and okra. In 2 years of investigation, viz., 1989 and 1990, an average of 24 and 22% greater leaf CO_2-exchange rates (CER) were found in the MD 65-11genotype. They attributed these higher values of leaf CO_2 to the presence of greater specific leaf weight (SLW) and to more leaf chlorophyll concentration in the genotypes possessing the super-okra and okra leaf types. They also recorded higher water use efficiency. Furthermore, these had 42% thicker leaves. It resulted in more SLW. Rubisco concentrations were similar in all genotypes chloroplasts.

Lee et al. (2007) investigated the changes in gene expression and the early events during cotton fiber development. They observed that cotton fibers derived from seed coats of the genus *Gossypium* are seed trichomes. These have arisen from individual cells present in the epidermis of seed coat. Cotton fiber development had followed four distinct and overlapping developmental stages: fiber initiation, elongation, secondary wall biosynthesis, and maturation. Fiber cell development follows a complex process which involves many pathways, including various signal transduction and transcriptional regulation components. They have identified through several analyses by using expressed sequence tags and microarray transcripts that there is a preferential accumulation of these transcripts during fiber development. They asserted that trichome development from seed coat in cotton is regulated by MYB transcription factors. Apart from MYB transcription factors, several pathways of signaling and phytohormones have their role in bringing about fiber development. Simultaneously, auxin and gibberellins promoted early stages of fiber initiation; and also ethylene- and brassinosteroid-related genes were upregulated during the fiber elongation phase. Furthermore, there is upregulation of genes associated with calmodulin and calmodulin-binding proteins in fiber initials. They suggested that additional genomic data, mutant and functional analyses, and genome mapping studies may reveal the critical factors mediating cotton fiber cell development.

Kakani et al. (2003) in their study on the effects of Ultraviolet B radiation on morphology and anatomy of cotton (*G. hirsutum* L.) found that leaves turned chlorotic initially and later developed necrotic patches on continuous exposure to UV B radiation. They observed a drastic reduction in both vegetative and reproductive growth with high UV B radiation and that the plants produced a smaller canopy indicated its sensitivity. They further observed an increase in the epicuticular wax content on adaxial leaf surfaces. There was also an increase in stomatal index on both leaf surfaces, while there was a drastic decrease in leaf thickness. The decrease in leaf thickness owes to a decrease in the thickness of the tissues, namely, the palisade and mesophyll tissue. There was no change in thickness of epidermis. High levels of UV radiation exhibited their influence on vegetative characteristics, while ambient and high levels of radiation had a strong influence on reproductive traits/parameters.

Egan et al. (2014) undertook a study on meta-analysis of herbicide drift of 2,4-D and Dicamba in soybean and cotton, and observed that commercial introduction of cultivars of cotton and soybean genetically modified with resistance to the synthetic auxin herbicides dicamba and 2,4-D though permitted the compounds to be used with greater flexibility, they could also let them become susceptible to non-target herbicide drifts. A meta-analysis approach of simulated drift experimental data led to the generation of global dose–response curves. Conversely, it was observed that cotton exhibited much tolerance to dicamba rather than 2, 4-D at vegetative and pre-flowering squaring stages. Visual injury symptoms of vegetative stages could not give an indication of yield losses. These global dose–response curves that were generated helps in offering proper guidance of herbicide applications to all users.

5.10 CONCLUSION

Cotton plants exhibit variations in their morphology and anatomy. Several researchers have identified the variations in the branching pattern, leaf types, root system, and utilized some of these anatomical and morphological traits in developing cotton genotypes tolerant to abiotic and biotic stresses.

KEYWORDS

- **botany**
- **taxonomy**
- **morphology**
- **anatomy**
- **cotton**

REFERENCES

Egan, F. J.; Kathryn, M. B.; David, A. M. A Meta-Analysis on the Effects of 2,4-D and Dicamba Drift on Soybean and Cotton. *Weed Sci.* **2014,** *62*, 193–206.

Jensen, W. A. The Ultrastructure and Histochemistry of the Synergids of Cotton. *Am. J. Bot.* **1965,** *52*, 238–256.

Kakani, V. G.; Reddy, K. R.; Zhao, D.; Mohammed A. R. Effects of Ultraviolet B Radiation on Cotton (*Gossypium hirsutum* L.) Morphology and Anatomy. *Ann. Bot.* **2003,** *91*, 817–826.

Lee, J. J.; Woodward, A. W.; Chen, Z. J. Gene Expression Changes and Early Events in Cotton Fiber Development. *Ann. Bot.* **2007,** *100*, 1391–1401.

Longstreth, D. J.; Nobel, P. S. Salinity Effects on Leaf Anatomy. Consequences for Photosynthesis. *Am. Soc. Plant Biol.* **1979,** *63*, 700–703.

Pettigrew, W. T.; Heitholt, J. J.; Vaughn K. C. Gas Exchange Differences and Comparative Anatomy among Cotton Leaf-type Isolines. *Crop Sci.* **1993,** *6*, 1295–1299.

Seelanan, T.; Schnabel A.; Wendel, J. F. Congruence and Consensus in the Cotton Tribe (Malvaceae). *Systematic Bot.* **1997,** *22*, 259–290.

Wright, J. Cotton Tree Facts Updated September 21, 2017.

CHAPTER 6

Physiological Basis of Cotton Growth and Productivity

ABSTRACT

This chapter discusses significant research advances till 2018 on physiological basis of cotton growth and productivity. It discusses various factors affecting cotton growth and productivity starting from germination and seedling establishment, vegetative growth, and flowering, fruiting and yields. Besides, it discusses various manipulation techniques used to determine their influence on cotton growth and productivity.

6.1 INTRODUCTION

Cotton (*Gossypium hirsutum*) is a fiber crop of varied importance. Its cultivation is taken up across several regions of the world under varied climatic conditions. The growth and development, the sympodial branch production, yield and fiber development, and fiber quality are influenced by several factors. The physiology behind these processes is discussed in brief in the chapter.

6.2 GERMINATION AND SEEDLING ESTABLISHMENT

Good germination and seedling establishment is required for good growth and productivity of a crop. Several studies are available on cotton germination, vigor, and seedling establishment.

6.2.1 SEEDLING EMERGENCE FROM SOIL CRUST

Emergence of seedling from the soil and establishing itself to put forth good growth and development is very essential for obtaining better yields. Uniform emergence of the healthy seedling is much required. Crust formation of the soils act as a hindrance to the emerging seedlings. A technique for seedling emergence of the cotton through the soil crust was developed by Bennet (1964). Alteration of the geometric characteristics of the soil was obtained by the use of evaporation suppressing agents, soil conditioners, and some chemicals. They led to a decrease in the crust strength. The soil top 3 inches moisture content had a positive correlative influence on the emergence of the cotton seedlings from the soil crust. Similarly, they observed a negative correlation between the crust strength and the emergence of cotton seedlings.

6.2.2 EFFECT OF SOIL TEMPERATURE ON THE SEEDLING EMERGENCE

Seedling emergence of cotton is influenced by day and night temperatures and also the soil temperatures. The time of planting also influences the emergence of seedlings from the soil. These affect the seedling rate of development. The cotton cultivars GC-510, SJ-2 seeds planted at 0800 and 1600 h planted at 3 weeks interval from beginning of February to April and were evaluated by Steiner and Jacobsen (1992) for the effect of soil temperature on seedling emergence and also seedling development rate index. Their study indicated that during the initial 100 h of planting of cotton seeds the seedling emergence was found to be more sensitive to the cool temperature stress. The two cultivars exhibited variations to the cool temperatures and in the seedling emergence. The seedling emergence in these genotypes was found to be sensitive to cool soil temperatures even at 160 h of planting. In both the cultivars the seedling emergence was not affected by day of planting heat units, so was also the effect of time of planting on the seedling development rate index. Although no relationship was found between seedling emergence and seedling development rate index, the cultivars exhibited variation to soil cool temperatures of planting.

6.2.3 RHYMICITY OF ETHYLENE PRODUCTION IN COTTON SEEDLINGS

Good germination and seedling establishment is required for good growth and productivity of a crop shown in Figures 6.1 and 6.2. Several studies have indicated that during the biosynthetic pathway of ethylene, light has an influence on the transformation of 1-aminocyclopropane-1-carboxylic acid into ethylene and an earlier stem is found to be controlled by endogenous rhythm within seedlings. An experimental study was conducted in cotton seedlings (*Gossypium hirsutum* L.) cotyledons by Rikin et al. (1984) to study the rhythmicity in ethylene production in these seedlings grown under a photoperiod of 12 h darkness and 12 h light. They found that ethylene production has shown an increase toward the end of dark period. Furthermore, during the first third of the light period, a maximum production was achieved which later showed a decrease. In young, mature, and old cotyledons (7–21 days old), these oscillations in ethylene evolution were found to occur. As these were found even after invertion of photoperiod, these oscillations in ethylene evolution appeared to be controlled endogenously. These oscillations exhibited rhymicity where there was continuous evolution of ethylene in continuous light, while in continuous darkness they appeared for some time and vanished for some period. They observed that always, a decline by light and an increase by darkness in the transformation of applied 1-aminocyclopropane-1-carboxylic acid into ethylene.

6.2.4 IN SITU EMBRYO CULTURE IN COTTON

Stewart et al. (1977) developed a phytohormone-supplemented medium (BTP) upon which in situ embryo culture and seedling development of cotton (*Gossypium hirsutum* L.) was taken up. They observed a good ovule growth on BTP, but none of embryos could develop to maturity on this medium. However, they observed that supplementation of NH_4^+ on the medium led to the production of mature embryos by more than 50% of the ovules and these ovules exhibited precocious germination after 8–10 weeks of culture. A separate medium to give balanced root and shoot growth was formed and after germination these were transferred on that and seedlings got established.

FIGURE 6.1 Germination.

FIGURE 6.2 Seedling establishment.

6.2.5 TILLAGE FERTILIZATION EFFECTS ON SEEDLING EMERGENCE

The dry land cotton production systems are on the increase in the US cotton belt. Correspondingly, the safe disposal of poultry litter is becoming a serious environmental problem in these regions. Nyakatawa et al. (2000) studied the tillage, cropping system, and N source effects on emergence of cotton seedling in Northern Alabama, on a Decatur silt loam soil for 2 years. Under no-till 40–150% higher cotton seedling counts were recorded at 1 and 2 days of seedling emergence. Similarly, 14–50% higher seedling counts were produced in Cotton–winter rye cropping system, rather than in cotton–winter fallow cropping in the first 4 days of emergence. A higher yield of 17–50% cotton seedling counts was recorded in poultry litter that was used as a source of N. In drier year, they found that the seedling counts correlated well with the growth parameters and lint yield. They attributed that these practices led to conservation of soil moisture during seedling emergence and resulted in bringing beneficial effects on seedling counts and growth parameters. Furthermore, their results revealed that conservation tillage promoted cotton germination, emergence, dry matter, and lint yield, and hence for attaining early cotton seedling emergence and growth no-till with winter rye cover cropping and poultry litter can be suggested in these soils.

6.2.6 IMPROVEMENT OF SEEDLING EMERGENCE

Inoculation of seeds with nitrogen fixing bacteria in several studies has revealed an improvement in seedling emergence, nutrient uptake, and growth of crops. There are certain strains of Brady *Rhizobium* and *Azotobacter* which have the ability to produce growth hormone indole acetic acid. Hafeez et al. (2004) inoculated indole acetic acid producing Brady *Rhizobium* and *Azotobacter* strains to four cultivars of cotton (*Gossypium hirsutum* L.) to study the effectiveness of these strains on seedling emergence, nutrients uptake, and growth. The results revealed that seedling emergence was increased by 3–9%, similarly an increase of 48%, 75%, and 57% was recorded in shoot dry weight, biomass, and N uptake, respectively. These beneficial effects were observed with inoculation of *Rhizobium leguminosarum* bv. *trifolii* E11 and *Azotobacter* sp. S8. Likewise, strain E11 led to an increase of root dry weight, root length, and area by 248, 332, and 283%, correspondingly, and the K^+ and Ca^{2+} uptake by 2–21% and 9–14%. Their results revealed that cotton growth was stimulated by (*Brady*) *rhizobium* strains owing to effective uptake of nutrients, attributed to an increase in root growth and production ability of these strains for IAA. However, they presumed growth promotion by *Azotobacter* sp. S8, as well as 4-indole-3-acetic acid production. Furthermore, during its growth at some stage it might also include biological N_2 fixation by this rhizobacterial strain.

6.2.7 SALINITY EFFECT ON EARLY SEEDLING GROWTH

Salinity has an effect of the seed germination and on seedling establishment in many crops. Kent and Lauchli (1985) in their study observed that there was a delay in germination and a reduction in germination by 200 mol m^{-3} NaCl in the presence of a complete nutrient medium. There was no improvement in germination with the supplemental addition of Ca^{2+} (10 mol m^{-3} as SO_4^{2-} or Cl^-) to the medium, however, it had led to a decrease in root growth and roots were infected by microbes. The beneficial effect of calcium cation on the cotton seedlings growth has been confirmed by the addition of $MgSO_4$ or KCl to the treatments of NaCl. Though the supplemental addition of Ca^{2+} partially hindered the influence for K^4 in the roots and for Ca^{2+} in both roots and shoots, it did not exhibit any effect on sodium contents. They concluded that in saline environments the high Ca^{2+} concentrations had a beneficial effect of increase of cotton seedlings root growth. This might be attributed to maintenance of K/Na-selectivity and also sufficient Ca status in the root.

6.2.8 ACCELERATED AGING AND EMERGENCE

The vigor of seeds is decreased on aging. Accelerated aging decrease the vigor of seeds. The impact of accelerated aging on germination and emergence of cotton seedlings was evaluated in two seed lots of cotton (*Gossypium hirsutum*) by Bishnoi and Delouche (1980). They had an average germination of 82.7%. In non-deteriorated seeds, a good relation existed between the standard germination and cold test. They obtained a correlation between seedling establishment and vigor tests of the lab except the conductivity tests. Furthermore, they observed that there is a rapid emergence and seedling establishment in the seeds which were of high quality rather than the deteriorated seeds.

6.2.9 IMPROVEMENT SEED VIGOR IN AGED COTTON SEEDS

Improvement in vigor of aged cotton seeds was investigated by YinFa (2017) by using high voltage pulsed electric field at very low frequency. They have used high voltage pulse power supply and arc electrode to treat aged cotton seeds with different voltages and frequencies where the treatment was imposed for 40 s; at an environment temperature set to 25°C, the relative humidity to 30%, the voltage of vertically downward electric field was 12–20 kV, the pulse frequency was 5–15 Hz, the electrode was arc, and the distance between electrode plates was 50 mm. The results revealed an increase in germination potential by 44.2%, the germination rate by 56.8%, the germination index by 64.3%, and the vigor index 81.8%, respectively, under optimal conditions.

6.2.10 COTTON SEEDLING TRAITS UNDER DROUGHT

In limited conditions of soil water availability, root growth and distribution in the soil are some of the traits required for adaptation to these conditions. Cook et al. (1992) studied the traits of seedling and first bloom plant characteristics in six cotton (*Gossypium hirsutum* L.) genotypes and influence of drought. They found that seedlings of cotton genotypes CD3H and CABUCS had higher levels of seedling vigor, rapid development of root system and establishment, and a lower root-to-shoot ratio. The regression analyses revealed that at first bloom the drought-induced boll abscission was positively associated ($R^2 = 0.47$, $P < 0.01$) with root-to-shoot ratio, and a dryland lint yield was negatively related ($R^2 = 0.30$, $P < 0.01$) with first-bloom root-to-shoot ratio. They suggested that improvement to drought tolerance and lint yields in cotton can be achieved when the cotton germplasms are selected for and combination of enhanced seedling vigor, rapid root-system establishment, and lower root-to-shoot ratios, particularly for regions which are subjected to limited distribution of rainfall conditions.

6.2.11 MONITORING AND QUANTIFICATION OF COTTON GERMINATION PROCESS

RuiZhi et al. (2018) explored the possibility of utilizing unmanned aircraft system (UAS)-based visible-band images in monitoring and quantifying the process of cotton germination. It had a consumer-grade red, green, and blue camera which was stabilized by a built-in gimbal system. They obtained an ultra-high image resolution of the germination stage, when the UAS platform was flown at an altitude of 15–20 m above ground. From the orthomosaic images and leaf polygons they developed a method for the calculation of the mean plant size and the number of germinated cotton plants with an average efficiency of 88.6%.

6.2.12 ALLELOPATHIC EFFECT OF COTTON STALK ON SEEDLING EMERGENCE

The cotton growth was found to be obstructed to large extent due to the cotton stalk returned to cotton field in Xinjiang province, China. Li YanBin et al. (2016) reported the naturally and microbially decomposed cotton stalks products allelopathic effects on the growth of cotton seedling. The outcomes revealed that when the seedlings were treated with the 30-day microbially decomposed cotton stalks, a better growth and physiological characteristics was exhibited by the seedlings rather than whose which were treated with the products of 30-day naturally decomposed cotton stalks. A low concentration of treated microbially decomposed cotton stalks, that is, 5 kg^{-1} increased the seedling dry weight and plant height. The gas chromatography–mass spectrometry analysis revealed that the naturally decomposed cotton stalk extracts contains higher concentrations of dibutyl phthalate (DBP) and diisobutyl phthalate (DIBP) compound. These were responsible for the inhibition of the cotton seedling growth in bioassay and might be the autotoxicity to cotton growth induces due to these allelochemicals.

6.2.13 TECHNIQUE OF THE FACTORY MUTUAL AID SEEDLING RISING IN COTTON

DongLin et al. (2013) by using the technique of factory mutual aid seedling raising of two crops, and also the machinery transplanting, studied the characteristics of growth and development and yield of cotton after harvest of winter wheat under field conditions. The results indicated that there was no difference in the yield and the pre-frost yield of seed cotton of the spring cotton using nutritional bowl seedling raising and transplanting after wheat harvest. They were identical to the spring cotton and the summer cotton using the technique of factory mutual aid seedling rising with wheat in same holes and machinery transplanting after wheat harvest. As the aforementioned, there was a delay in the growth process in the treatment of spring cotton interplanting in wheat field. Its seed cotton yield was decreased by 4.2–4.7%, and its rate of cotton yield before frost was reduced by 0.9–1.3% points. The technique of the factory mutual aid seedling rising of cotton with wheat in same holes has realized specialization, commercialization, and scale business. It also improves the cotton machine transplanting. Thus, their research has revealed that the cotton transplanting of traditional nutritional bowl cotton seedlings after winter wheat harvest may be replaced by the machinery transplanting of factory mutual aid cotton seedlings after winter wheat harvest.

6.2.14 SEED COATING TO RELIEVE LOW TEMPERATURE DAMAGE TO SEEDLINGS

Xing Xing et al. (2018) investigated the relieving effect of seed coating agent Xinluzao 57 on low temperature damage of cotton seedlings. Their results revealed an improvement in the cotton seedling chlorophyll a, b, and carotenoid contents under low temperature stress and also an increase in net photosynthetic rate (Pn), transpiration rate (Tr), stomatal conductance (Gs) augmented to 33.71%, 52.00%, and 96.63% and 38.57%, 55.17% and 103.42%, respectively. However, there was a decrease in intercellular carbon dioxide concentration (Ci) by 15.15% and 11.12%, respectively. Thus, their results revealed that an improvement in photosynthetic pigments in cotton seedlings, regulation of stomatal opening, enhancement in the fixation and utilization of CO_2, etc., can be obtained by seed coating agent which contributes to the reduction of the damaging effects of low temperature stress.

6.3 GROWTH AND DEVELOPMENT

6.3.1 VEGETATIVE GROWTH

Several factors influence the vegetative growth of cotton (Fig. 6.3). Several modeling approaches are taken up to study the growth, development, and productivity of cotton in various environments.

FIGURE 6.3 Vegetative growth and flowering.

6.3.1.1 PLANTING DATES AND GROWTH

Farid et al. (2017) observed through their studies at Sahiwal, Punjab (Pakistan) that planting dates have an effect on the growth of cotton. It was observed that cotton that was sown on 15th May had given 45% higher yield compared with the late planting of cotton on 15th June. They concluded that planting dates in the cotton cultivars affected the sympodial branches per plant and average boll weight of the cotton crop. They also observed that the cultivar FH-142 was suitable for early planting of cotton at 15th may it was observed that cultivar (FH-142) was a suitable cultivar for early planting of cotton at 15th for increased seed cotton yield under semi-arid conditions of Sahiwal.

In seven cotton varieties (CIM-600, CIM-616, CIM-622, CRIS-641, DNH-105, DNH-40, and DNH-57), of Dera Ismail Khan, Pakistan earlier sowing produced additional vegetative growth instead of seed cotton yield. At cold temperature, the flowering and boll formation was stimulated by late planting. It had an adverse effect on cotton yield (Khan et al., 2017). Furthermore, their study has shown that DNH-105 of Dera Ismail Khan, Pakistan, planted on April 1 is well suited to that area and correspondingly in irrigated conditions had the ability to enhance cotton yield and quality.

The yield and phenological responses of Bt cotton cultivar MNH-886 in semi-arid climates to different sowing dates (Shah et al., 2017) indicated that early sown (planted on February 14 and March 1) crop caused more delay in days to start emergence, to complete 50% emergence. The less time to germinate and a better and uniform seedling establishment were observed on late sown (30th April and 15th May). A positive correlation was found between phenological events such as days to start squaring, flowering, boll formation and boll opening, and seed cotton and lint yield of Bt cotton. Similarly cotton yield and boll weight showed a positive association with fiber quality traits. The study indicated that 16th March is the best option of planting Bt cotton in arid to semi-arid conditions of South Punjab, Pakistan, for a harvest of higher seed cotton and better quality fiber from lint yield.

YouSheng et al. (2018) evaluated 68 cotton varieties on their response to growth and yield to different planting dates. They measured 17 indexes, including plant height, stem diameter, leaf area, chlorophyll fluorescence parameters, yield, and so on. It was observed that the Y (NO) and conductivity indexes showed a greater variation coefficient in cotton varieties (lines). These also had significant difference among cotton varieties (lines) to drought stress. In a function value method, they evaluated the drought resistance of cotton varieties (lines) and from among the 68 materials, 7 were selected as high resistant materials, 20 as medium resistant materials, and 41 as the weak resistant materials. Furthermore, they classified the 17 indexes into three kind of factors using principal component analysis, (1) Fv/Fm, Y(NPQ), and ΦPSl, (2) stem diameter and conductivity, (3) were stomatal number on the back of leaf and stomatal area on the front of leaf kind, which had more load. These variables may be useful in identifying the drought resistance of cotton during bud stage to flowering stage. They concluded that as selection criteria to the evaluation of the drought resistance of cotton during bud stage to flowering stage, the Fv/Fm, Y (NPQ), ΦPS II stem diameter, conductivity, stomatal number on the back of leaf, and stomatal area on the front of leaf could be used.

6.3.1.2 EFFECT OF SOWING CONDITIONS AND CHANGING MICROCLIMATE

Premdeep et al. (2017) considered the different sowing conditions and changing microclimate effect on the HD 123, H 1098, and RASI 134 cotton cultivars. Different growth stages of these treatments were computed by radiation and thermal use efficiency (TUE). The results showed that the highest radiation and thermal indices were utilized by the crop sown during April 2nd week and compared with other development stages at days to 50% flowering weather parameters were found to be positively correlated with crop parameters. Negative association was found between morning and evening relative humidity and seed cotton, cotton seed, cotton lint, and bolls per plant during vegetative, flowering stage, and boll opening stage, whereas during boll opening stage Tmax, Tmin, and vapor pressure deficit exhibited a strong positive association with seed cotton, cotton seed, cotton lint, and bolls per plant. Furthermore, the agrometeorological indices, photothermal unit, heat unit next to heat use efficiency were highly interrelated with seed cotton, cotton seed, cotton lint, and bolls per plant and the most important weather parameter affecting the cotton yield up to 82% during boll opening were maximum temperature, morning relative humidity, vapor pressure deficit, while during flowering stage sunshine hours was most effective.

6.3.1.3 ROW SPACING

It is stated that the water efficiency increases effectually with the row spacing expansion of cotton. It reduces the number of plants per hectare together with increasing the micro-catchment for water; assist in facilitation of heavy harvesting machinery controlled traffic conversion. Bartimote et al. (2017) assessed the effects of 1.5 m row spacing on cotton yield, fiber quality and water use efficiency (WUE). The results revealed a better performance of cotton in 1.5 m row spacing system in terms of WUE and machine traffic impact. The WUE was greater with higher gross margin in the 1.5 m system cotton, though the cotton matured more slowly in this system of planting, it had a stronger and longer better quality fiber. There was also an increase in the gross margin potential in the 1.5 m system. Where heavy machinery had not been used widely it could completely offset the cost of controlled traffic conversion, within one season for a field.

6.3.1.4 EFFECT OF PLANT DENSITY ON VEGETATIVE AND REPRODUCTIVE DEVELOPMENT

Both planting density and root zone volume effects the vegetative and reproductive development of cotton (*Gossypium hirsutum* L.). Carmi (1986) in their study on plant density and root zone volume effect on the vegetative as well as the reproductive development in cotton found that there was a reduction in both the shoot and root growth per plant with 6–10 per m^2 increase in plant density or with 10 to 2 L decrease of root zone volume in pots. A drastic decrease in growth was observed when there was a decrease in root zone volume. This decrease in growth was not that significant with increase in planting density. At a water stress of 0.0–0.15 MPa which was maintained in the pots along with the limitation of roots to 2 L, it was observed that this has led to the production of compact plants with short branches. However, these plants had high boll density. There was no variation in the flowering rate in these plants and other normal plants grown in larger pots at 60–100 DAS. However, in plants which were grown in the smaller root zone volume flowering and boll opening earliness was observed. Furthermore, it was observed that though there was a decline in the average yield (seed cotton) per plant by an increase in plant density, there was an enhanced yield per unit soil area in both root zone volumes. Even though there was a reduction in the vegetative growth, a greater proportion of allocation of the total assimilate produced to the bolls was seen in smaller root zone volume plants. Their research study has concluded that if the root zone volume is restricted there would be an enhancement in the flowering and an increased translocation of assimilates preferentially to bolls rather than to the vegetative growth.

Plant growth and yield of cotton are affected by the plant density and spacing. Jost et al. (2000) in their study on effect of ultra-narrow row spacing on cotton growth and yield conducted at Brazos botton college station, TX, observed that in the cotton grown in the 19 cm row spacing the plant height and node counts at crop maturity were reduced. Compared with wider row spacings, closure of canopy was rapid in the 19 and 38.1 cm row spacings. In a relatively wet growing season, though there was no effect on yield, an increase in yield was observed in 19 and 38.1 cm row spacings in a dry growing season. A more number of harvestable bolls of 84.6% were obtained in a 19 cm row spacing at the first fruiting position. Furthermore, on nodes 6 through 10, 76.1% of the bolls were produced and there was a decrease in fiber length in narrow row spacing. The study recommended ultra-narrow row spacing as a viable option of maintaining yields by reducing the costs of production.

6.3.1.5 ROOT DEVELOPMENT

Hao et al. (2017) considered different irrigation methods to study the soil water movement stimulation and transplanted cotton root development. In the North China Plain (NCP), after the winter wheat harvest, the cotton is often transplanted. But, before transplanting, owing to the seedling aperture disk restrictions, the transplanted cotton root system is deformed. Therefore, to explore the distorted roots of transplanted cotton and movement of soil water a field experiment and a simulation study were conducted using border irrigation (BI) and surface drip irrigation (SDI) during 2013–2015. The results revealed that in the shallow root zone (0–30 cm) SDI was favorable to root growth while in the deeper root zone (below 30 cm), the BI was favorable to root growth. For the distorted root system of transplanted cotton, SDI is well suited for producing the optimum soil water distribution pattern. Well-developed root system occurred under SDI. HYDRUS-2D model was found to be more effective in describing the soil water content (SWC) by comparing the experimental data and model simulations under different irrigation methods. The research indicated that these are much helpful in designing an optimal irrigation plan for BI and SDI in transplanted cotton fields. Furthermore, from this for cotton transplantation the wide use of planting pattern can be promoted.

6.3.1.6 GROUNDWATER ON COTTON GROWTH

Under arid conditions, generally the soil–plant–atmosphere system water balance and sustainable development of cotton is influenced by the groundwater. In the region of Xinjiang, northwest of China, the major water resource for cotton is the groundwater. At the site of Aksu water balance station, the average 1.84 m groundwater depth has contributed to 23% of crop transpiration. Ming et al. (2015) evaluated the impact of groundwater on cotton growth and root zone water balance by using Hydrus-1D coupled with a crop growth model and determined that the groundwater table level had high influence on cotton growth and the soil water balance components and different groundwater table positions showed positive and negative effects on cotton growth, where the capillary rise from ground water exhibits a significant impact on the growth of cotton.

6.3.1.7 GAMMA IRRADIATION IN COTTON

Aslam et al. (2018) stated that in cotton gamma irradiation of 10 Gray is effective to increase the agronomical characters and tolerance to disease. To generate new genetic variability and develop new cotton mutants with desirable characters, two cotton cultivars NIAB-78 and REBA-288 were crossed by the use of pollen irradiated with 10 Gray (Gy) of gamma rays. In subsequent generations, the plants exhibited significant variations from control/ parents and from M_2 population a desirable mutant NIAB-777 having higher yield, early maturity, high tolerance to cotton leaf curl virus (CLCuV) disease, and insect pests and suitable for high density planting was selected.

6.3.1.8 PHOTOSYNTHESIS

The association between the photosynthetic rate and growth and fruiting of cotton and other crops was studied by Mauney (1976) and cotton (*Gossypium hirsutum* L.) plants subjected to an atmosphere enriched with CO_2 to 630 ppm (v:v) ($HiCO_2$) during daylight hours in a glasshouse, where with an average increase in CO_2 exchange rate (CER) to 15%, the temperatures were trolled continuously to produce a daily maximum of 35°C and a minimum of 21°C, were able to continue their larger size and improved absolute growth.

6.3.1.9 CARBON AND NITROGEN CONTENTS IN SUBTENDING LEAF

XinHua et al. (2010) studied the variations of carbon and nitrogen contents in the cotton boll subtending leaf in two cotton cultivars (KC-1 and AC-33B) in Nanjing and Xuzhou, Jiangsu Province, China in 2005, in response to application of nitrogen at 0, 240, and 480 kg/ha. The changes of total carbon and total nitrogen contents in the subtending leaf could be simulated with quadratic function $Y \equiv at^2 + bt + c$ [Y stands for the total carbon or total nitrogen content (%), t stands for boll age (d)]. The change of C:N ratio could be simulated with a logistic equation. When the date of reaching the minimum content of the total carbon was earlier and the minimum content of the total carbon was higher, the average rate of cotton boll, seed, and fiber biomass during the speedy accumulation period increased. Meanwhile, the duration of cotton boll and fiber biomass was shortened and the duration of the seed biomass speedy accumulation was prolonged. The period of cotton boll and fiber biomass speedy accumulation was prolonged when the date of reaching the total nitrogen minimum content was later. The minimum content of total nitrogen was high, resulting in prolonging the cotton seed biomass speedy accumulation period. The duration for cotton boll and fiber biomass speedy accumulation was prolonged by lengthening the speedy increasing duration and slowing the average rate of C:N ratio increase during the speedy increasing period. Therefore, the changes of carbon and nitrogen in the subtending leaf of cotton boll were closely related to the cotton boll biomass, boll development could be improved by adjusting the contents of carbon and nitrogen and the C:N ratio in the subtending leaf.

6.3.1.10 FERTILIZATION AND PRODUCTIVITY ON GROWTH AND DEVELOPMENT

An adequate plant population is required for good productivity of cotton. Investigation was carried out by Sawan et al. (2009) in Egyptian cotton (*Gossypium barbadense* cv. Giza 86s) to study nitrogen fertilization direct and residual and potassium and plant growth retardant (Mepiquat chloride) foliar application effects on Egyptian cotton growth, seed yield, seed viability, and seedling vigor. In the next crop season, cotton genotypes exhibited improved seed yield along with increased seed viability and seedling vigor with the combined application of *N* at a rate of 142.8 kg ha^{-1} and spraying with K_2O at 1.15 kg ha^{-1} and mepiquat chloride at 0.048 + 0.024 kg ha^{-1}.

6.3.1.11 WATER AND NITROGEN

PeiLing et al. (2010) studied the physiological indices influenced by coupling effect of water and nitrogen under alternative furrow (AF) irrigation and a single-factor effect of water or nitrogen on cotton physiological indices, cotton yield, cotton biomass was observed. The results of nitrogen single-factor effect showed that the population physiological indices, yield, biomass, and nitrogen application rate were significantly and positively correlated when the nitrogen rate of 56.2–122.8 kg hm^{-2} was applied and under AF irrigation the cotton growth and development can be coordinated by utilizing the water and nitrogen coupling effect on cotton physiological indicators, biomass and yield, consequently increasing cotton yield and water and nitrogen use efficiency.

6.3.1.12 POULTRY LITTER

Poultry litter is commonly recommended for cotton because it is an effective N fertilizer and has considerable amounts of K and Mg but whether the sufficient amount of K and Mg needed by cotton is present in poultry litter has not been documented. So to determine whether the commonly recommended litter rate of 4.5 Mg ha^{-1} supplies sufficient K and Mg obtained from the same improves Mg nutrition in cotton. Tewolde et al. (2010) conducted an experiment with broiler in Mississippi at Coffeeville and Cruger which had contrasting soil K and Mg levels. The results revealed that 4.5 Mg ha^{-1} litter supplies sufficient K and Mg nutrition is dependent on the sufficient N fertilization in cotton, therefore the external N supply might be more important to cotton Mg nutrition than the external supply of Mg.

6.3.1.13 IMPACT OF CROP ROTATION ON GROWTH

A study has been undertaken by Pettigrew et al. (2016) in Mid-South United States (US) on cotton growth and agronomic performance when grown in rotation with soybean. The results revealed that cotton plants were 13% taller in the years when it was grown following soybean, owing to interception of 6% more sunlight. These cotton plants were found to contain 13% higher leaf Chl concentrations in comparison to those cotton plants that were grown continuously without crop rotation. Rotation of cotton after soybean also produced increased yields, though there was no effect of the rotation sequences on fiber quality. They attributed the increase in yield of cotton to the increased soil N via N fixation from the prior soybean crop and/or due to altered soil microbial populations favorable to the subsequent cotton crop.

6.3.1.14 CONTINUOUS CROPPING EFFECT ON SOIL MICROFLORA

Cotton is an important cash crop. It is usually cropped continuously for a long term. This often results in the imbalance of soil microbial ecology, increase in the incidences of soil-borne diseases, decrease in yield and quality. This prevents the cotton industry. Several research studies have shown that biochar application can increase soil microbe activity. Biochar is effective because of its higher porosity, large specific surface area, and effective nutrient absorption ability. Most of these were effective in modifying soil environment, controlling soil-borne diseases, and in alleviating continuous cropping obstacles. YunYan et al. (2017) investigated the effect of cotton stalk biochar application on the soil microflora in a continuous cotton cropping sequence under usage of antagonistic actinomycetes. The results showed that biochar application brought an increase in number of bacteria, actinomycetes, fungi, and antagonistic actinomycetes in the soil. The antagonistic actinomycetes exhibited a change in number and proportion of dominant microorganisms in the soil. Furthermore, there was also an improvement in the number and proportion of *Bacillus* and *Streptomyces.* The number and proportion of *Micromonospora* were lower, though there was an increase in the number of *Aspergillus oryzae*, *Aspergillus niger*, and *Trichoderma koningii*, there was a decrease in their respective proportions. Therefore, their research studies have shown that the application of antagonistic actinomycetes and cotton stalk biochar can bring about an increase in the number of biocontrol agents, enhance disease-controlling and growth-promoting abilities of antagonistic actinomycete, and also improve the structure of soil microbial community. Thus, it had a potential to alleviate the adverse effects of continuous cropping of cotton.

6.3.1.15 NO-TILL/COVER CROP

Li et al. (2013) reported the wheat or rye cover crop allelopathic effect on growth and yield of no-till cotton. Cover crops suppressed cotton growth and reduced its lint and seed yield. The magnitude of suppression varied over years. Cotton planted into cover crops had lower leaf chlorophyll content. In soils where cover crops were grown, there were elevated ($P < 0.05$) concentrations of three allelochemicals, namely, [2,4-dihydroxy-7-methoxy-(2H)-1, 4-benzoxazin-3(4H)-one (DIMBOA); 2,4-dihydroxy-(2H)- 1,4-benzoxazin-3 (4H)-one (DIBOA); 2-benzoxazolinone (BOA)]. Thus, their results suggested that a reduction in cotton growth and yield when grown with cover crops might partly be attributed to the release of the allelochemicals by the cover crop.

6.3.1.16 FILM MULCHING ON CARBON FLUX OF COTTON

Environmental factors and human activities influence the carbon fixation and emissions that were produced from croplands. Li et al. (2018) conducted a study in Wulanwusu, northern Xinjiang, an arid region of Northwest China and recorded −304.8 g C m^{-2} (a strong sink) cumulative net carbon flux (NEE) by considering the associations among carbon fluxes and environmental factors in a drip-irrigated, film-mulched cotton field in arid region. The net carbon flux was found to be higher in the cotton cropland devoid of plastic film mulching and drip irrigation. Furthermore, the gross primary production (GPP) associations to air temperature (T_{air}), net solar radiation (R_n), and soil water content (SWC) was found to be greater when time is increased from a half-hour to a month. This was attributed to plastic film mulching protection and resistance and flexibility of ecosystem. At time scales from minutes to days, a strong correlation was noted with the GPP and R_n than T_{air}. SWC and vapor pressure deficit (VPD) showed a weakly correlation with GPP at all time scales. The soil had sufficient water because of plastic film mulching and regular drip irrigation.

6.3.1.17 MONITORING OF COTTON GROWTH

Chao et al. (2014) used normalized differential vegetation index (NDVI) in growing season data on FY-3/MERSI data for monitoring cotton growth during growing season. The FY-3/MERSI data enabled in monitoring growth in large cotton areas while it could not predict growth effectively particularly in the small patch. They mentioned that the normalization weight method for monitoring the cotton growth is better than mean value method.

6.3.1.18 GROWTH MODELS

Jones et al. (1974) used the SIMCOT II, simulation model for analyzing the nitrogen requirements of cotton (*Gossypium hirsutum* L.). The amount of daily N absorbed by the crop depends on its requirement for the growth of new flush and N available in the soil pool. However, during shortages of N periods, the N requirements of the new growth are met from the supply of N compounds from the breakdown of some reserve compounds. Based on the maximum and minimum N concentration for new growth of leaves, stems, roots, burrs, and seed of cotton, the daily N requirements were calculated for these plant parts. It was observed that during the period of N shortages there was a reduction in the new growth. Furthermore, it has also resulted in increased abscission of leaves, bolls, and squares. The SIMCOT II model run on the basis of N balance during the boll setting period has indicated that shortages of N would lead to limitation of growth and abortion of reproductive organs. The yield curves and simulated characteristics of fruiting were identical to that of real plant.

Wang et al. (1977) has studied a cotton–herbivore interaction through the development of a population model for plant growth and development. They have used this model in Acala SJ-II cotton to simulate its growth and development in California. They gave a mathematical framework to couple the plants and herbivores interactions as well as the biological inferences of their harm to the plant.

COTCO2 is a cotton growth simulation model developed for understanding the response of growth of cotton to varying atmospheric CO_2 concentrations. Wall et al. (1994) has utilized this model for understanding the cotton growth by this model which is capable of simulating the major plant processes such as photosynthesis, photorespiration, stomatal conductance, and its role in leaf energy balance which are likely to be influenced by CO_2. Furthermore, C_3 plants at the carboxylation and oxygenation level the impact of atmospheric CO_2 concentration on photosynthesis and photorespiration can be simulated with this model. They have utilized this model for stimulating the growth of different organs, namely, leaf blade, stem segment, taproot and lateral roots, and fruit including the squares and bolls and by measuring the carbohydrate and nitrogen requirement the potential growth was calculated. Furthermore, they calculated the actual growth on the basis of substrate availability, the potential growth, and water stress. They were mainly interested in describing the model general structure, its current position, and developmental strategies of the future.

Zhong et al. (2011) maximized the original GOSSYM model for coupling it with the regional Climate–Weather Research Forecasting model (CWRF) and developed a geographically distributed cotton growth model. The model comprised the physics, redesign, and software. They developed it keeping in view that there is a need of the coupling model system that would represent the climate and cotton interactions and predict the cotton (*Gossypium hirsutum* L.) production under a varying climatic conditions.

6.3.1.19 COSIM MODEL

Xinjiang is the largest cotton producing area in China accounting for more than 50% of its total cotton production. XueJiao et al. (2017) by using COSIM model in Xinjiang developed a dynamic prediction method for cotton yield. The measured climatic data in the recent 50, 30, 20, 10, and 5 years was substituted for the unknown climatic data from forecasting day to harvest day. Using the results, they validated the best scheme for yield prediction that could predict the accuracy reaching to 81.3–99.6%. They specified that for prediction of the regional cotton yield, consideration should be given to sowing time.

Crop growth modeling enables the assessment and prediction of the optimum nutrient requisites of different crops and has arisen as an efficient tool. As phosphorus (P) is one of the most important major nutrient and in commercially important arable crops like cotton (*Gossypium hirsutum* L.), there is a need to develop strategies which can optimize phosphorous use. Thus, Amin et al. (2016) undertook a phosphorus use study in cotton. For semi-arid climate in Vehari-Punjab, Pakistan, the CSM-CROPGRO-cotton model was used. In this study, under the semi-arid climate of southern Punjab, Pakistan, two cotton cultivars were grown by applying P at three different rates and the observed and simulated P use in cotton was estimated by evaluating Decision Support System for Agrotechnology Transfer (DSSAT) sub-model CSM-CROPGRO-Cotton-P. It is concluded that to forecast the cotton yield in the semi-arid climate of Southern Punjab, cotton production system under different levels of P the CROPGRO-Cotton-P would be a suitable tool.

In North Xinjiang under mulched drip irrigation, Jun et al. (2016) studied modeling response of cotton yield and water productivity by calibrating and validating a two-dimensional soil water transport and crop growth coupled model. In Urumqi, Xinjiang Uygur Autonomous Region, China, during 2010 and 2011, cotton growing seasons cotton yield was estimated by applying three irrigation levels, namely, 50%, 75%, and 100% of full irrigation when the average soil moisture within the root zone (40 cm for the squaring stage and 60 cm for the bloom stage) was depleted to 60% and 70% of the field capacity until the soil water content reached to 85% and 95% of the field capacity for the squaring and bloom stage. Irrigation schedule was determined by measuring weekly soil water content with Trime-FM probe to 100 cm depth. At squaring, bloom, and boll-forming stages, leaf area index and aboveground biomass of cotton plant data was recorded. With CHAIN 2D and the crop growth model of EPIC integrated program subroutines and functions, the coupled model was coded and written in FORTRAN 90 for Windows system. To study the interaction between root water uptake and crop growth in this coupled model, the root water uptake model of Vrugt was coupled with the root depth growth model. Furthermore, the response of cotton yield and water productivity to irrigation amount was studied by calibrating and validating the two-dimensional soil water transport and crop growth coupled model by soil water content dynamic, crop growth indexes, and seed cotton yield obtained from the field experiments. The results revealed that with increasing amount of irrigation the seed cotton yield and water productivity were increased and decreased through quadratic functions and indicated that in the North Xinjiang region for cotton 280–307 mm is the suitable mulched drip irrigation amounts in view of yield and water productivity and in predicting the soil moisture, above ground biomass, seed cotton yield and total water use percentage the coupled model performed well.

To predict the responses to various situations of climates and irrigation strategies, a model of crop growth AquaCrop is used. Shuai et al. (2018) evaluated the performance of AquaCrop by conducting 4-year irrigation experiment in 2012, 2013, 2015, and 2016 growing seasons for cotton with irrigation treatments of covering full (100%), over (115% and 145% of full), and deficit (55–90% of full) in a saline region of southern Xinjiang of China with film-mulched drip irrigation for two typical soils and the suitable irrigation amounts for cotton under many scenarios of initial soil–water content (SWC) and soil salinity were determined. Based on the suggested parameters for cotton in AquaCrop manual, the model was calibrated using 2016 data sets and validated with other year data sets. With coefficient of determination r^2 >0.77 canopy cover, soil water storage of the root zone and above ground biomass simulations was found to be fitted well with the field observations but the yield was slightly underestimated with the index of agreement d >0.92. Simulation for less than 80% of full irrigation treatments was given as reliable for soil salinity, whereas it was underestimated for over 80% of full irrigation treatments.

In crop management, estimation of yield is a crucial task. But the traditional methods of estimation are costly, time-consuming, and difficult for extension to large areas, while remote sensing can provide quick coverage over a field at any scale. So for large-scale earth observation satellite remote sensing is used. Huang et al. (2016) used very high-resolution digital images (2.7 cm pixel-1) obtained from an inexpensive small multi-rotor UAV for estimating cotton yield by two methods: (1) to estimate cotton plant height, three-dimensional point cloud data got from multiple digital images of the cotton field was used and henceforth estimate yield and (2) from the background of the digital images of the defoliated cotton field just before harvest cotton boll signatures were segmented and then by estimating cotton plot unit coverage yield was estimated. The results showed that by estimating plant height (R^2 = 0.43, compared with R^2 = 0.42 for yield estimation through manually measured plant height) the cotton yield can be estimated accurately using low-altitude remote sensing with an inexpensive small UAV. Furthermore, by estimating cotton boll coverage in each plot the method can offer consistent cotton yield estimation with Laplacian image processing. This study possibly will assist in estimation of cotton yield.

6.3.2 FLOWERING AND FRUITING

Cotton (*Gossypium hirsutum* L.) lint yields were increased by effective plant breeding efforts. Wells and Meredith (1984) undertook a study to give an insight of reproductive dry matter partitioning on cotton yield in Obsolete and Modern Cotton Cultivar by monitoring 12 cultivars at different times during the season from Stoneville and Deltapine background for the reproductive growth by numbers and dry weights of squares, immature bolls, and mature bolls. They observed that modern cultivars of cotton produced flowers much earlier and squares in greater proportion compared with obsolete cultivars. In the newer cultivars, the production of early squares and flowers in greater proportion has contributed to large amounts of early bolls. The abscission of bolls that occurred between 117 and 142 DAP was similar in most of the cotton cultivars. "STV 213,"

"STV 825," "DPL 41," and "DPL 16" modern cultivars produced higher reproductive to vegetative ratios 0.95, 0.86, 0.84, and 0.71 kg reproductive dry weight/kg vegetative dry weight at 117 DAP. A large proportion of lint yield was produced in modern cultivars mainly because of a greater apportioning ability of dry matter to reproductive organs and also to the augmented reproductive development. This was a resultant of the maximum production of leaf mass and leaf area.

Vistro et al. (2017) conducted a study at the experimental fields of Cotton Section, Agriculture Research Institute, Tandojam during the year 2013–2014 to investigate the impact of plant growth regulators on the growth and yield of cotton, in sindh-1 variety of cotton, various plant growth regulators significantly affected the growth and yield character at ($P < 0.05$). Pix applied at 1000 mL/500 L of water at bud formation resulted in the maximum plant height (137 cm), monopodial branches per plant (1.9), symonopodial branches per plant (23.0), opened bolls per plant (30.1), unopened bolls per plant (4.0), seed cotton yield per plant (97.4 g), seed cotton yield (3074.0 kg ha^{-1}) and G.O.T (34.5%). It is concluded that at 5% probability level all the characters in cotton variety sindh-1 were significant except G.O.T%. Their study indicated that yield of cotton crop can be enhanced by applying plant growth hormones effectively. An application of Pix at bud formation at 1500 mL/500 L of water would be more effective in obtaining not only more number of bolls per plant and maximum seed cotton yield per plots but also for obtaining maximum seed cotton yield kg ha^{-1}.

6.3.2.1 *CARBON DIOXIDE ENRICHMENT*

Mauney et al. (1994) investigated response of cotton (*Gossypium hirsutum* L.) growth and yield to a free-air carbon dioxide enrichment (FACE of 550 ymol mol^{-2} CO_2) in an open field near Maricopa, Arizona for three growing seasons. In a season in 1991, CO_2 concentration by 48% resulted in 37% increase in the biomass, while there was an increase of 43% in harvestable yield. They attributed this to an increase in early leaf area, more profuse flowering, and a longer period of fruit retention. The WUE was increased to the equal extent in both irrigated and water stressed plots by FACE treatment. An increase in biomass production, rather than a decrease of consumptive use caused an increase in WUE.

Different factors influence development and productivity of flowering and fruiting stage.

6.3.2.2 PLANT DENSITY

Planting density has an influence on yield, yield components, as well as fiber properties of cotton. Sadik et al. (2017) in Aydın Province observed that planting density studied in altered rows 3, 6, 9, 12, 15, 18, 21, and 24 cm under the wheat/cotton double crop showed a significant effect on characteristics such as seed cotton yield, number of days to open the first boll, the number of monopodial and sympodial branches, number of bolls, 1st position boll number, and also on the seed cotton weight per boll. However, it had an insignificant impact on other characteristics. They concluded that the most appropriate distance on row is 18 and 24 cm for secondary product cotton production.

Ali et al. (2017) investigated the 28 upland cotton genotypes by environment and GGE-biplot analyses for seed cotton yield in in three locations. In genotypes (G) across environments, significant ($P < 0.01$) differences were observed for seed cotton yield. Over different years and locations for genotypes an average mean performance were observed. Due to G × Y × L, the interaction effects were also significant ($P < 0.01$) and the involvement of genotypes, environments (years, locations), and their interactions in total sum of squares varied from 3.01 to 37.90%. Significant variations were found among different genotypes in different locations. Seed cotton yield was significantly and positively correlated ($P < 0.01$) with earliness, morphological, and yield traits, whereas seed cotton yield was negatively correlated with majority of the fiber quality traits. Genotypes NIBGE-4 and IR-NIBGE-2620 were distinguished as best cultivars with high stability and seed cotton yield based on stability analysis, GEI, and GG-biplot analysis.

In the Mississippi for cotton production the possibility of using modern transgenic cotton cultivars in ultra-narrow rows (<38 cm) and its effect on growth, lint yield, and fiber quality were investigated by Nichols et al. (2004). They compared the cotton grown in ultra-narrow row spacings with cotton grown in 101 cm spacing and ultra-narrow row crop exhibited a decrease in plant height and number of sympodia, total nodes, and total bolls per plant with equal to or higher lint yields than 101 cm spacing crop. Significant interactions occurred between spacing and cultivar, which affected the mean lint yield. In ultra-narrow rows the glyphosate-resistant transgenic cultivars produced better or equal yields in 2 of 3 years than conventional cultivars. However, there was no improvement in the lint yield of the okra-leaf cultivars even in the narrow row spacing. Thus, they could not make any conclusions about the effect of plant stature on lint yield, where row spacing had negligible effect on fiber quality and ultra-narrow row spacing appears as a viable agronomic parameter affecting cotton growth and development.

6.3.2.3 EFFECTS OF TEMPERATURE

Pettigrew (2008) in Mississippi Delta undertook a study on two cotton (*Gossypium hirsutum* L.) genotypes (SureGrow 125 and SureGrow 125BR) grown under a warm temperature regime (about 1°C warmer) and an ambient temperature control to record the variances in agronomic and physiological performance. There was no variation in the genotypes in their response to the temperature regimes. Lower nodes above white bloom (NAWB) were observed under warmer temperatures which indicate a somewhat advanced crop maturity. In the warm regime, the production of a 6% smaller boll mass, with 7% less seed produced per boll were observed which resulted in 10% lower lint yield in 2 out of 3 years. This reduction was found to be resultant of the warm regime. But the warm temperature regime produced a 3% stronger fiber. Ovule fertilization may get affected when temperatures become too hot, leading to reductions in seeds produced per boll, boll masses, and finally, lint yield.

6.3.2.4 COTTON BOLL DEVELOPMENT AND PRODUCTION

Several interesting studies were undertaken on cotton boll development, its production, and fiber properties (Fig. 6.4).These studies were directed to relate the productivity of cotton, boll, and fiber properties with carbon assimilation by leaves and its translocations to bolls which were confirmed by adopting experimental techniques.

6.3.2.5 STEM NODE COUNTS IN GROWTH MONITORING

The strategies that can be adjusted in response to expected or realized changes in plant growth and development need to be established to improve the management efficiency in cotton (*Gossypium hirsutum* L.). For plant development, practical and active measuring tools are the successive counts of main-stem nodes bearing sympodia with a preflower fruiting structure in the first position (Bourland et al., 1992). Using main-stem

FIGURE 6.4 Fruiting and fiber development.

node counts, the concept for monitoring the growth and development of cotton plants was proposed. They suggested a technique of monitoring the growth by a count of number of main-stem nodes above the sympodial branch bearing a white flower in the first position from the main axis (NAWF). By associating NAWF to the crop growth and yield parameters measurements, the physiological basis of NAWF was established. Then, by retaining, size and number of seeds related with first-position bolls NAWF critical value was assessed. The effects of cultivars, plant densities, irrigation, and two insect pests on NAWF were evaluated in separate tests. The number of NAWF was found to be closely associated to deviation in canopy photosynthesis, signifying that crop growth activity can be determined using NAWF. The potential economic value of flowers calculated on the basis of individual boll measurements showed a rapid decrease as NAWF reached to 5.0 and the number of days to NAWF of 5.0 (representing the last effective flower population) was calculated from regression equations of NAWF by days from planting. Using this management factors influences on plant development were differentiated.

It is suggested that the foremost advantage of this monitoring technique could be offered with increased precision and confidence in end-of-season management decisions.

6.3.2.6 PHOTOSYNTHETIC CARBON PRODUCTION

Wullschleger and Oosterhuis (1990) investigated the developing of cotton leaves and bolls for photosynthetic carbon production and its use. They assert that for defining the yield productivity, the estimation of leaves photosynthetic C contribution to vegetative and reproductive processes is very much essential. In cotton (*Gossypium hirsutum* L.), the canopy stands morphological complexity creates an important barrier in determining specific yield relationships. In a 2-year study undertaken by Wullschleger and Oosterhuis (1990) it was observed that just before subtended flower anthesis the sympodial leaves attained maximum photosynthesis and afterward it exhibited a decline throughout the boll-filling period. Furthermore, the rare synchronization between C production and C utilization was revealed from the evaluation of the individual sympodial leaves and their subtended bolls at mainstem Nodes 8, 10, and 12 carbon budgets. At mainstem Node 10 Carbon import requirements for the first three fruiting positions were 50, 37, and 21%, respectively. For the bolls at mainstem Node 8 C import of >60% was found to be essential to sustain the growth during the season. It was observed that the subtended bolls can acquire the total C needed efficiently only from leaves at mainstem Node 12. Furthermore, the need of considerable translocation of photosynthate from adjacent leaves and leaves outside the mainstem node was indicated by the deficiencies of carbon found at mainstem Nodes 8 and 10. By studying the movement of 14C-assimilate from leaves to developing cotton bolls for short term this hypothesis was verified. Therefore, the assessment of avenues for increased yields via crop genetics were simulated by breeding objective for enhancement of C production by leaves. These simulations offered a benefit of improved leaf longevity in the continuance of C production.

6.3.2.7 TEMPERATURE AND CARBON DIOXIDE EFFECT ON GROWTH AND DEVELOPMENT

Reddy et al. (1999) reported that at Mississippi State the boll and fiber growth parameters of cotton (*Gossypium hirsutum* L.) (cv. DPL-51) gets affected with the temperature and atmospheric carbon dioxide concentration [CO_2]. Boll size and maturation periods were declined with an increase in temperature. Though the boll growth exhibited an increase with an increase in temperature to 25°C, it showed a decline in growth at the highest temperature. The rates of boll growth, size, or the maturation periods were not influenced with the atmospheric [CO_2]. The most sensitive period of cotton development was found to be the boll retention period. At high temperatures, that is, 1995 plus 5 or 7°C the continuous production of squares and bolls was observed but almost no bolls were retained to maturity. So, for cotton boll survival the upper limit is 32°C, or 5°C warmer than the 1995 US Mid-South ambient temperatures. Furthermore, 40% more squares and bolls were produced at 720 $\mu L\ L^{-1}$ atmospheric CO_2 through temperatures. At less than optimal temperatures for boll growth (25°C) the fibers that were developed were longer, however, with an increase in temperature, more uniform fiber length distributions were observed. On the other hand, the increase in temperature up to 26°C increased the fiber fineness and maturity linearly, which further exhibited a reduction at a temperature of 32°C. A linear decrease in short-fiber content was observed as the temperature increased from 17°C to 26°C though it was greater at higher temperature. On any of the fiber parameters they could not record an elevated atmospheric [CO_2] effect, although tremendous effects on boll set and properties of fiber were noted from the changes in temperature. Thus, they suggested that under optimal water and nutrient conditions, the necessary functional parameters to construct fiber models may be obtained from the relations among temperature, boll growth, developmental rate functions, and fiber properties.

Similarly, a study on carbon dioxide and nitrogen nutrition effects on growth, development, yield, and fiber quality in cotton cultivar NuCOTN 33B by Reddy et al. (2004) revealed that an increase of [CO_2] under both N treatments led to a decline in N concentration of leaf. With these low N concentrations of leaf the elevated [CO_2] effect in the production of higher lint yields was not influenced. Plants grown at elevated [CO_2] and N+ conditions exhibited a greater response.The [CO_2] exhibited nonsignificant effect on fiber quality. However, the N concentrations of leaf, which showed variation with [CO_2], affected most of the fiber quality parameters either positively or negatively. During boll maturation period significant positive correlations were observed between Leaf N and mean fiber length ($r^2 = 0.63$), fine fiber fraction ($r^2 = 0.67$), and immature fiber fraction ($r^2 = 0.65$), while Leaf N was found to be negatively correlated with mean fiber diameter ($r^2 = 0.61$), short fiber content ($r^2 = 0.50$), fiber cross-sectional area ($r^2 = 0.76$), average circularity ($r^2 = 0.74$), and micronafis ($r^2 = 0.65$). It is concluded that if N is optimum the fiber quality and yield get affected with future elevated [CO_2]. They suggested that the incorporation of these developed algorithms into process-level crop model will be much helpful in optimizing the production and fiber quality of cotton.

6.3.2.8 EFFECT OF ROOT ON COTTON GROWTH

Carmi and Shalhevet (1983) studied the effects of roots on cotton growth. They mentioned that root zone volume strongly influence the cotton (*Gossypium hirsutum* L.) plants growth and assimilates distribution among their vegetative and reproductive organs. Plants grown in small pots exhibited a decrease in root growth, along with a significant reduction in vegetative growth and an increase in the reproductive organs proportionate dry weight accumulation. Thus, they determined that the differences in root system development as affected by the pot capacity caused differences in cotton vegetative and reproductive growth rate and it is not affected with the deficiency of N, P, K, or water stress.

6.3.2.9 EFFECT OF POLYMER COMPOUNDS ON SOIL AGGREGATES

YuLian et al. (2017) conducted an experiment at Regiment 146, Shihezi, Xinjiang Uygur Autonomous Region to analyze the influence of different polymer compounds on soil aggregates and to identify the appropriate polymer compound to increase cotton yield under salt conditions using four treatments, namely, M1 (poly acrylic acid salt treatment, K-PAM), M2 (poly acrylamide treatment, PAM), M3 (cellulose treatment, HEC), and control treatment (no application of compound, CK). To evaluate the soil pH and conductivity of soil, samples from different growth stages of cotton were collected. The results reveal that compared with CK the soil conductivity under M1 and M2 treatments in different soil layers was low at flowering and boll development stage, while in M3 soil conductivity was found to high in 0–20 cm and low in 20–40 cm soil layer. At boll development stage in 20–40 cm soil layer compared with CK the soil pH was low in both M2 and M3 treatments, whereas in other growth stages of M2 treatment the soil pH was not affected, indicating the polyacrylic acid salt is most outstanding application and improves the soil structure increasing cotton yields under salt conditions.

Various biotic and abiotic stress factors affect reproductive growth, flowering, fruiting, boll formation, and productivity of cotton.

6.3.3 FACTORS AFFECTING GROWTH AND YIELD

6.3.3.1 ABIOTIC STRESS

6.3.3.1.1 UV Radiation on Cotton

Reddy et al. (2005) observed that the cotton plants grown in sunlight controlled-environment chambers under optimum water, nutrient, and temperature conditions exhibited decrease in dry matter production, plant height, leaf area on exposing to higher UV-B radiation levels 16 kJ m^{-2} d^{-1} and in these genotypes certain changes were induced by UV-B radiation. They classified the cotton genotypes, DP 458B/RR, NuCOTN 33B, and DP 5415RR as tolerant; Pima S7 and FM 832 B as intermediate; and SG 521 B and Tamcot HQ95 as sensitive on the basis of Total Response Index (TRI).

6.3.3.1.2 Increased UV-B Radiation

Hong et al. (2017) reported the increased UV-B radiation reduces cotton photosynthetic leaf area, chlorophyll content, and net photosynthetic rate due to the destruction of PS II reaction center. To verify this under field condition all the growth stages of cotton were exposed to 20% and 40% increased UV-B radiation using ultraviolet lamp and dry matter accumulation, photosynthetic pigment content and seed cotton yield were studied. The results revealed that the increasing UV-B radiation inhibits the growth of cotton stems, leaves, dry matter accumulation, and seed cotton yield with more pronounced effect at seedling stage. Under 20% and above ambient UV-B radiation the contents of chlorophyll a (Chla) and chlorophyll b (Chlb) found to be increased, while under 40% ambient UV-B radiation Chla, Chlb, and Chla/Chlb decreased with no change in stomatal conductance (Gs) and transpiration rate (Tr) and increased intercellular CO_2 concentration (Ci) showing that the non-stomatal limitation factors causes the reduction in photosynthesis. With enhancing UV-B radiation the decreased maximum quantum efficiency (Fv/Fm), operating efficiency (ΦPS II), linear electron transport rate (ETR), and photochemical quenching (qP) of PSII and increased non-photochemical quenching (NPQ) were confirmed by chlorophyll fluorescence parameters analysis, indicating that all the chlorophyll fluorescence parameters were associated with Pn changes which decreases under increased UV-B radiation.

Jin et al. (2010) conducted a field trial with cotton cv. Xinluzao 13 in Wulanwusu Agrometeorological Experiment Station, Xinjiang, China in 2006. They studied the effect of enhanced ultraviolet radiation (UV-B, 280–320 nm) at 0, 0.5, 1, and 1.5 W/m^2 on physiological indices, yield, and quality of cotton fiber. There was an increase in the chlorophyll content and soluble protein content initially which later exhibited a decline with an increase in UV-B radiation dose. Furthermore, there was also an increase in the contents of proline, while there was a decrease in the yield and quality of cotton. Thus, their results indicated that an enhancement of UV-B radiation results in the damage to cotton and a greater injury arises when this radiation exceeds the threshold of the cotton plant, resulting in a drastic decline in the quality and yield of cotton.

6.3.3.1.3 High Temperature

In cotton, the yield gets affected due to the high temperature during the reproductive stage. Rahman (2004) evaluated the upland cotton for cellular membrane thermostability (CMT) during the fruiting stage by quantifying the electrolyte leakage after heat treatment. Except cotton in several crop species the heat tolerance and higher yields were owing to higher CMT. There was variation in CMT for both upland cotton cultivars and hybrids. With changing temperature regimes the cultivars and hybrids relative ranking get modified, however the heat-tolerant and susceptible groups stayed moderately stable. FH-634, CIM-448, HR109-RT, and CIM-443 cultivars appeared as heat susceptible, while cultivars FH-900, MNH-552, CRIS-19, and Karishma were comparatively heat tolerant (thermostable). Among cotton cultivars a greater relationship was observed among CMT and SCY than hybrids. In the presence of heat stress a higher SCY due to higher CMT was indicated by the regression analysis. Furthermore, under supra-optimum greenhouse conditions CMT was found to be positively associated with SCY. The positive relation was also observed under early and late field regimes. However, under optimum (non-stressed) greenhouse conditions CMT showed a negative association with SCY. The results revealed that in upland cottons both susceptible cultivars and hybrids gave higher yields in the environment lacking heat stress, indicating that these two traits were not dependent on each other and their association is determined by the presence or absence of heat stress. Under heat-stressed conditions the difference in the ability of cotton cultivars and hybrids to regulate CMT reflects its physiological adjustment to heat stress or heat hardening in upland cotton. It was recommended that for distinguishing heat-tolerant and susceptible cottons, CMT may be used as a useful technique.

6.3.3.1.4 Quantitative Trait Locus Mapping

Drought and salt tolerances being quantitative traits are controlled by multiple genes, environmental factors, and their interactions. In upland cotton (*Gossypium hirsutum* L.) the drought and salt stresses can result in above 50% yield loss. *G. barbadense* L. (the source of Pima cotton) possesses desired traits for abiotic and biotic stress tolerance accompanied by high fiber quality. Owing to interspecific hybrid breakdown the introgression of drought and salt tolerance from Pima to upland cotton has been a challenge (Abdelraheem et al., 2018). Under the greenhouse and field conditions, Abdelraheem et al. (2018) undertook drought and salt tolerance quantitative trait locus mapping in upland cotton introgressed recombinant inbred line population. For this study, TM-1/NM24016, an upland recombinant inbred line population with stable introgression from Pima cotton was used. Using total of 1004 polymorphic DNA marker loci comprising RGA-AFLP, SSR, and GBS-SNP markers genetic map of 2221.28 cM span was constructed. The morphological, physiological, yield, and fiber quality traits of the population and its two parents were estimated. The results revealed that the decline in cotton plant growth at the seedling stage and reduction in lint yield and fiber quality traits in the field were due to drought under greenhouse and field conditions and salt stress in the greenhouse. Out of total 165 QTLs for salt and drought tolerance in cotton most of the chromosomes exhibited the phenotypic variation of 5.98–21.43%. For salt and drought tolerance a common QTL from among these 165 QTL were detected under both the greenhouse and field conditions. From multiple tests in the greenhouse and the field, this study reported the consistent abiotic stress tolerance QTL for the first time which can be used in understanding the genetic basis of drought and salt tolerance and to develop cotton cultivars with abiotic stress tolerance by means of molecular marker-assisted selection.

6.3.3.2 BIOTIC STRESS

6.3.3.2.1 Temperature Effects on Insects of Cotton

Saad et al. (2014) conducted a study in determining the temperature influence on the phytoseiid mites' population dynamics on cotton cultivars; Bt (Bt555, Bt-Tarzan-1, and Bt-3701) and one non-Bt (Anmol) in Sargodha. With simple correlation and regression methods the data was analyzed which revealed that population of these mites was significantly positively affected by temperature.

6.3.3.2.2 Climate Change Effect on Growth and Yield

Reddy et al. (1997) undertook a comparative study of the global climate change effect on growth and yield of cotton at ambient and twice ambient atmospheric concentrations of CO_2 and at five temperatures using Mississippi 1995 temperature as a reference. At 23°C, 4–6 times higher leaf area and at 20 days after emergence dry weight as at average temperature (1995 ambient) were observed when grown at 28°C. An increase in temperature declined the number of days to first square, flower, and open boll; however, there was no effect on the above developmental rates with an increase in double the atmospheric CO_2. Temperatures higher than 28°C or 1995 average whole-season temperatures influenced the boll retention and growth in mid- and late-season and at 1995 ambient temperature plus 5 or 7°C the fruits were not retained to maturity. At 5–7°C above the 1995 ambient conditions vegetative growth of the entire season was not declined significantly through temperature but vegetative dry matter accumulation was increased by 40% across temperatures with twice ambient CO_2. In a distinct experiment, based on long-term average US mid-south July temperatures cotton grown at a range of temperatures showed comparable outcomes on fruiting.

Ye (2000) analyzed the flowering and fruiting genetic behavior at different development stages by employing an additive-dominant genetic model with genotype by environment interaction with 2-year data from 4 × 4 diallel crosses. The variance analysis outcomes revealed that at early period dominant effects mainly controls the behaviors of flowering and fruiting, but by additive effects afterward. On the flowering and fruiting behavior the insignificant effects of GE interaction were found when compared with genetic main effects. Before August 1st the nonsignificant positive dominant correlation and negative or zero additive correlation between average flower number, boll number at different development stages, and total number of bolls were unveiled from the covariance analysis. However, the contrary results were found later on. Different gene action intensity was found at different development stages from additional conditional variances analysis. The highest gene action level was observed in last 10 days of July and the first-middle 10 days of August and during the flowering and fruiting period time interval of study played a significant role on exploring the gene action. While choosing the time interval of investigation the consideration of the experiment, environmental condition, investigating traits, and developing period is essential.

6.3.3.2.3 High Temperature

Crop production mainly gets affected due to the extreme global climate change and global warming, high temperature stress. Extreme heat stress frequently occur in July and August in the cotton (*Gossypium hirsutum* L.) growing areas of China, especially in the Yangtze River valley, during the peak time of cotton flowering and boll loading, leading to lower boll set and lint yield. So to stabilize yield in the current and future warmer weather conditions there is a need to screen the cotton germplasm and cultivars for high temperature tolerance. Lie et al. (2006) commenced screening of 14 cotton cultivars for high temperature tolerance (10–50°C at 5°C intervals) for in vitro pollen germination and pollen tube growth. In different cotton genotypes in vitro pollen germination and pollen tube length changes in response to different temperature were observed. Among the 14 cultivars the average cardinal temperatures (T_{min}, T_{opt}, and T_{max}) showed variations and were 11.8, 27.3, and 42.7°C for pollen germination and 11.8, 27.8, and 44.1°C for maximum pollen tube length. The maximum pollen germination and pollen tube length varied from 25.2% to 56.2% and from 414 to 682 μm, respectively. Boll retention and boll numbers per plant also exhibited variations corresponding to the weather conditions. On the basis of principle component analysis of the combination of pollen characteristics in an in vitro experiment and boll retention testing in the field environment by screening method 14 cotton cultivars were categorized as tolerant, moderately tolerant, moderately susceptible, and susceptible to high temperature.

6.3.3.2.4 Herbicide Effect

Snipes and Byrd (1994) investigated the fluometuron and MSMA postemergence topical applications impacts on cotton yield and fruiting at Delta Branch Experiment Station, Stoneville for 4 years. In the cotyledon to 1-leaf growth stage these treatments of herbicide were applied to cotton. There was a visual injury of cotton after these treatments at 14 days after treatment. It was 14–28% for fluometuron, 9–26% for MSMA, and 22–34% for the combination. All herbicide treatments led to a decrease in the first harvest seed cotton yield. At later harvests, the herbicide treatments resulted in equal to or greater cotton yield compared with untreated control. Averaged over 4 experiment years, it was observed that there was no reduction in total bolls per plant; percentage of bolls in first, second, or outer positions; highest sympodium with two bolls; or number of sympodium with bolls in the first or second position by the herbicide treatments. But, node number of the first sympodia by one and 1.5 positions were increased by MSMA and the combination treatment indicating the delay in maturity.

6.3.3.2.5 Water Productivity

Jalota et al. (2006) using simulation analysis investigated the crop water productivity of cotton (*Gossypium hirsutum* L.)–wheat (*Triticum aestivum* L.) systems. The simulated results revealed the reduction in both yield and ET of cotton with the reduction in amount of irrigation water input below the economic optima. There was a decrease in CWP consequently to variable amounts depending upon soil texture, precipitation, and irrigation regimes. A higher decrease in CWP was not observed in cotton with a reduction in the post-sowing irrigation. Precipitation increased the CWP. The crop growth stages flowering to boll formation in cotton was identified as most sensitive to water stress.

6.3.3.2.6 Mineral Nutrition

Nutrient availability in the soil has a strong influence on the vegetative development and attainment of reproductive stage and yield. Gardner et al. (1967) in his study on effects of nitrogen in cotton on its vegetative and fruiting characteristics observed that at early growth stages in peak flowering areas there was a reduction in the vegetative branches development, elongation of internode, and fruiting by N deficiency. On the other hand, when there is availability of adequate N, in longer seasons or "two-peak" flowering areas an increase in fruiting was observed and late season following earlier N deficiency also increased fruiting. Furthermore, it was also noted that even when there was limitation of N, the yield and fruiting were increased by N applications. N nutrition and yield and fruiting relationships were shown by analysis of petioles.

6.3.3.2.7 *Shade*

Zhao and Oosterhuis (1998) studied the variation in nonstructural carbohydrate composition in cotton at different growth stages of cotton in response to shade. Shade of an 8-day period (of 63% reduction in photosynthetic photon flux density) was imposed at four growth stages [i.e., pinhead square (PHS), first flower (FF), peak flower (PF), and boll development (BD)] and the effects were studied. The leaf photosynthesis, chlorophyll concentration, and nonstructural carbohydrate (hexose, sucrose, and starch) concentrations in leaves, floral bracts, and floral buds were determined to understand the effect of shade in these field-grown cotton plants. At all the four growth stages by shade a 43–55% reduction in leaf photosynthetic rate, and a 14 (on a leaf area basis) or 73% (on a dry weight basis) increase in total concentration of chlorophyll were observed, though there was not much influence on the leaf dark respiration rate. Furthermore, there was a sharp decline in concentration of starch in leaves and floral bracts, with minor changes in hexose concentration. Shade at the BD stage decreased the total nonstructural carbohydrate (TNC) concentration in 20-days-old floral buds. Among the four growth stages, the maximum reduction in leaf TNC concentration was caused by the shade at the PF (1994) or BD (1993) stage. Furthermore, a greater reduction in the bracts and floral buds TNC concentrations was observed due to shade at the BD stage. Shade during plant reproductive growth led to a significant decrease in leaf photosynthesis and TNC concentrations.

6.3.3.2.8 *Film Mulch*

Li et al. (2018) analyzed the associations between carbon fluxes and environmental factors in a drip-irrigated, film-mulched cotton field in Wulanwusu, northern Xinjiang, an arid region of Northwest China using Eddy covariance measurements. Their results revealed a very high [−304.8 g C m^{-2} (a strong sink) cumulative net carbon flux (NEE)]. This was still higher in the cotton cropland deprived of plastic film mulching and drip irrigation.

At time scales from minutes to days a strong association was seen between the gross primary product GPP and R_n than T_{air}, whereas at time scales from days to weeks it reversed. This result is largely determined by the biochemical characteristics of photosynthesis. As plastic film mulching and drip irrigation permit soil to uphold sufficient water weak correlations were seen between GPP and SWC, vapor pressure deficit (VPD) at all time scales.

6.3.3.2.9 Seed Quality

In a study, on 136 cotton seed samples (*Gossypium hirsutum* L.) from the 1973 National Cotton Variety Tests oil content, protein content, seed index, and percent immature seed were verified. The results revealed that the level of oil and seed maturity was influenced by the environment far more than the cultivars. Cultivars had more effect on protein content comparable as environmental conditions, however no interaction was observed. On the basis of the results it is suggested that breeders might find it more practical to screen for protein content than oil content, unless higher diversity for oil content can be witnessed in cotton germplasm (Turner et al. 1976).

6.3.3.2.10 Climatic Factor on Cotton Boll Production

Sawan (2009) studied the different climatic factors effects on flower and boll production in Egyptian cotton. Evaporation, sunshine duration, humidity, surface soil temperature at 1800 h, and maximum air temperature were assessed as the important climatic factors affecting flower and boll production. Furthermore, evaporation, minimum humidity, and sunshine duration were the most effective climatic factors during previous and later periods of boll production and retention. The flower and boll production showed a negative association with either evaporation or sunshine duration, whereas with minimum humidity that association was positive. They specified that the fourth quarter period of the production stage was the most suitable and practical time for collection of the data, to define the effectual prediction equations for cotton production. Thus, during this fourth quarter, evaporation, humidity, and temperature were the major climatic factors governing the production of cotton flower and boll. The 5-day interval was observed to be effectively and sensibly related to yield parameters. The most significant climatic variable influencing flower and boll production was found to be evaporation, followed by humidity.

Thus, they mentioned than an advance forecast of 5–7 days of accurate weather forecast would offer an opportunity to evade adverse impacts of climatic factors on cotton production and cultural practices can be used appropriately.

6.3.3.2.11 Floral Bud Loss, Photosynthesis, and Carbon Partitioning

Photosynthesis contributes to the productivity of a crop. In cotton, floral bud losses occur because of insect damage. Holman et al. (1999) studied photosynthesis and carbon partitioning in response to this floral bud loss caused by plant bugs (*Lygus lineolaris* Palisot de Beauvois) and bollworms (*Helicoverpa zea* Boddie). In the infested plants, 33% square abscission was observed at the first sympodial fruiting position while it was only 5% in control insecticide treated plot. Insect infestation resulted in a decrease in yield by 21%. It was observed that insect treatment increased the light penetration through the canopy by 4%, which lead to 17% increase in photosynthesis of the eighth main-stem leaf from the terminal leaf. At 4 weeks after the initiation of flowering, 21% higher canopy photosynthesis was found in the infested plants. CO_2 labeling studies in infested plants indicated a higher recovery of ^{14}C in the terminal node (terminal leaf plus main stem above the terminal leaf). Less of ^{14}C was found to remain in the branch at the same node as the source leaf. Though there was no increase in node number, there was an increase in plant height. Their study indicated that insect-induced abscission treatments also have alike effects as manual fruit removal treatments reported by others. They specified that by the use of either approaches future studies need to be justified. In the US mid-south variations in carbon exchange and allocation causes early fruit loss, but poor late-season growing conditions often avoids yield advantage.

Intensive research activities have been undertaken to determine the effects of various factors on flowering and fruiting stages cited below.

6.3.3.2.12 Water Potential Effect

Grimes and Yamada (1982) studied the correlations between minimum or midday leaf water potential (ψ1) and cotton (*Gossypium hirsutum* L.) growth parameters, mainstem elongation, and fiber growth in an experiment conducted over a 3-year period in the San Joaquin Valley of California to assess the suitability of using minimum (ψ1) as an index for irrigation scheduling and found that mainstem elongation (E_L) was highest after stress was lessened by irrigation. A minimum reduction (ψ1) (bars) there was a linear decline in E_L (cm/day). From the available observations they derived the function $EL = 5.08 + 0.200\,(\psi 1)$. A minimum (ψ1) drop in water potential to −24 bars ceased the elongation of mainstem. Fiber growth (elongation and weight increase) reduced markedly when (ψ1) reached −27 to −28 bars, indicating that fiber growth maybe a favored sink at high stress levels. A linear reduction in minimum (ψ1) reduced was observed with time after irrigation. Furthermore, they obtained high yields when minimum (ψ1) was permitted to drop to about −19 bars before irrigating.

6.3.3.2.13 Normalized Difference Vegetation Index for Nitrogen Stress in Cotton

Plant et al. (2000) mentioned that on plant growth and development relatively low cost set of detailed, spatially distributed data can be obtained effectively from remotely sensed electromagnetic reflectance data. The normalized difference vegetation index, or NDVI, is one such normally used vegetation indices. In California, in Acala, cotton field experiments with treatments of water or nitrogen stress level aerial photographs were taken. NDVI integrated over time showed a significant association with lint yield, this was evident in those experiments and there was a significant stress effect on yield. The stress factors were reflected from spatiotemporal pattern of NDVI. The NDVI stress patterns were also equivalent with the beginning of measurable water stress. The presence of nitrogen stress was indicated by NDVI even in those cases where the stress has not resulted in yield reduction. NDVI was associated with nodes above white flower and was strongly associated with nodes above cracked boll in a study of the correlation of NDVI with late season plant mapping indices. Thus, their studies have indicated NDVI is a better indicator of nitrogen stress than an alternate vegetation index, the relative nitrogen vegetation index.

6.3.3.2.14 UV-B Irradiance on Growth and Development

UV-B enhancements in many of the agricultural crops cause decrease in yield, change in species competition, decline in photosynthetic activity, susceptibility to disease, and variations in structure and pigmentation.

Gao et al. (2003) studied effect of higher UV-B irradiance under field conditions on cotton growth, development, yield, and qualities. Increased irradiance of UV-B 9.5% all through the growing season had a negative impact on cotton growth. There was a 14% reduction in height, 29% in leaf area, and overall 34% reduction in total biomass. Furthermore, it had also led to a decrease in fiber and a drop of 72% in economic yield and 58% reduction in economic coefficient.

6.3.3.2.15 Intercropping

The cultivation of more than one crop species on the same piece of land simultaneously is known as intercropping. Mostly it is considered as the practical application of basic ecological principles such as diversity, competition, and facilitation.

In northern China in Anyang, Henan Province wheat and cotton intercropping is practiced at large scales which varies in the number of wheat and cotton rows in the alternating strips of either crop and were categorized as 3:1, 3:2, 4:2, and 6:2. Zhang et al. (2007) determined dry matter accumulation, yield, land equivalence ratio (LER), and lint quality. To periodic harvest data expolinear growth equations were fixed to categorized the cotton growth patterns in monocultures and intercrops. Compared with cotton monoculture fixed parameters showed a growth delay of 11.8 days in the 3:1 system, 6.3 days in the 3:2 system, 6.9 days in the 4:2 system, and 5.6 days in the 6:2 system. In the linear growth phase, the growth rate was greatest in the monoculture (8.9 g m^{-2} d^{-1}) but not significantly different from 3:2 and 4:2 systems, it was higher in the 3:1 (7.0 g m^{-2} d^{-1}), 4:2 (7.7 g m^{-2} d^{-1}), and 3:2 (8.4 g m^{-2} d^{-1}) systems, whereas lowest in the 6:2 system (5.9 g m^{-2} d^{-1}). During the seedling phase of cotton these findings are inferred in terms of the competitive effect of wheat. In the 3:1 system this competitive effect was strongest causing a long growth delay. In the 6:2 systems the presence of large distance between cotton rows reduces the interception of radiation by the cotton leaf canopy after wheat harvest. It has resulted in a relatively low rate of linear growth. Their results specified that the intercropping effects on cotton quality were below detection thresholds.

Shopan et al. (2012) evaluated cotton-based intercropping for Northern Region of Bangladesh. At Cotton Research Centre the experiment on the cotton cost benefit analyses based intercropping was conducted, following

Randomized Complete Block Design with three replications. Compared with other treatments the (Cotton + Red amaranth + Potato + Maize + Sunhemp) treatment has given more benefits (Cost benefit ration 2.38) and in seed cotton yield, also for potato and wheat yield statistically significant and profitable result was found among six treatments.

6.3.3.2.16 Plant Density

Bednarz et al. (2005) made a Cotton Yield stability analysis across population densities on aTifton loamy sand (Fine-loamy, kaolinitic, thermic Plinthic Kandiudults in 1997 and 1998), where cotton was planted on 91 cm row widths at seeding rates ranging from 3.5 to 25.1 seeds m^{-}. There was production of more mainstem nodes and monopodial branches with better fruit retention in lower population densities. It has led to greater fruit production per plant. An inverse association was observed between boll size and population density and a similar inverse association was also observed between the mean net assimilation rate from first flower to peak bloom and population density. Furthermore, it was observed that with an increase in population density there was also an increase in the mainstem node of peak boll set and in the first sympodial position a greater fruit production on a ground area basis was seen, while that in third positions and monopodial branches was more with a decrease in population density. Similarly, increased population density has also resulted in an increase of accumulative seed cotton from sympodial branches. However, there was no influence of population density on the total fruit number and seed cotton yield per area. Through the manipulation of boll occurrence and weight across population densities the yield stability was attained.

6.3.3.2.17 Conservation Tillage and Irrigation Methods on Yield and Water Use Efficiency

In Darad area of Fars province, Afzalinia et al. (2014) evaluated the conservation tillage and irrigation method effects on the cotton yield and WUE. The cotton yield and WUE found to be significantly affected by the conservation tillage methods and irrigation. The highest cotton yield and WUE was found with the reduced tillage treatment, while the surface irrigation method has given the minimum cotton yield and WUE.

6.3.3.2.18 Root Depth and Water Consumption

The root distribution and its concentration in the soil surface layer have a profound influence on the water consumption and growth of cotton plants. Hao et al. (2014) studied the cotton root distribution and water consumption influenced by dripper discharge. At each growth stage the cotton root distribution and water consumption were measured under five different dripper discharge (0.5, 1.0, 1.5, 2.0, 2.5 L/h). The findings indicated that cotton roots were concentrated in the surface layer, 10–30 cm in the soil depth direction. However, with an increase in dripper discharge cotton root length concentrated place of root length was found be present in the shallow soil layer. As a negative exponential function of soil depth, they described the vertical distribution of cotton root weight, where the largest root weight found to be obtained from the soil surface layer. Furthermore, there was a decline in the ratio of deep root weight, with an increase in the discharge of the dripper. There was a gradual reduction in the cotton root length and root weight from the cotton stem to both sides on the horizontal direction. However, with an increase in the dripper discharge, though there was a decrease in the ratio of root length (or root weight density) below the cotton stem, there was an increase in the ratio of root length (or root weight density) in both sides. The study indicated that roots of cotton root changed their structure from a compact form to a lengthy structure. During the entire growth period the cotton field water consumption showed a bimodal type curve. There was an increase in the water consumption intensity at seedling stage, with an increase in dripper discharge. There was a delay in the moment of water consumption depth moving down, and postponement in the period of maximum water consumption.

6.3.3.2.19 Irrigation Regimes

Dagdelen et al. (2009) reported the effect of different drip irrigation regimes in N-84 cotton variety in the Aegean region of Turkey on cotton yield, WUE, and fiber quality. With an increase in water use an increase in leaf area index (LAI) and dry matter yields (DM) were found to exhibit an increase. Drip irrigation levels affected the fiber qualities. The results revealed that for the semi-arid climatic conditions, well-irrigated treatments (T_{100}) could be more useful and irrigation of cotton with drip irrigation method at 75%

level (T_{75}) would exhibit a significant benefit both in terms of large WUE and saving of irrigation water. Under conditions of limited water supply this would have a benefit of deficit irrigation. In an economic perspective, they mentioned that a saving in irrigation water (T_{75}) by 25.0% caused 34.0% decrease in the net income, though the net income of the T_{100} treatment was found to be reasonable in those areas that had no water shortage.

6.3.3.2.20 Supplemental Irrigation on Yield

Sui et al. (2014) studied the cotton yield and fiber quality influenced by application of supplemental irrigation (based on soil–water sensor measurements) and graded levels of nitrogen (N) (0, 39, 67, 101, 135, and 168 kg ha^{-1}) in the Mississippi Delta during 2011 and 2012. N uptake of cotton plants was reduced by soil–water stress. A positive linear correlation existed between yield and leaf N content under irrigated and non-irrigated plots. There was an improvement in the fiber quality, including all fiber length parameters with irrigation. Under irrigated and non-irrigated conditions, though there was a decrease in micronaire there was an increase in neps and yellowness with the rise of leaf N. Thus, the research findings have indicated that application of excessive amounts of nitrogen may have a harmful effect on cotton fibers properties.

6.3.3.2.21 Irrigation Frequencies and Picking Timings

Deho et al. (2017) used Sadori, Chandi-95, and Malmal, three promising cotton varieties to assess the irrigation frequencies and picking timings influence on seed cotton yield and its attributes. Four irrigation frequencies (five, six, seven, and eight irrigations) and four cotton pickings timings on the basis of percent boll openings (30%, 50%, 70%, and 90% boll opening) were used for estimation. From the study six irrigations frequencies and cotton picking at 50% boll opening verified as best for improving cotton yield and fiber quality.

6.3.3.2.22 Suitable Restructuring Tilth Layer

ShuLin et al. (2017) conducted a study on the soil physical and chemical properties, cotton development traits, weeds, disease, and presanility indices using four treatments, that is, T1 (exchanged a soil layer of 0–15 cm for that of 15–30 cm), T2 (exchanged a soil layer of 0–20 cm for that of 20–40 cm, loosed 40–55 cm layer), T3 (exchanged a soil layer of 0–20 cm for that of 20–40 cm, loosed 40–70 cm layer), and CK (rotary tillage in the depth of 15 cm). All the indices displayed variations among treatments. In treatments T2, T3, and at 20–40 cm soil layer in T1 the total N, available P, available K content was higher compared with CK. Restructuring tilth layers was efficient in controlling weeds and decreasing diseases and presanility. In T2, the disease and presanility indices (DPI) were decreased by 41.7% and 31.9%, leading to 6.1% and 10.2% increased lint yield in 2014 and 2015 compared with CK. Furthermore, before flowering stage restructuring tilth layers increased cotton roots decreasing the above ground dry accumulation but after flowering stage the dry matter accumulation, boll number per plant, boll weight, and lint yield was found to be higher in restructuring tilth layers compared with CK, indicating suitable restructuring tilth layer (T2) is most efficient measure to improve cotton yields and to solve the constraints of continuous cotton cropping fields.

6.3.3.2.23 Different Amounts of Water and Fertilizers

To obtain the optimal combinations of different amounts of water and fertilizers and to analyze the combined effects of seed cotton yield, water and fertilizer use efficiency, and economic benefits in northern Xinjiang of Northwest China. HaiDong et al. (2018) performed a field experiment where five N-P2O5-K2O fertilization levels (150-60-30, 200-80-40, 250-100-50, 300-120-60, and 350-140-70 kg ha^{-1}) were assigned to main plots and the sub-plots were allotted with three drip irrigation levels, termed as full irrigation (1.0 ETc, ETc is the crop evapotranspiration), medium irrigation (0.8 ETc), and low irrigation (0.6 ETc). The result showed that increasing irrigation water improved the LAI, dry matter accumulation, seed cotton yield, partial factor productivity (PFP), and the economic benefits for the same level of fertilizer application. Furthermore, at 362.3–462.5 mm irrigation period and 212.5-85-42.5 to 367.5-147-73.5 kg ha^{-1} fertilizer intervals (N-P2O5-K2O) the seed cotton yield, economic benefits, and WUE attained ≥90% of their maximum values as confirmed by a multi-objective optimization model via binary quadratic regression analysis.

6.3.3.2.24 Deficit Irrigation and Nitrogen Effect

Singh et al. (2010) conducted a field study for 2 years in cotton variety Ankur-651 Bt. They investigated the deficit irrigation (Etc 1.0, 0.9, 0.8, 0.7, 0.6, and 0.5), nitrogen (80, 120, 160, and 200 kg N ha^{-1}) and plant growth minerals effects on seed cotton yield, water productivity, and yield response factor. Along with 43.1% higher seed cotton yield 26.9% water was saved at 1.0 Etc drip irrigation. To the tune of 9.3% of the maximum yield imposing irrigation scarcity of 0.8 Etc caused a decrease in seed cotton yield. With drip Etc 0.6-0.5, N up to 160 kg ha^{-1} provided the highest yield, afterward it showed a decline. The seed cotton yield was increased by 14.1% with the foliar spray of plant growth mineral (PGM). At various irrigation and N levels the productivity of water ranged from 0.331 to 0.491 kg m^{-3}. On combined basis, at 20% irrigation deficit the crop yield response factor of 0.87 was calculated.

6.3.3.2.25 Effective Underground Water Levels

YaPing (2016) stated the major physiological indicators like SOD, POD, and MDA gets affected with low moisture in cotton. About 30 cm underground water levels were found to be appropriate for early development period of cotton, whereas for middle and late stages 50–70 cm was appropriate. In young cotton plants the superoxide radical level is very low so before flowering stage at different water levels the activity of SOD was approximately same. Under both high and low water levels MDA content was higher but in cotton 50–60 cm level was more suitable. The POD activity was low at depleted ground water level.

6.3.3.2.26 Above-Ground Dry Matter Accumulation

A field trial was conducted with three cotton cultivars (Bt variety 33B, conventional variety CCRI 12, and Bt hybrid CCRI 46) in a farm of Cotton Research Institute, Chinese Academy of Agricultural Sciences, China in 2007. ShaoDong et al. (2010) studied the cotton genotypes above-ground dry matter accumulation under five nitrogen application levels (0, 90, 180, 270, and 360 kg/ha). They simulated the above-ground dry matter with the expolinear growth equation, and described the nitrogen dilution effect with a power function. The expolinear growth equation was well fitted to the growth of cotton above-ground dry matter. The parameters could well reflect the effects of cotton genetics and nitrogen application on the accumulation of cotton above-ground dry matter. The power function of dry matter and nitrogen content showed the declination of cotton above-ground nitrogen content with the dry matter accumulation, and the parameter could be used to explore the ability of nitrogen uptake. The analysis of above-ground dry matter accumulation and nitrogen dilution effect showed that the optimal nitrogen application rate was 180 kg/ha and excessive nitrogen application had no effect with cotton growth and nitrogen uptake.

6.3.3.3 FERTILIZATION EFFECT ON COTTON GROWTH AND DEVELOPMENT

6.3.3.3.1 Phosphorus

WenXuan et al. (2018) investigated the P fertilizer treatments (0, 75, 150, and 300 kg P_2O_5 ha^{-1}) effects on root/mycorrhizal processes and P uptake by cotton plants to study the chance of gaining high P fertilizer use efficiency and high yield, by optimizing P fertilizer application. The enhanced root length, hyphal density, and also the increased spatial distribution of cotton roots in the soil were resulted at low P application rate (75 kg P_2O_5 ha^{-1}). This has also increased the apparent phosphorus recovery (APR). But, with 150 kg P_2O_5 ha^{-1} a greater P uptake and cotton yield was observed. Thus, this study indicated that it is hard to maximize APR and cotton yield at the same time.

6.3.3.3.2 Ammonium and Phosphorous Fertilization

Soil salinity in many arid regions of Central Asia affects crop production. In cotton (*Gossypium hirsutum* L) it inhibts root growth and the capacity to scavenge phosphorus (P) from soil. Xin Xin et al. (2018) reported that localized ammonium and phosphorus fertilization can increase the cotton lint yields by bringing about a reduction in soil pH of rhizosphere and salinity. They hypothesized that acidification of the rhizosphere pH would decrease salinity in the rhizosphere soil and increase plant growth and P nutrition. In a 2-year experimental study performed in a drip-irrigated cotton field in Xinjiang, China, with six combinations of localized nitrogen (N) and P applications, drip irrigated every 10–15 days after the early bud stage (60 days after sowing) indicated that the localized supply of superphosphate and ammonium sulfate as starter fertilizers reduced the pH from 7.6 to 7.4. At the seedling stage the salt content was also decreased from 2.1 to 1.7 g/kg in the cotton rhizosphere soil. There was an increase in the P uptake and above-ground biomass. Furthermore, rhizosphere soil pH was decreased and P uptake, above-ground biomass were increased by the fertigation with ammonium sulfate at the flowering stage. There was an increase of lint yield by 15% of the limited supply of superphosphate and ammonium sulfate and fertilizer management. It is concluded that to increase cotton production in highly saline soils in dry regions the localized ammonium sulfate and superphosphate may be an effectual soil amelioration measure.

6.3.3.4 MANIPULATION TECHNIQUES USED ON DEVELOPMENT OF FLOWERING AND FRUITING

Overcrowding of leaves, flowers, and fruits may affect the productivity of cotton. Types of flowering habits and developmental stages of cotton flower are shown in Figures 6.5 and 6.6, different manipulation techniques such as defoliation, removal of flowers, etc. are used to see their effects on cotton productivity.

Different practices are used for possible effects on cotton productivity.

6.3.3.4.1 Growth Regulator in Managing Plant Height and Population

Mepiquat chloride

The excessive vegetative growth offset the full-season cotton (*Gossypium hirsutum* L.) varieties increased yield potential. This often causes undesirable fruit shed and boll rot. Siebert and Stewart (2006) studied response of cotton plant density to the Mepiquat Chloride application for managing plant height. Single application at 12 nodes (15.2 g a.i. ha^{-1}) or early bloom (45.8 g a.i. ha^{-1}), sequential applications at 12 nodes (15.2 g a.i. ha^{-1}) and early bloom (30.6 g a.i. ha^{-1}), or the modified early bloom schedule (a plant growth regulator application decision aid that recommends rates and timing based on plant growth parameters) were applied to the plant populations of 152,883; 101,929; and 50,958 plants ha^{-1}. The results revealed that final plant height was reduced at least 15 cm by application

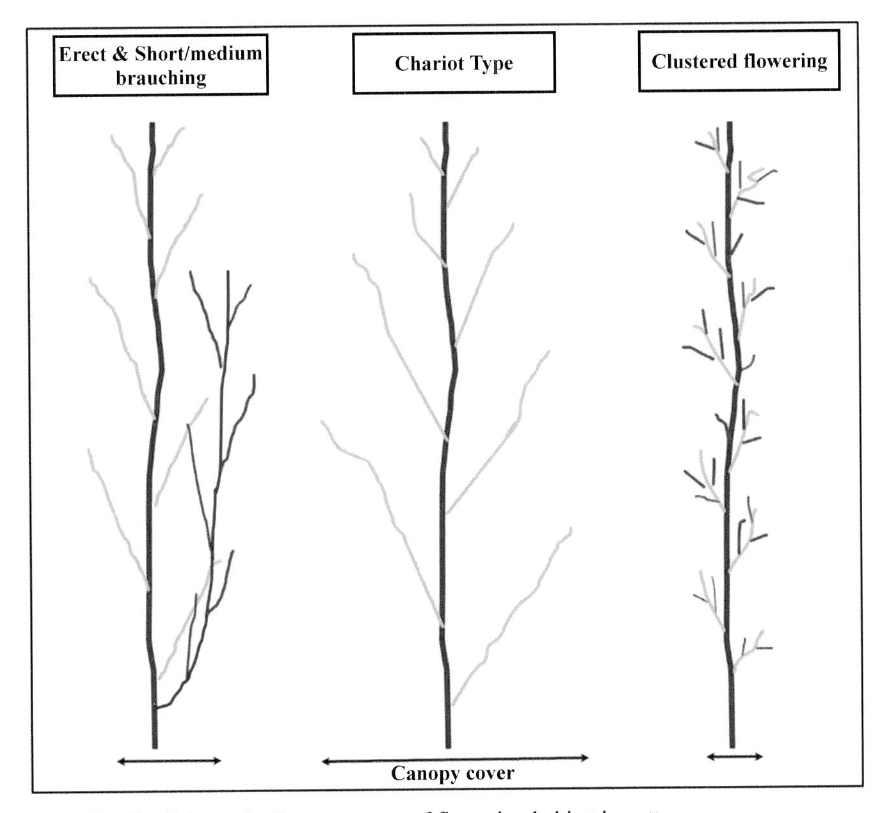

FIGURE 6.5 Schematic diagram: types of flowering habitat in cotton.

FIGURE 6.6 Developmental stages of cotton flower.

of Mepiquat, it has led to a reduction in the number of main stem nodes. Furthermore, they observed that there was an inverse relationship between the lint yield and plant population. With the modified early bloom application strategy with mepiquat chloride a significant yield response was found to occur. In 1 of 2 year Mepiquat chloride application improved lint yield, it was found to be suitable in decreasing plant height, irrespective of the plant population. Thus, their study has indicated that decreasing plant population would have no harmful effects on lint yield or fiber properties and by the use of plant growth regulator it may be valuable in attaining a desired plant stature with less intensive management.

DongMei et al. (2010) conducted a multi-site field trial in Linqing City, Xiajin County and Huimin County with cotton (*Gossypium hirsutum*) cv. Lumianyan 28 in the Shandong Province of China in 2008. They studied the interaction of 3.00, 5.25, 7.50, and 9.75 plants/m^2 plant densities and retention of vegetative branches (VB) on yield, yield components, earliness, and economic index of cotton. A significant interaction was observed among plant density and plant pruning and these affected the yield components. Boll weight was higher under VB removal. With increasing plant density the number of boll/m^2 was increased. Earliness under VB removal was higher. Furthermore, the earliness at 5.25–7.50 plants/m^2 was better than at 3.00 and 9.75 plants/m^2. The biological yield, economic index, and yield components were either individually or interactively affected by plant density and plant pruning or resulted in significant interaction on economic yield. It was concluded that VB retention can compensate for yield loss due to lower population under low plant density, while VB removal under middle and high plant density in Shandong Province was still beneficial to yield increase.

6.3.3.4.2 Removal of Flower Bud

Pettigrew et al. (1992) mentioned that cotton (*Gossypium hirsutum* L.) has an ability to partly compensate for the loss of floral buds (squares). They investigated on the removal of early season floral bud and its influence on cotton genotypes "DPL 50," a normal leaf type, and three leaf type isolines of "MD 65-11" (normal, okra, and super-okra) growth, yield, and fiber properties. They have removed the early developing squares either by hand, or were induced to abscise by application of ethephon [(2-chloroethyl) phosphonic acid]. Early after application there was a reduction in plant height by as much as 11% by ethephon application in 1989 and in 1990 these surpassed the check by 5% in height. Overall results showed that there were no significant differences among treatments for plant variables, yield, and quality. This study demonstrated that for early square cotton has potential to compensate loss, however, did not recommend that early square removal constantly leads to enhanced yields or fiber quality.

Similarly, Bednarz and Roberts (2001) studied the early season bud removal effects on the spatial distribution. The early season floral bud removal was started from the second week of squaring and continued for 1, 2, or 3 consecutive weeks. These were removed by hand. At 90 days after planting, in their estimations on plant height, LAI, main stem node number, fruit present, total fruiting positions, and dry weights it was observed that there were no variances in spatial yield distribution with the lowest level removal of floral bud. On the other hand, they observed that there was a decline in the probability of harvesting a mature boll in the lower canopy and an increase in upper canopy, with an increase of early-season floral bud.The removal of floral buds produced less of first sympodial position fruit and more of third sympodial position fruit at harvest. Therefore, their results have indicated that early-season floral buds removal contributes to additional production of seed cotton on more apical and distal fruiting positions. These changes in spatial yield distribution effectively replaced those floral buds removed early in the season. This was attributed to, because at crop maturity the total seed cotton yield did not vary among the treatments.

Compensatory growth

Pest damages cause loss of fruits. Cotton has capacity of yield compensation. Sadras (1995) undertook a review on compensatory growth in cotton after loss of reproductive organs. He made (1) the present hypotheses summary about cotton yield compensation for pest damage affecting reproductive organs, (2) the nutritional and hormonal hypotheses about growth synchronization existence on the hypothesis that fruit loss strongly affects the carbon and nitrogen economies and plant-growth regulators balance, (3) presents the hypothesis that the combined effects of fruit loss on yield potential (as affected by changes in acquisition and partitioning of carbon and nitrogen) and on growing conditions (as affected by changes in the spatial pattern on the plant and seasonal timing of fruiting) are represented by actual yield responses. This assumption could form a basis for yield compensation experimental and simulation studies in cotton.

6.3.3.4.3 Removal of Fruiting Branch

YiZhen et al. (2018) investigated the removal of early fruiting branches impacts on leaf senescence and yield by changing the sink/source ratio of field-grown cotton. The previous studies in cotton have shown that removal of early fruiting branches (FB) changes the sink/source ratio. It also delays the leaf senescence. However, the regulation of the changed sink/source ratio on leaf senescence and yield formation is not yet clearly known. From two near isogenic cotton lines, late- and early-senescence lines 2 or 4 early FB were removed and plants with intact FB used as control in this study. The late-senescence line had higher leaf chlorophyll content (Chl) and photosynthetic (Pn) rate than the early-senescence line but in the late-senescence line the malondialdehyde (MDA) accumulation and boll load per leaf area (BLLA) were lower than those in the early-senescence line. In both cotton lines the Pn rate, Chl concentration, and the GhLHCB gene expression were enhanced by the removal of FB but reduced the BLLA and MDA accumulation, indicating that the removal of FB reduces sink/source ratio and suppresses leaf senescence. Though removal of FB increased the iP + iPA content it had little effects on Z + ZR or DHZ + DHZR contents and by reducing the ABA and JA related genes and increasing the expression of ABA catabolic related genes it decreased the ABA and JA contents. The results indicate that the increased iP + iPA and reduced ABA and JA contents, along with the differentially expressed biosynthesis and catabolic linked genes associated with FB removal may attribute to the delayed leaf senescence following the reduced sink/source ratio. The overall findings proposed that in cotton removal of early fruiting branches would practically control the sink/source association and suppress leaf senescence. Delaying leaf senescence would not essentially raise cotton yield except when earliness is not affected or when normal maturity is attained.

6.3.3.4.4 Flower Removal

In Arizona, of differing experimental designs during 1986–1987, Terry et al. (1992) studied the early-season cotton, *Gossypium hirsutum* L., effects on square loss and insect control (primarily thrips, *Frankliniella occidentalis* (Pergande), on the capacity of cotton plants to replace lost fruit and also maintain lint quality in several field trials. Some degree of compensation for damage to fruiting structures and yield was revealed from all trials depending upon the timing and plant growth stage. However, removal of squares leads to insignificant differences in yield at 4 weeks after square production initiation. Among treatments most of the cotton quality measures were comparable. For any test, differences were not calculated in fiber strength, elongation, or any color or trash index. Compared with the squares removed early or no removals the shorter fiber lengths were found in treatment with squares removed at 4 weeks after square initiation. The aldicarb-treated "Deltapine 90" cotton at pinhead square produced fibers of higher micronaire than aldicarb-treated "Deltapine 77;" however, there were no differences between treated and untreated plots. Higher uniformity was observed in treated "Deltapine 90" than Deltapine 44, in the same way the plots which were treated with phorate at planting also had higher uniformity ratios.

Jones et al. (1996) during various stages of reproductive development measured the flower removal effects on subsequent yield development, boll development, and fiber properties in cotton two cultivars at Clayton, NC. They imposed various flower removal treatments and studied the impacts of early-, mid-, and late-season flower loss on reproductive development of cotton. Early flower removal treatments (3rd week and earlier) delayed boll development. At the seasons end it did not result in any significant decrease in yield. The total fiber yields were reduced by 13–33% by later flower removals (4th week and later). A positive correlation was seen between the boll weight and fiber properties. This is related to the amount of competition among developing bolls. From the treatments having largest negative effect on yield the largest bolls were obtained. The Micronaire was the only fiber variable affected by flower removal. Its increased values appear to be related with later removal. Data indicated that with significant yield drops occurring from even the least severe, late-season removal treatment of 6th week and later flowers late-season the flower losses (4th week and later) were the most harmful to yield. If the season permits adequate time for compensatory reproductive growth and if additional losses because of plant stress or insect damage are controlled early loss (3rd week and earlier) of fruit can be tolerated.

Initial flower buds or bolls abscission or abortion in the cotton (*Gossypium hirsutum* L.) increases the significance of retaining fruit on adjacent or succeeding fruiting positions. The plant growth regulator mepiquat chloride (MC) (*N*, *N* dimethylpiperidinium chloride) increases boll retention on lower reproductive branches (sympodia) and decreases vegetative growth, and in some cases lessens upper sympodial productivity. Cook et al. (2000) studied early flower bud loss and the Mepiquat Chloride effects on yield distribution of cotton. Within-plant yield distribution they verified the effect of different levels of early flower bud removal and doses and application timings of MC. At two field locations in 1992 before applying MC they removed zero or an average of two or four flower buds located at first positions on lower sympodia from each plant. This bud loss had contributed to a more retention at second positions on lower sympodia and on slightly higher sympodia. On lower sympodial second positions the compensation and yield was improved by applying two biweekly doses of 24.5 g ha^{-1} of MC at early bloom. By the MC treatments when fruit retention on lower sympodia was high the number of fruiting positions and yield on higher sympodia were found to be reduced. The biological responses made by early bud loss and MC treatments interacted positively and loss of early buds brings about small increases in vegetative growth and ameliorated this effect. After early bud loss, the negative effects on upper sympodia were not found with the improvement of compensation on lower sympodia in plants treated with MC.

In cotton (*Gossypium hirsutum* L.) the compensatory growth often occurs due to the loss of reproductive organs. It results in the changes in the morphological, physiological, and photosynthetic patterns. Wells (2000) investigated the removal of early flower buds for the first 2 weeks of flowering effect on leaf pigment, namely, leaf chlorophyll (Chl) and anthocyanin and canopy photosynthesis. There was an extension in the flowering period with the flower bud removal, except in the years where high temperatures prevailed. There was variation in the canopy photosynthetic rates every year of their study. In 1997, there was 15% larger area under the seasonal photosynthetic curve which was seen on four dates and plants with flower bud removal. After 100 days of planting, higher chlorophyll concentrations were recorded in leaves of those plants with flower removal. Similarly, these also had higher anthocyanin levels. A negative relationship was found between the anthocyanin levels and either Chl a/b ratio or Chl concentration. The variations in the pigment contents may be related with delayed senescence in the plant ontogeny. The pigment levels late in development were found to be related with canopy photosynthesis. However, there was no Chl loss in all the treatments. In 2 years, prior to variances in Chl concentration, there were significant differences in canopy photosynthesis. This has given an insight that in response to early fruit loss, plants exhibit morphological and physiological adaptations.

Defoliation

Overcrowding of leaves affect uniform capture of solar radiation and distribution of assimilates in bolls.

Different studies are undertaken to investigate effect of defoliation of cotton productivity.

Timing of defoliation and fruiting gap may cause a shift in benchmarks for timing various agronomic practices that are aimed at maximization of yields of cotton and in optimization of quality. In 1999, 2000, and 2001, Faircloth et al. (2004) investigated a timing of cotton defoliation to (1) determine the creation of a fruiting gap effect on defoliation timing and (2) to compare the usage of the open boll percentage at defoliation (OBPD), nodes above cracked boll (NACB), and micronaire readings at defoliation as tools for timing defoliation. By removing fruit from several continuous nodes on plants physically and defoliating the plots on the basis of various OBPD values each year, they formed a fruiting gap in late July or early August. After defoliation, they recorded the OBPD and NACB, at defoliation with the retention of lint, they determined the micronaire. In both 1999 and 2000, by delaying defoliation beyond 60 OBPD they observed a yield advantage particularly in those treatments that contained a gap. In both the years 1999 and 2000, a direct association was observed among OBPD and both yield and micronaire. Thus, the studies demonstrated that without any sacrifice in the yields it might be possible to terminate cotton growth before the 60% boll open in years where no fruiting gaps occur. Thus, under circumstances where less risk of quality-based reductions are involved, farmers may shift to defoliation, and hence harvest. The micronaire at defoliation can be used as an effective technique for timing defoliation.

In cotton (*Gossypium hirsutum* L.) lint yield and quality gets affected with defoliation timing. In Georgia, Louisiana, Tennessee, and Texas during a 3-year period based on heat-unit accumulation Gwathmey et al. (2004) evaluated a timing method, prior to defoliation in diverse environments for its effects on yield and fiber quality. In all the locations after five nodes above white flower (NAWF=5) defoliation timing treatments of 361, 417, 472, and 528°–days (DD, base 15.6°C) were applied. Defoliation at 472 (DD15) after NAWF = 5 was imposed as control treatment.

There was variation in the crop maturity which was measured on the basis of timing of boll opening that occurred in the control treatment plots. From the maximum yield in 9 of 16 site-years yields were not significantly different in these plots, however, a yield loss which averaged 18.5% was observed in four environments, particularly when the control treatment was too early and the control treatment was very late, this lost time averaged 4 days to harvest. The heat-units after NAWF = 5 that were necessary to attain the earliest maximum yield, across the environments, were found to be related with yield level. Furthermore, each increment of 100 kg lint ha^{-1} was found to be related with 12 more DD15 and found that fiber properties were less sensitive to defoliation compared with yield.

Similarly, the effects of defoliation timings were studied with reference to the net revenues in ultra-narrow-row cotton (UNRC) (*Gossypium hirsutum*) by Larson et al. (2005) at Milan, TN, in 2001, 2002, and 2003 cotton cv. PM 1218 BG/RR was planted in 25.4 cm rows and using heat-unit accumulation after node above white flower (NAWF) two defoliation timing criteria were imposed. The NAWF = 5 plus 472 DD (base 15.6°C) standard defoliation criterion were assigned to the main plot and 14 ± 1 day after each of the three desiccation treatments (paraquat, sodium chlorate, or no desiccant) the subplot were allotted with the NAWF = 2 plus 472 DD, defoliation criterion. A finger-type stripper 8 ± 1 day was employed to harvest the plots after each desiccation treatment. By the use of fiber quality data obtained from the experiment and price quotations of North Delta spot they calculated the price differences for fiber quality and estimated the net revenues with lint yields, price differences, and desiccation costs. An increase in lint yield and net revenue by 9% was observed in those treatments where defoliation until NAWF = 2 plus 472 DD was delayed. They did not observe any adverse effect on the lint yield, price difference, or net revenue with the application of a desiccant after defoliation, though it has somewhat improved gin turnout, which may be an economical parameter.

Boll population

In Burleson Country, on a Weswood silt loam soil, Bynum and Cothren (2007) determined the proper defoliation timing to optimize lint yield and fiber quality by calculating heat unit (HU; DD15.5) accumulation beyond cutout, defined as five nodes above white flower (NAWF = 5). Three nodal positions (NAWF = 3, 4, and 5) were allotted in main plots and the HUs (361, 417, 472, 528, and 583) accumulated beyond the corresponding nodal positions (NAWF) were assigned to five subplots. At this time they applied a tank-mix as a harvest aid. Immediately before defoliation for percent open boll (POB) and nodes above cracked boll (NACB) they evaluated 10 plants $plot^{-1}$. In south central Texas lint yield was adversely impacted when defoliation was started at 472 HUs beyond NAWF = 5. They observed that a 29% more lint yield was recorded in defoliation at 60 POB rather than from defoliation based on the present benchmark of NAWF = 5 plus 472 HUs. Their result suggested that in all production locations, NAWF = 5 was not a sign of the last effective flower population. Furthermore, it was also not a suitable guide for timing cutout, particularly when the end-of-season management decisions are based in south central Texas.

6.3.4 COTTON FIBER DEVELOPMENT AND ELONGATION

In Figure 6.7 different methods of crossing in cotton to develop commercially important cotton hybrids are shown. For plant cell elongation and cell wall biogenesis studies, cotton fiber is considered as best model. Cotton fibers are unicellular, so cell elongation can be assessed separately from cell division. Kim et al. (2001) conducted plant and in vitro studies on growth of cotton fiber, considering it model for plant cell elongation and cell wall biogenesis. This is much extended in their size or structure as cotton fibers, accurately their much extended structure and unique chemical make-up.

Growers and cotton processors are interested of fiber quality and production. Physiologists and agronomists in most of their cotton production studies concentrated to improve the yields. Bradow and Davidonis (2000) undertook fiber quality and the cotton production measurements from a perspective of physiologist. They mentioned that for producing the cotton with high yield and standard fiber quality without increased production costs the producers have inadequate production practices. So the processors of fiber are interested to get the utmost quality cotton at the lowermost price. They make attempts to encounter the processing requirements by mixing bales with diverse average fiber properties.

The development and elongation of cotton fiber (Shown in Fig. 6.8) is a function of structural, physiological, biochemical, and molecular changes.

FIGURE 6.7 Method of crossing in cotton.

Sufficient research inputs are directed to study the mystery of fiber cell development and elongation of cotton fiber is cited below.

Cotton being a fiber of high commercial importance, concerted research activities are directed on the mystery of cotton fiber development, its elongation at cytological and at molecular level.

Beasley et al. (1974) estimated cotton fiber growth adopting a quantitative procedure and fibers were stained selectively with Toluidine blue 0 dye consequently, in an acid–alcohol solution it was destained. The fiber development was measured from absorbance of the dye containing destaining solution. It is then stated in terms of total fiber units (TFU). At 624 nm one OD unit was given one TFU value. The most favorable conditions and procedure for staining and destaining times and also the ovule ratios solution, namely, (1) in 80 mL 0.018% toluidine blue O 20 ovules with related fibers stained for 15 s, (2) by 60 s wash non-absorbed dye removed, (3) in 100 mL glacial acetic acid–ethanol–water (10:95:5), ovules destained (4) after 1 h of destaining absorbance determined. For the total fiber units expression determination this procedure is considered perfect and accurate.

FIGURE 6.8 Fully opened cotton boll.

6.3.4.1 INVOLVEMENT OF CYTOSKELETON IN COTTON FIBER DEVELOPMENT

Seagull (1993) reported about the involvement of cytoskeleton in cotton fiber growth and development. It is mentioned that the wall physical properties and characteristics of cell expansion are influenced by the cellulose microfibrils organization in plant cell walls. An excellent model system for the analysis of the biological regulation of cell wall patterns is observed from the developing cotton fibers. Present research mentions that the cytoskeleton performs most important role in developing cotton fibers. In many plant systems and the developing cotton fibers the deposition and organization of cellulose microfibrils in the cell walls is directed by the cytoskeleton. Regulation of the changes in microfibril patterns during fiber development are governed by both microtubules and microfilaments. The variations in the cytoplasmic microtubules orientation can be changed by the polylamellate architecture of the fiber wall. These in turn seem to guide the microfibril deposition orientation in each consecutive fiber wall layer. In a swirled pattern of bundled microfibrils, cellulose is deposited in the fiber wall in the drug-induced absence of microtubules. The interactions that occur among adjoining microfibrils may affect the cell wall organization on a localized level. Microfilaments seemed to be indirectly engaged in the cellulose microfibrils deposition, in contrast to the direct involvement of microtubules on wall organization. Recent evidence revealed that microfilaments control variations in microtubules patterns to have an effect on wall organization. In many plant systems the manipulation of cytoskeletal components genetically is one way in the direction of future direct manipulation of cell expansion characteristic and may bring about enhancements in the cotton fibers textile qualities.

In the cotton fiber development the research of Schubert et al. (1994) indicated that the elongation and secondary wall thickening phases during the fiber development were not alienated in time. However, before fiber elongation stops a considerable fraction of secondary wall thickening occurs. Schubert et al. (1994) investigated the kinetics of cell elongation and secondary wall thickening of the cotton fiber development in field-grown Stoneville 2013 cotton (*Gossypium hirsutum* L.). The lint fiber development was investigated from anthesis to maturity. The data was recorded on elongation of lint fiber, dry lint weight per seed, and dry lint weight per seed per unit of length and were plotted against boll age. Using computer curvilinear regression analysis this was fitted to proper best-fit curves. The mathematical equations were established from curve-fitting procedures. These equations enabled in precise growth rate curves differentiation. The elongation rate curve demonstrated that at 27 days after anthesis, the lint fiber elongation stopped. It was found that during this period it was found that there was accumulation of 40% weight of final weight values by dry fiber weight and dry fiber weight per unit length. At 16–19 days after anthesis the constant fiber dry weight per unit length was observed from the data analysis after spinning; hypothetically, at this age, thickening of secondary wall begins. These results put forward that during the fiber development the elongation and secondary wall thickening phases were not separated in time.

Seagull et al. (2000) studied cell diameter and wall birefringence changes during cotton fiber development. They mentioned that the biological mechanisms controlling the expansion characteristic of the cell wall regulate the perimeter of the fiber and establish cell diameter. Four varieties of cotton (*Gossypium hirsutum*, MD51ne, DP50, and DP90; and *G. barbadense* L.) studied the changes in fiber diameter that occurred during the elongation and secondary-wall synthesis phases during the development of fiber. In the first 30 days of development, the varieties exhibited an increase in diameter. Therefore, they specified that a dynamic fiber property established during a long period of development is the perimeter of fiber, rather than during a short stage early in development.

6.3.4.2 PHYSIOCHEMICAL STRUCTURAL CHANGES

Beasley and Ting (1974) studied the in vitro development of fiber in fertilized ovules of cotton grown in nutrient medium containing different plant growth substances. Forty-eight hours after anthesis from ovaries ovules were isolated aseptically and these were floated on the liquid medium surface. The medium was supplemented with the only source of nitrogen KNO_3 and the main source of carbohydrate glucose. Based on the intensity of destaining solution using colorimeter the quantity of total fiber produced from the ovule surface was ascertained. Staining solution used was toluidine blue O, where the ovules were immersed for about 15 s. A marked stimulation of fiber development was found in solution of gibberellic acid. From fertilized cotton ovules a marked fiber production inhibition was found in kinetin and abscisic acid contained medium. It was found that gibberellic acid was able to control and overcome the total fiber production inhibition, which is caused by both kinetin and abscisic acid. Though indoleacetic acid could overwhelm the inhibition caused by abscisic acid to a great extent, it could not control the total fiber production inhibition caused by kinetin.

6.3.4.3 BIOCHEMICAL CHANGES DURING FIBER DEVELOPMENT

Meinert et al. (1977) studied the changes that occurred in the biochemical composition of the fibers. This was analyzed in the fiber cell wall composition during its development starting from 5 days of postanthesis, the elongation early stages through the period of secondary wall formation. Both from fibers developing on the plant and from fibers excised, cultured ovules obtained cell walls were used. It was observed that the elongating cotton fiber cell wall had a dynamic structure, which was expressed as a weight percent of the total cell wall. During the period of fiber elongation remarkable changes were observed in the contents of cellulose, neutral sugars (rhamnose, fucose, arabinose, mannose, galactose, and noncellulosic glucose), uronic acids, and total protein. The analysis of absolute changes in the walls with time was expressed as grams component per millimeter of fiber length. The thickness of cell remained constant until about 12 days postanthesis. Later it exhibited a marked increase in thickness which was attributed to the completion of cellulose deposition in the secondary wall. Between 12 and 16 days of postanthesis total wall increase per millimeter fiber length was contributed by increase in all components. They observed that secondary wall cellulose deposition started at around 16 days postanthesis (at least 5 days before to the elongation cessation) and it lasted till 32 days of postanthesis. A sharp decline in protein and uronic acid content were observed at the time of the beginning of secondary wall cellulose deposition. During development some of the individual neutral sugars showed changes in their content, the most important change occurred just before the onset of secondary wall cellulose deposition, namely, a large increase in noncellulosic glucose.

From methylation analyses during fiber development significant change in amino acid composition of cell wall is not observed in neutral sugars, while the glucose was found to be 3-linked at least in part.

6.3.4.4 MOLECULAR CHANGES DURING FIBER DEVELOPMENT

Cotton which is one of the most important textile crops has long cellulose-enriched mature fibers and on the role of gene for fiber development, elongation, and seed development adequate information is not available. Ling et al. (2003) reported that the initiation of cotton fiber cell gets repressed due to the sucrose synthase gene expression suppression. Single-celled hairs that are initiated at anthesis from the ovule epidermis are cotton fibers. The role of the sucrose synthase gene (*Sus*) in cotton fiber and seed is tested by the transformation of cotton with *Sus* suppression constructs. For early fiber development 0–3 days after anthesis (DAA) and when the fiber and seed reached maximum in size 25 DAA were focused for analysis. In the ovule epidermis the fiberless phenotype was found to be contributed by 70% or more suppression of *Sus* activity where in these ovules the fiber initials were less and shrunken or collapsed. The level of *Sus* suppression showed a strong correlation with the degree of inhibition of fiber initiation and elongation, this was attributed to the hexoses reduction. By 25 DAA in the seed coat fibers and transfer cells the presence of *Sus* suppression but not in the endosperm and embryo was demonstrated from a portion of the seeds in the fruit. While *Sus* suppression both in the seed coat and in the endosperm and embryo was unveiled from the remaining seeds in the fruit. Except for considerably reduced fiber growth these transgenic seeds were comparable to wild-type seeds and with the transfer cells loss these seeds were shriveled and were <5% of wild-type seed weight. The results revealed that in the single-celled fibers initiation and elongation *Sus* plays a restrictive role and the fiber development get repressed with the *Sus* suppression only in the maternal seed tissue without having any effect on embryo development and seed size, whereas in the endosperm and embryo further suppression inhibits their own development blocking adjacent seed coat transfer cells formation and stops the development of seed completely.

The epidermal cells from the outer integuments of the ovule develop into cotton fibers. At the same time, Ji (2003) carried out the isolation and analyses of genes which were specially expressed during early cotton fiber development using subtractive PCR and cDNA array. For identification of genes involved in cotton fiber elongation using cDNA made from 10 days postanthesis (d.p.a.) wild-type cotton fiber as tester and cDNA from a fuzzless-lintless (*fl*) mutant as driver subtractive PCR was performed. With most of the earlier published cotton fiber-related genes 280 independent cDNA fragments were recovered. Their cDNA macroarrays revealed that in elongating cotton fibers 172 genes were highly upregulated. In the descriptive cases this was established by *in situ* hybridization. In 10 d.p.a. fiber cells 50-fold more accumulation of 29 cDNAs, which comprised a putative vacuolar (H^+)-ATPase catalytic subunit, a kinesin-like calmodulin binding protein, several arabinogalactan proteins and key enzymes which were involved in long chain fatty acid biosynthesis was observed when compared with that in 0 d.p.a. ovules. Auxin signal transduction, the MAPK pathway and profiling upstream pathways were observed during this process and during the fast fiber elongation period expansin-induced cell wall loosening were also activated.

Arpat et al. (2004) described the cell elongation functional genomics in developing cotton fibers. They stated that the cell expansion rate and duration, fiber morphology, and other significant agronomic traits are regulated by several factors. In cotton fiber they performed rapid cell elongation genetic characterization. In a cultivated diploid species (*Gossypium arboreum* L.) from developmentally staged fiber cDNAs nearly 14,000 unique genes were found to be assembled from 46,603 expressed sequence tags (ESTs). In the cotton genome 35–40% of the genes were represented by the fiber transcriptome. In *silico* expression analysis indicated that rapidly elongating fiber cells determine important metabolic activity, with the majority of gene transcripts, signified by three major functional groups—cell wall structure and biogenesis, the cytoskeleton and energy/carbohydrate metabolism. Among biogenesis of primary and secondary cell wall, dynamic changes in gene expression were shown by oligonucleotide microarrays unveiling that for cell expansion dbEST fiber genes are highly stage-specific—from the fiber dbEST the lack of known secondary cell wall-specific genes supported this assumption. 2553 "expansion-associated" fiber genes exhibited downregulation significantly during the developmental switch from primary to secondary cell wall syntheses and during secondary cell wall synthesis 81 genes were found to be upregulated significantly. In developing fibers during the modification of secondary cell wall these genes were involved in biogenesis of cell wall and energy/carbohydrate metabolism which is constant with the cellulose synthesis stage. In plant biology for studying the basic biological processes with applications in agricultural biotechnology the first detailed observation of an expanding cell transcriptome genetic complexity is provided through this work.

Subsequently, Yong et al. (2006) on the basis of sequencing of a cotton fiber upland cotton (*Gossypium hirsutum*) cDNA library and consequent microarray analysis observed that during fiber elongation biosynthesis of ethylene functions as one of the most significantly upregulated biochemical pathways. During this growth stage a greater level of expression of the *1-Aminocyclopropane-1-Carboxylic Acid Oxidase1-3* (*ACO1-3*) genes were found to be responsible for ethylene production. The amount of ethylene released from cultured ovules exhibited association with *ACO* expression and the fiber growth rate. The robust fiber cell expansion improved with exogenous application of ethylene, but the fiber growth was inhibited due to its biosynthetic inhibitor L-(2-aminoethoxyvinyl)-glycine (AVG). A modest upregulation during this growth stage was observed in the brassinosteroid (BR) biosynthetic pathway. BR or its biosynthetic inhibitor brassinazole (BRZ) treatment similarly increased or inhibited the fiber growth. The BR treatment effect was much less than that of ethylene and by ethylene the ethylene inhibitory effect on fiber cells could be overwhelmed, but the BR reversed the AVG effect to much less extent. These outcomes indicated that ethylene performs a key role in stimulating cotton fiber elongation. Moreover, ethylene might cause improvement in cell elongation by augmenting the sucrose synthase, tubulin, and expansin genes expression.

Yang et al. (2006) investigated the accumulation of genome-specific transcripts, transcription factors, and few phytohormonal regulators during the early stages of fiber cell development in allotetrploid cotton. From *Gossypium hirsutum* L. Texas Marker-1 (TM-1) immature ovules (GH_TMO) 32,789 high-quality ESTs were obtained and their computational and expression analyses was reported, then these ESTs were gathered into 8540 unique sequences including 4036 tentative consensus sequences (TCs) and 4504 singletons. Nearly 15% of the unique sequences in the cotton EST collection were represented in this collection and about 178,000 existing ESTs were found to be obtained from elongating fibers and non-fiber tissues. Furthermore, a significant increase in the percentage of MYB and WRKY genes encoding putative transcription factors and genes encoding predicted proteins involved in auxin, brassinosteroid (BR), gibberellic acid (GA), abscisic acid (ABA), and ethylene signaling pathways was observed in GH_TMO ESTs. During fiber cell initiation cotton homologs associated to *MIXTA*, *MYB5*, *GL2*, and eight genes in the auxin, BR, GA, and ethylene pathways were induced but suppressed in the naked seed mutant (*N1N1*) that is impaired in fiber formation. Furthermore, the phytohormonal pathway-related genes were induced former to the activation of *MYB*-like genes, suggesting the important role of phytohormones in cell fate determination. It is suggested that in *G. hirsutum* L. all functional classifications comprising cell-cycle control and transcription factor activity AA subgenome ESTs were selectively enriched. In cotton allopolyploids during the early stages of fiber cell development the general roles for genome-specific, phytohormonal, and transcriptional gene regulation were determined from these results.

Cotton fiber development is defined into four distinct and overlapping developmental stages: fiber initiation, elongation, secondary wall biosynthesis, and maturation. Lee et al. (2007) investigated the early events of cotton fiber development for gene expression changes. They mentioned that development of fiber cell is a composite process involving several pathways. The different pathways of signal transduction and transcriptional regulation components were included in this. Transcripts accumulated preferentially during fiber development were identified from several analyses with expressed sequence tags and microarray. In Arabidopsis the MYB transcription factors which specially control the development of leaf trichome may also control the development of seed trichome in cotton. Furthermore, several phytohormones and other signaling pathways mediating the cotton fiber development were made known from transcript profiling and ovule culture experiments. It was found that the auxin and gibberellins promote early stages of fiber initiation and during the fiber elongation phase the ethylene- and brassinosteroid-related genes gets upregulated while in fiber initials the genes related with calmodulin and calmodulin-binding proteins were found to be upregulated. The key factors mediating the development of cotton fiber cell were revealed from the additional genomic data, mutant and functional analyses, and genome mapping studies.

Pu et al. (2008) reported that for the development of cotton (*Gossypium hirsutum* L.) fiber the R2R3 MYB Transcription Factor GhMYB109 is required. In development of cotton fiber they confirmed the role of the cotton fiber-specific R2R3 MYB gene *GhMYB109* and to confirm its fiber-specific expression a 2-kb *GhMYB109* promoter was sufficient as demonstrated from the analysis of transgenic reporter gene. A considerable decrease in fiber length was caused due to the antisense-mediated suppression of *GhMYB109*. They also stated that in transgenic cotton many genes associated to the fiber growth of cotton were significantly reduced. Their results unveiled that for the development of cotton fiber *GhMYB109* is necessary and a highly conserved R2R3 MYB transcription factor mechanism in cell fate determination in plants was revealed.

Kwak et al. (2009) investigated the global expression and complexity of small RNAs during cotton fiber initiation and development by adopting Solexa (Illumina Inc.) developed deep sequencing approach and from wild type (WT) and fuzz/lintless (*fl* Mutant in the WT background) cotton ovules two small RNA libraries were constructed, then more than 6–7 million short sequences were generated by sequencing these libraries separately. Out of more than 13 million sequence reads generated at least 22 conserved candidate miRNA families including 111 members were identified out of these families in developing cotton ovules vast majority of expressed miRNAs were build up with seven families. In cotton ovules two additional cell-type-specific novel miRNA candidates were identified and for most of conserved miRNAs, total 120 unique target genes were predicted. Therefore, in this study, during the cotton fiber development the enrichment of a set of microRNAs is reported and the significant variances in expression abundance of miRNAs among the wild-type and mutant is demonstrated suggesting that these differentially expressed miRNAs possibly regulate the transcripts distinctly engaged in development of cotton fiber.

In cotton ovule epidermal cells development into the elongated seed fibers is regulated by MYB transcription factors and R2R3 MYB, *GhMYB25-like* performs a main function in the very early stages of fiber cell differentiation which is detected from its reduced expression in a fiberless mutant of cotton (Xu142 *fl*). In this study, Walford et al. (2011) mentioned that a *GhMYB25-like* promoter–GUS construct was expressed mostly before anthesis (−3 days postanthesis, d.p.a.) in the epidermal layers of cotton ovules, which increases in expression in 0-d.p.a. ovules, mainly in those epidermal cells expanding into fibers and then at +3 d.p.a. in elongating fibers and decreases afterward. This was constant with large quantity of *GhMYB25-like* transcript during fiber development. Phenocopying the Xu142 *fl*mutant the cotton plants with fiberless seeds, but normal trichomes in another place were resulted because of *GhMYB25-like* RNA interference suppression. Similar to the Xu142 *fl* these plants had low expression of the fiber-expressed MYBs, *GhMYB25,* and *GhMYB109*, signifying that *GhMYB25-like* is upstream from those MYBs. In *GhMYB25* and *GhMYB109*-silenced transgenic lines the absence of variation in transcript level of *GhMYB25-like* supported this hierarchy. The elevated expression of *GhMYB25-like* is seen in ovules of transgenic cotton having added copies of the native gene, but no observable increase in fiber initials was seen, signifying that GhMYB25-like may interact with other factors to differentiate epidermal cells into fiber cells.

6.3.4.5 PLANT GROWTH REGULATORS IN FIBER GROWTH AND DEVELOPMENT

An ideal textile raw material which is suitable for environment and human health is the naturally colored cotton. Meng et al. (2017) investigated the plant growth regulator effects on fiber growth and development of colored cotton. The ovule of cotton varieties Z1-61, Lyumian CC28, and RT-baixu (CK) were cultured in the media containing different concentrations of plant growth regulators (MeJA, SHAM, BR, BRz, FLD, ETH, $CoCl_2$, PAL inhibitor, 4CL inhibitor, Urea, and Chl). Duncan's new multiple range method was applied for determination of fiber color, fiber length, ovule fresh weight, ovule dry weight, and fiber dry weight after 30 days. Light colored cotton fiber was noted in Z1-61 and Lyumian CC28, particularly in treatments of salicylhydroxamic, phenylalanine ammonia solution enzyme inhibitor and 4CL inhibitor. Higher concentration of the salicylhydroxamic acid produced a lighter color fiber. Brown cotton fiber lighter was produced by FLD and $CoCl_2$, while in $CoCl_2$ the green color cotton fiber was little influenced. It also resulted in inhibition of callus formation also. The fiber length, ovule fresh weight, fiber dry weight, and ovule dry weight were decreased in salicylhydroxamic acid, BRz, FLD, $CoCl_2$, phenylalanine ammonia solution enzyme inhibitor or 4CL inhibitor. Furthermore, it was found that for the brown cotton fiber pigment synthesis and addition of 5 g L^{-1} urea or 1 mg L^{-1} Chl was found to be advantageous.

6.3.5 FACTORS AFFECTING FIBER QUALITY AND YIELD

Different factors during the growth of cotton influence the fiber quality and fiber yield.

6.3.5.1 WATER STRESS

Pettigrew et al. (2004) with eight different cotton genotypes grown under both dryland and irrigated conditions carried out a field study to analyze the moisture deficit stress effects on reproductive growth, lint yield, yield components, boll distribution, and fiber quality. On white bloom counts, nodes above white bloom, lint yield, components of yield, end-of-season plant mapping, and fiber quality weekly data was recorded. The genotypes from the two soil moisture regimes showed a similar response for all the parameters. Irrigation delayed cutout, vegetative growth, owing to strong assimilate demand (average of 6 days) reproductive growth. Later in the growing season flowering was sustained due to the delayed maturity of these plants. Mainly due to 19% reduction in number of bolls, a 25% decrease in dryland plants lint yield was observed during the sufficient moisture deficits years. In irrigated plants at higher plant nodes (>Node 10) and at the more distal positions (≥2) on the sympodial branches more bolls were produced. Though irrigation had no effect on most fiber traits, but nearly 2% longer fiber was produced in 3 out of 4 years of irrigation. With irrigation increase in the plant yield because of more bolls production and extra fruiting branch delay shows that these are the areas on the plant where high yields should be stabilized.

6.3.5.2 NUTRITION

6.3.5.2.1 Potassium

Cassman et al. (1990) evaluated different levels of potassium on the lint yield and fiber quality of Acala cotton on a vermuculitic soil under irrigation with four levels of potassium nutrition, namely, 0, 120, 240, or 480 kg K ha^{-1}. Every year a significant seed–cotton yield response was obtained with the quantity of K applied. The quantity of lint yield was higher compared with the seed yield. This increase was seen with an increase in the potassium supply. Thus, there was a large increase in lint percentage as plant was supplied with K. The micronaire index indicated a higher percentage of lint that had an increased fiber length and thickened secondary walls. This lint percentage was found to be more in those plants which received the application of potassium fertilizer. In regression analyses, they observed a positive relationship between the dependent variables, namely, the fiber length, micronaire index, fiber strength and percent elongation, and fiber length uniformity ratio (dependent variables) and independent variables, namely, fiber K concentration at maturity, leaf K concentration at early bloom, and index of soil K availability, in both the cultivars. The regressions between the cultivars indicated that "Acala GC510" had a better fiber quality than that of "Acala SJ2." Thus, their research concluded that an important determinant under field conditions about the cotton fiber quality is the supply of K.

6.3.5.2.2 Nitrogen and Potassium

Read et al. (2006) conducted outdoor studies in upland cotton (*Gossypium hirsutum* L.) for determining the separate effects of nitrogen (N) and potassium (K) stress at flowering stage on lint yield and fiber quality. They observed that stress could not alter the length of fiber consistently. Their results suggested that the plant status of N and K contents indirectly affected the fiber development in the early stages. There was a reduction in the yield by deficiency of nitrogen. It resulted in the early termination of the reproductive growth on the plant. The environment also influenced the crop response to N stress. Furthermore, it was observed that there was a decrease in yield and lint weight $boll^{-1}$ and micronaire values of 3.7 under severe K deficiency. Their results supported the evidence that N stress have an indirect effect on cotton growth and also supported the evidence that K deficiency in cotton also adversely affect the reproductive growth, boll weight, and translocation of sugar.

6.3.5.3 BA AND ABA APPLICATION ON YIELD AND QUALITY

JingRan et al. (2013) studied two cotton cultivars Kemian 1 and NuCOTN 33B in the lower reaches of Yangtze River (Nanjing) and in the Yellow River Valley (Anyang), China, the 6-BA and ABA applications effects on yield, quality, and photosynthate contents in the cotton subtending leaf. The results showed that there was an increase in the contents of sucrose and starch and also the transformation rate of sucrose on application of 6-BA in the subtending leaf, while the application of ABA has resulted only in the regulation of the balance of endogenous hormones. Furthermore, with different planting dates there was an increase in boll weight and fiber qualities in cotton plants, those which were given the application of 6-BA. Whereas in Anyang, the decrease ranges of cotton yield and quality was less for the application of ABA with planting date of May 25th in cotton plants. It was attributed that an enhancement in the cotton yield and quality by the application of 6-BA may be due to an enhancement in the photosynthate contents and sucrose transformation rates. However, the application of ABA in improvement of cotton yield and quality might be attributed to an increase in the stress resistance of cotton plants.

6.3.5.4 EFFECT OF PIG MANURE

An experiment was carried out by DaBing et al. (2009), the seedlings of cotton were treated with water (control), PMC (pig manure compost) solution extracted by water (W), and PMC solution extracted by 0.5 mol/L potassium sulphate (K), respectively, and the effects of PMC extracts on growth and nutrient utilization of cotton plant were studied. The results revealed that K and W treatments increased the plant height, numbers of fruit branch, and bud and boll of cotton plant. Similarly, they brought an increase by 2.30 and 1.86 times in the yields of seed cotton. They also caused a rise in cotton plant biomass of cotton plant by 71.4% and 52.9% in budding stage, 114.2% and 83.5% in flowering-boll stage, and 106.1%

and 90.5% in opening ball stage. It was concluded that PMC extracts can be used as a kind of liquid effective organic fertilizer, effective in improving growth and yields of cotton.

6.3.5.5 TRANSGENIC COTTON

6.3.5.5.1 Injury by Glyphosate Application in Transgenic Cotton

XiaoYan et al. (2013) studied the effects of glyphosate application applied at different rates in transgenic cotton at the various stages, namely, cotyledon stage, 3–4 true leaf stage and blooming and boll-forming stage. The results showed that in the early days after spraying glyphosate applied during different cotton growth stages with 1640–9840 g/hm generally caused cotton injury though there was no effect on cotton fiber quality. In conclusion, the cotton seedling period is the optimal application time of glyphosate in the glyphosate insect-resistant cotton field.

6.3.5.5.2 Enzyme Activity and Nutrient Content of Transgenic Cotton

Chun et al (2013) in Meichang county of Tianjin, China in 2011, studied the effect of transgenic double-gene cotton on enzyme activity and nutrient content in rhizosphere soil. The study was carried out in transgenic cotton sGK321 (Cry1Ac+CPTI), transgenic cotton with double insect-resistant genes (Cry1Ac+Cry2Ab), transgenic cotton with insect-resistant and herbicide-resistant genes (Cry1Ac+Epsps), and non-transgenic cotton cv. Shiyuan 321. A lower content of available phosphorus and ammonium nitrogen were recorded in transgenic cotton with double insect-resistant genes than Shiyuan 321, where nitrate nitrogen showed no difference between them. The available phosphorus and nitrate nitrogen of transgenic cotton with insect-resistant and herbicide-resistant genes were also higher than Shiyuan 321, but nitrate nitrogen was significantly lower than Shiyuan 321. Significant differences are not found between transgenic cotton sGK321 and Shiyuan 321 regarding the activities of urease, alkaline phosphatase, and catalase in the rhizosphere soil. Further significant differences were not found in the soil nutrient content (except nitrate nitrogen) and soil enzyme activity in the rhizosphere soil of transgenic cotton sGK321 and Shiyuan 321.

6.3.5.5.3 *Effect of Mycorrhizal Fungi on Bt Cotton*

Song and Feng (2013) analyzed the Bt cotton and incubation of arbuscular mycorrhizal fungi (AMF) effects on soil nutrients content and soil enzyme activity in plant rhizospere in pot trial. The soil quality and the key soil enzymes activities in the rhizosphere soil were high in Bt cotton than non-Bt cotton. In the rhizosphere of Bt cotton, the content of soil total nitrogen, total phosphorus, and organic matter were 25.90, 3.82, and 15.80% higher. These outcomes indicated that Bt cotton planting exerted no harmful effect on soil quality or soil enzyme activities, but undermined the potential positive effect of AMF on the interaction between the plant and the soil. Furthermore, the nitrogen content in the plant should not be used as a good predictor measuring the potential effects of AMF on the nutrient uptake of plants.

XiaoGang et al. (2015) made comparison of the physiological characteristics of transgenic insect-resistant cotton and conventional lines. The results showed that the difference in genetic backgrounds is the main factor responsible for the effects on biochemical characteristics of transgenic cotton when incubating with cotton *Fusarium oxysporum*. However, genetic modification had a significantly greater influence on the stomatal structure of transgenic cotton than the effects of cotton genotypes. The results revealed that the differences in genetic background and/or genetic modifications may introduce some variations in physiological characteristics and these should be considered to explore the potential unexpected ecological effects of transgenic cotton.

6.3.5.5.4 *Nutrient Uptake*

Jat et al. (2014) investigated the Bt cotton (*Gossypium hirsutum* L) productivity and nutrient uptake at Hisar. Closer spacing (100 × 40 cm) lead to highest seed cotton yield (kg/ha). N and K uptake were influenced positively by plant spacing, while it had a nonsignificant effect on P uptake. After harvest of crop it also exhibited a significant effect on available N, P, and K contents of soil. There was a maximum N, P, and K content in soil with 100 × 60 cm spacing. The N application with 2 split doses resulted in maximum available N, P, and K in soil. After harvest of crop high levels of available N, P, and K were found in soil with 125% RDF.

6.3.5.5.5 Physiological Characteristics of Transgenic Insect-Resistant Cotton

XiaoGang et al. (2015) compared the physiological characteristics of transgenic insect-resistant cotton with conventional lines of cotton. The results indicated that the genetic differences within the cultivars were responsible for bringing about the biochemical characteristics of transgenic cotton affected by incubation of cotton with *Fusarium oxysporum*. But the stomatal structure of transgenic cotton was influenced by genetic modification. The results revealed that in physiological characteristics the variations can be introduced with the differences in genetic background and/or genetic modifications. These are to be studied for exploring the possibility of unexpected ecological effects of transgenic cotton.

6.3.5.5.6 Bt Cotton on Soil Microbial Activities

Bt-toxins (Cry proteins) are produced by transgenic Bt-cotton and due to their binding ability on soil components these toxins may get accumulated and persist in soil. Yasin et al. (2016) studied two transgenic Bt-cotton genotypes (CIM-602 and CIM-599) with cry1 Ac gene and two non-Bt cotton genotypes (CIM-573 and CIM-591) to analyze the effect of Bt-cotton on soil microbiological and biochemical characteristics. Results revealed that Bt-cotton had no harmful effect on microbial population (viable counts) and enzymatic activity of rhizosphere soil. Different biochemical components such as phosphatase, dehydrogenase, and oxidative metabolism of rhizosphere soil and cation exchange capacity were higher with Bt cotton. Similarly, the rhizosphere of Bt-cotton genotypes had higher total nitrogen, extractable phosphorous, extractable potassium, active carbon, Fe and Zn contents. It can be resolved that microbiological activities and nutrient dynamics of soils do not get affected with the cultivation of Bt-cotton expressing cry1 Ac.

6.3.5.5.7 Correlation and Path Analysis

Khan et al. (2014) conducted an experiment at Khanewal Pakistan using path coefficient analysis to observe the direct and indirect effects of fiber traits on seed cotton yield in 48 cotton (*Gossypium hirsutum*) genotypes (i.e., 12 parents and their 36 F_1 hybrids) at two locations. Results revealed that at Khanewal location GOT percentage had positive direct effect (0.734) while negative direct effect (−0.03) on seed cotton yield at Multan. At both Multan and Khanewal locations the GOT percentage had indirect positive effects for fiber strength (1.785) and fiber uniformity index (0.187), whereas negative indirect effects were observed for upper half mean length. Toward seed cotton yield fiber length showed positive direct contribution (3.272) for Khanewal, however fiber length had negative direct effect (−0.218) for seed cotton yield at Multan. At Multan, fiber strength exhibited positive direct effects (0.798) for seed cotton yield, whereas negative at Khanewal location. The negative direct effects (−0.421) were depicted by micronaire value for seed cotton yield at Khanewal while positive (0.265) at Multan. Micronaire value showed positive indirect effects for fiber strength and fiber uniformity index at Khanewal, whereas at Multan it is positive for fiber strength and negative for FU index. The upper half means length exhibited positive direct effects (1.207) on seed cotton yield at Khanewal and negative direct effects (−0.529) at Multan location.

YanLi et al. (2015) conducted the correlation and path analysis between seed cotton weight and main traits. This was carried out to have a provision of the theoretical basis of selection of traits that were related to yield and quality for breeding high-yield and high-quality cotton variety. The correlation analysis showed that positive correlation existed between the single boll weight with boll number and seed cotton weight ($P < 0.01$). In addition, the node number of main stem exhibited a negative correlation with seed cotton weight, and the corresponding correlation coefficient was −0.4110. The path analysis showed that the single boll weight ($P < 0.05$), boll number, and plant height had large, direct, and positive contributions to seed cotton weight. The corresponding direct path coefficients were 0.5859, 0.3222, and 0.3024, respectively. There was little positive contribution by the stem diameter to seed cotton weight, while the lint percent and node number of main stem had direct negative effect on seed cotton weight. This has given an indication that low lint percent would be helpful for improving the seed cotton yield.

6.3.5.6 CLIMATE CHANGE IMPACTS

6.3.5.6.1 Modeling

QunYing et al. (2015) reported effectiveness of agronomic practices in the Australian cotton industry to overcome the climate change impacts.

Climate change is impending cotton production. In almost all the world's leading cotton regions, cotton production is affected. They adopted a system modeling approach, using CSIRO Conformal Cubic Atmospheric Model driven by four general circulation models (GCMs) to study the climate change effects on cotton production in different regions of Australia under different climate change and management situations. For measuring the effect of varying planting times and irrigation schedules that are expected to trigger the cotton lint yield, water use, and WUE in 2030 a process-oriented cotton model (CSIRO OZCOT) was applied. Their simulation results indicated that the response of cotton to climate change is region specific and variable. For each condition diverse management strategies are required.

6.3.5.6.2 Effect of N, P, and K

ChengSong et al. (2009) studied the of N, P, and K fertilizer application effects in saline soil in Yellow River Delta on growth of Bt cotton cultivar SCRC28. The experiments were conducted with in coastal saline soil in the Yellow River Delta, Shandong province. The results showed that in low, middle and high levels of salinity field, nutrient uptake of NP, NPK of Bt cotton was high, while there was low $Na+$ uptake of cotton. Furthermore, nutrient use efficiency in agronomy (NUEa) of cotton, the biomass yield and lint yield were highest. It was concluded that the effective way to alleviate nutrition difficulties of saline soil was application of rational fertilizer, based on classification of soil salinity. In coastal saline fields it improves the cotton nutrition and augments the NUEa and cotton yield.

6.3.5.6.3 Enzyme Activity Bt Cotton

Mina et al. (2011) investigated the Bt cotton effect on enzymes activity and microorganisms in rhizosphere. Under Indian subtropical conditions *Bacillus thuringiensis* (Bt) transgenic cotton (var. Mech 162) and its isogenic non-Bt counterpart were assessed for the risks of transgenic crop on the soil ecosystem for 3 successive years. At different growth stages, that is, seedling, vegetative, flowering, bolling, and harvesting stages the activities of dehydrogenase, alkaline phosphatase, nitrate reductase, and urease soil enzymes were assayed to study the effect of Bt cotton on soil biochemical properties. At different growth stages in Bt and non-Bt cotton plants they also observed Bt cotton effect on soil microorganisms, number of nematodes, collembola and ants representing micro, meso, and macrofauna, respectively, in rhizosphere. Results revealed during crop growth period between Bt and non-Bt cotton rhizosphere there was no significant difference ($P < 0.05$) in alkaline phosphatase, nitrate reductase, and urease activity. But dehydrogenase activity was significantly high ($P < 0.05$) in the Bt cotton rhizosphere. At most of the growth stages numbers of micro, meso, and macrofauna were more in Bt cotton rhizosphere. In number of nematodes, collembola, and ants, significant temporal and spatial variations were seen between Bt and non-Bt cotton plants rhizosphere. Their study has indicated that the Bt cotton variety Mech 162 does not pose any threat to soil microorganisms or the soil biochemical properties.

6.3.5.6.4 Nitrogen

Rong et al. (2011) considered the influence of irrigation and nitrogen fertilizer in regulation of the net photosynthetic rates in the flowering and also in the boll-forming stages in cotton that is grown on sandy farmland in the marginal oasis, in the middle reaches of the Heihe River basin. An increase in the nitrogen fertilizer application rates resulted in an increase in the average daily net photosynthetic and transpiration rates of cotton leaves both in the flowering and also the boll-forming stages. However, there was a slight decline in the net photosynthetic and transpiration rates when application of N rate exceeded 300 kg/ha. Positive correlations were observed among the net photosynthetic rate in the flowering and boll-forming stages with seed cotton yield, straw biomass, boll number per plant, boll weight, and seed weight. Thus, these outcomes have given an insight of the influence of irrigation and nitrogen fertilizer application on net photosynthetic rate of cotton leaves during the flowering and boll-forming stages. Further, they also have an influence on the growth and yield of cotton.

Similarly, ZiSheng et al. (2011) in Liaoyang, Liaoning, China, studied the effect of nitrogen application rates (0, 240, and 480 kg/ha) on cotton biomass accumulation, nitrogen uptake, and nitrogen fertilization recovery rate in cotton in the extremely early-maturation region in two cotton cultivars Liaomian 19 and NuCOTN33B, during 2007–2008. A Logistic curve equation was used to describe the accumulation dynamics of biomass and nitrogen. It was observed that the biomass and nitrogen accumulation characteristics of cotton were changed with the nitrogen application rates and also affected the yield and quality of cotton. Nitrogen rapid accumulation began 10–12 days earlier than that of biomass accumulation. Furthermore, highest accumulation of biomass and nitrogen were observed with the nitrogen rate of 240 kg/ha. Similarly, most harmonious Eigen values of the dynamic accumulation model of cotton were obtained. Meanwhile, there was also the highest recovery rate of nitrogen and lint yield with a best quality fiber. Higher dose of nitrogen rate of 480 kg/ha reduced the biomass amount and accumulation rate as well as that of the nitrogen. It also reduced the distributive indices of biomass in reproductive organ and lint yield.

6.3.5.6.5 Mulching and Planting Techniques on Growth and Yield

Nalayini et al. (2011) studied the growth and yield performance of cotton (*Gossypium hirsutum*) expressing *Bacillus thuringiensis* var. Kurstaki and non-Btgenotypes of "RCH 20" in response to the polyethylene mulching and planting techniques. They compared three planting methods, namely, single row, triangular, and double row (paired row). A higher yield (4328) was obtained from Bt cotton "RCH 20.". Both-Bt and non-Bt cotton were found to be benefited by polymulching, a 56.4% higher yield was benefited by this mulching particularly in the Bt cotton. Under polymulching in 0–45 cm soil depth the available soil moisture was higher (24.9%), while it was only (19.8%) under conventional method. Polymulching has also resulted in increased soil temperature of 3.9–4.0°C, in the soil depth of 0–15 cm. They specified that this temperature was much favorable for a quick and faster mineralization and also for the nutrients mobilization in the soil.

6.3.5.6.6 Drymatter Accumulation

RenSong et al. (2011) studied in southern Xinjiang, in super high-yield cotton grown under the natural ecological conditions of the characteristics of dry matter accumulation and its distribution. They also studied the absorption and transfer of nutrients during different growth periods. The super high-yield cotton had a lint yield of above 3000 kg/hm^2. The results indicated that the dry matter accumulation is fast. A large quantity of dry matter accumulation occurs over a long period in this super high-yield cotton. Furthermore, they observed that before the flowering stage, this genotype distributed the photosynthetates to the stem and leaf organs; and exhibits a decline to these organs after flowering. An increase in the distribution of photosynthetates occurs after flowering to the boll and bud organs. This ensures a good yield in the latter period. They mentioned that the distribution of photosynthates to the stem and leaf of the cotton even after the flowering stage occurs in high, middle, and low-yield cotton, as a result of which their yields are affected to a large extent. Similarly, in the super high-yield cotton before flowering absorption and concentration of N, P_2O_5 and K_2O in the stem and leaf occurs and these concentrates reach a peak value between flower forming stage and flower peak stage, while the accumulation peak of the boll and bud appears at boll peaking stage. However, this peak value is remarkably much higher than that of high-yield and middle and low-yield cotton. The time for the nutrients absorption t1 is N > K_2O > P_2O_5 with relatively long Δt; big GT and it decreases with the decrease of the yield in super high-yield cotton.

6.3.5.6.7 Crop Stubbles

The incorporation of various crop stubbles in the soil rather incorporation of continuous cotton stubble as in monocropping in to the soil environment has a dramatic effect on the soil environment and nutrients availability in the soil. WenXiu et al. (2011) studied the crop stubbles effects on cotton yield and soil environment in a continuously cropped cotton field for 8 years. The results revealed that incorporation of different crop stubbles resulted in the higher amount of soil organic matter, soil available nutrients, and cotton yields rather than continuous cotton cropping treatment. Among the soil available phosphorus was increased with the processing of tomato stubble, whereas wheat and corn stubbles could increase the soil available potassium. With all crop stubbles, apparent increases in the microorganism mass in soil were observed compared with that in cotton.

6.3.5.6.8 Planting Pattern and Density

Stephenson et al. (2011) investigated the three cotton planting patterns (19 or 38 cm twin rows and 97 cm single rows) at five plant densities (7, 9, 11, 13, and 15 plants m^{-2}) effect on growth, yield, and fiber quality of new cotton (*Gossypium hirsutum*). There was no influence of the planting pattern on variables of plant structure or yield, seed cotton or lint yield, lint percentage, lint or seed index, and fiber quality. On the main axis, the first sympodial branch was also not influenced by plant density, however, with a decrease in plant density, the number of monopodial branches increased. There was also a reduction in the total bolls per plant with increasing plant density. Similarly, an inverse relationship was found to exist between the first position boll retention to plant densities. Plant density of 11 plants m^{-2} resulted in more of seed cotton and lint yields. Fiber length, micronaire, strength, or uniformity of fiber remained uninfluenced by plant density. The data indicated that adverse effect on cotton growth, yield, or fiber quality does not arise by seeding cotton in twin-row planting pattern.

6.3.5.6.9 Physiological Characteristics of Mainstem Functional Leaves

HaiPeng et al. (2012) analyzed in cotton varieties, Wanmian 38 (brown-fiber cotton) and Wanmian 39 (green-fiber cotton), and the parental material Wanmian 25 (white-cotton), the differences in physiological characteristics of main-stem functional leaves. There was no variation between the colored and white cotton in the change tendency of chlorophyll, soluble protein, and soluble sugar contents; however, these reached a peak at the full flowering stage in colored cotton. The three indices at the whole stage of the colored cotton varieties were lower and rapidly declined after the full flowering stage. The change tendency at the whole development stage of SOD, POD, and CAT activities were comparable in the two colored cotton varieties. At the full flowering stage, the SOD and POD activities reached their peak and after the bolls opening exhibited a decrease. The green-fiber cotton at the early flowering stage had higher SOD, POD, and CAT activities than the brown-fiber cotton. It had lower SOD, POD, and CAT activities at the other stages (bud stage, full bloom stage, and boll-opening stage). Similarly, brown-fiber cotton at the early flowering stage also had a higher content of MDA than green-fiber cotton and white cotton; however, at the boll opening stage green-fiber cotton had a higher MDA content. It was observed that damage to green-fiber cotton under stress, functional leaves resulted in increased activities of protective enzymes to lessen the damage to the cotton plant. However, the cell defense enzymes activities in green-fiber cotton reduced rapidly at the boll stage, resulting in the senescence of cotton.

Onanuga et al. (2012) in hydroponically grown cotton (*Gossypium*) studied the phosphorus and potassium nutrients residual level by applying low P, low K, and high PK nutrients or by spraying these with indole-3-acetic acid (IAA), gibberellic acid (GA3), zeatin (Z), and their combinations at high PK nutrient level. In the first experiment, nutrient solution level in the hydroponics affected the cotton planted in them, irrespective of varieties. However, the P and K nutrients residual levels were influenced by low P, low K, and high PK treatments, regardless of the two cotton varieties planted in the hydroponics pots. Higher nutrients solution level was observed in Zhong cotton variety than Xin cotton variety. In general, except at 43 DAT hormonal applications after every nutrients change had no effect on residual P and K in the nutrients solution. However, at 80 and 90 DAT for P and 74 and 90 DAT for K high residual nutrients in the nutrient solution grown with Xin cotton variety there were varietal differences. The research highlighted that in order to avoid wastage one has to take into consideration the wise usage of mineral fertilizers and synthetic plant hormones.

6.3.5.6.10 Soil and Foliar Applied Boron

Kumar et al. (2018) studied the B uptake of cotton (*Gossypium hirsutum* L.) and its influence on yield in B-deficient calcareous soil of south-west Punjab. A field experiment was planned using six levels of soil-applied B (0.0, 0.5, 1.0, 1.5, 2.0, and 2.5 mg B kg^{-1} soil) and two levels of foliar-applied B (0.1% and 0.2% borax and granubor solution) as treatments with three replications following RBD factorial design. The results showed that with the rising levels of soil applied boron up to 1.0 mg B kg^{-1} the seed cotton yield and its contributing characters and root biomass increased significantly and the mean soil B content and its uptake by seed cotton improved at 45, 75, 105, and 145 days after sowing considerably over the control, while in foliar method of B application it remained same at all growth stages. In cotton, 0.1% borax solution foliar solution was effective than 0.2% borax solution and in both sources the efficacy of borax and granubor was equal. Finally, for soil applied boron up to 1.0 mg B kg^{-1} uptake and for foliar spray 0.1% borax or gonubar solution was found to be effective in increasing the seed cotton yield.

6.3.5.6.11 Microelements

Boron

Jun et al. (2012) studied yield-promoting effects after chelated-B fertilizer in cotton crop in Xinjiang. They applied commercial B fertilizer and chelated B fertilizer to cotton plant by foliar application in the field experiment conducted. There was an increase in the number of bolls per plant, boll weight, and cotton yield with commercial B and chelated B. Increasing yield effects of 4.49–13.36% were found with the commercial B fertilizer, while chelated B fertilizer had yield-increasing effects (20.39%) on cotton yield. In the growth process of cotton, the dry matter weight of cotton increased by all treatments, though a significant increase in the dry matter occurred in dry matter weight of root, boll hull, and seed. The resulted concluded that chelated B fertilizer is more helpful for cotton reproductive organ's growth.

6.3.5.6.12 Remote Sensing in Crop Yield Estimations

Crop yield information is obtained from multitemporal remote sensing images to a large extent than the monotemporal images. From this multitemporal remote sensing data the information should be used to improve the precision of crop yield estimation. Zhong Ling et al. (2012) adopted the studying area. By integrating the concept of cotton growing area with time-series NDVI data similarity analysis they proposed a method of cotton yield estimation. Initially, they determined the NDVI as the dominant factor of cotton yield estimation via correlation analysis between vegetation index and cotton yield from all sampled plots. Then, according to cotton variety and soil condition the whole studying area was divided into several cotton growing areas. For each growing area the linear-fitting analyses was used for acquiring the yield model coefficient. Lastly, for each cotton pixel the multiple linear regression coefficients were determined with similarity analysis between NDVI vectors from unknown-yield cotton pixels and all known ones as the investigated yields. Thus, time-series NDVI data assisted in the realization of cotton yield estimation throughout the whole studying area. The analyses showed that between the estimated and investigated yield the coefficient of determination (R^2) can reach to 0.77, indicating that the method is reasonable and adjustable.

6.3.5.6.13 Cotton Water Stress Index

Qi et al. (2012) studied the correlations between cotton water stress index (CWSI) from infrared thermography and photosynthetic parameters during flowering and boll-forming stage of cotton to establish a linear regression model function between CWSI and photosynthetic parameters. The linear regression model established provided a potential tool for monitoring cotton water-stressed status by using CWSI in Xinjiang. At flowering and boll-forming stage of cotton they used Fluke infrared thermal camera and obtained the canopy infrared thermal images of two cotton cultivars grown under four levels water treatments. By the image processing technology, they applied this information for acquiring the temperature of canopy sunlit leaves, and add wet artificial reference surface (WARS) temperature to empirical formulation which was defined by Jones, the calculation of CWSI. Simultaneously, they obtained the net photosynthetic rate (Pn), stomatal conductance (Gs), and transpiration rate (Tr) of cotton with LI-6400 portable photosynthesis system. Exploration of the quantitative relationship between cotton CWSI and photosynthetic parameters showed that cotton CWSI increased with water stressed and correspondingly there was a reduction in Pn, Gs, and Tr. In four key growth periods of flowering and boll-forming stage of cotton, CWSI were negatively correlated with Pn, Gs, and Tr. The average of the correlation coefficient of CWSI between Pn, Gs, and Tr under different level water treatments were rCWSI-Pn = −0.882 3**, rCWSI-Gs = −0.907 3**, and rCWSI-Tr = −0.935 6**, respectively, during flowering and boll-forming stage of cotton. The study concluded that cotton CWSI and photosynthetic parameters Pn, Gs, and Tr can simultaneously reflect the water status during the flowering and boll-forming stage of cotton.

6.3.5.6.14 Water Use Efficiency

In Xinjiang autonomous region the cotton fields were extensively applied with drip irrigation under film. It is beneficial in saving water, raises soil temperature, decreases soil salinity in plant root zone, and increase yields. SongRui et al. (2013) studied the influence of this on the water use efficiency of drip-irrigated cotton in different planting modes under film in Xinjiang. According to the local conditions of heat, radiation, soil, and mechanization, an appropriate planting mode for drip-irrigated cotton under film has to be selected under the present field conditions of cotton production. The method chosen helps in regulating the distribution of soil water and salt in field, progressing cotton growth, increasing cotton yield and labor productivity, and increasing farmers' income. On the basis of three typical planting modes of drip-irrigated cotton experimental techniques, they suggested that the super wide film mode might be a good choice for efficient utilization of the agricultural water resource and increased income.

6.3.5.6.15 Rhizosphere Microbial Activity

Kaware et al. (2012) studied rhizosphere microbial activity in organic *G. arboreum* AKA-8 cotton grown on vertisol. The organic sources of nutrients for this rainfed *G. arboreum* cotton variety AKA-8 were FYM, vermicompost, castor cake, and in situ green manured with sunhemp. The results indicated that from flowering stage to boll bursting stage there was a decrease in the bacteria, fungi, and actinomycetes population in the rhizosphere. However, bulky organic manures, namely, FYM and vermicompost led to an increase in the populations of these in the rhizosphere. Highest seed cotton yield of 18.38 q ha^{-1} was acquired with the application of castor cake at 500 kg ha^{-1}. Thus, the research indicated that multiple benefits are attained by small doses of concentrated organic manure like castor cake rather than heavy doses of bulky organic manures like FYM, for organic *G. arboreum* cotton cultivation.

6.3.5.6.16 Crop Management Practices

Rajpoot et al. (2016) studied the water productivity, weed suppression, and nutrient dynamics in Bt-cotton (*Gossypium hirsutum*) based intercropping systems in a semi-arid Indo-Gangetic plains region, influenced by different crop management practices. Highest seed cotton yield (3.22 tonnes/ha), system productivity, IWUE, and IWP were higher in transplanted cotton (*Gossypium* sp.) than in direct-seeded cotton. There was suppression of weed population in transplanted cotton as well as in Cotton + cowpea intercropping system than in cotton + okra system and sole cotton. Planting geometry of 90 × 60 cm proved superior in seed cotton yield, system productivity, IWUE, and IWP than 120 × 45 cm. Highest seed cotton yield were attained from the sole cotton rather than from the intercropping systems. In transplanted cotton there was highest uptake of total N and P, though the K uptake was found to be highest under direct seeded cotton. There was a high removal of N and P by cotton + cowpea intercropping system, highest K uptake was highest in cotton + okra intercropping system. The results indicated that transplanted Bt-cotton with planting geometry of 90 × 60 cm under cotton + cowpea intercropping system contributed to highest seed cotton yield, system productivity, less weed infestation, and high IWUE and IWP. It has also comparatively built a positive N balance and less depletion of P and K than direct seeding in a semi-arid IGPR.

6.3.5.6.17 Crop Establishment Methods in Bt Cotton

Paul et al. (2016) reported that in Bt-cotton the system productivity, economic efficiency, and water productivity may be influenced by crop establishment methods and Zn nutrition. They estimated the direct effects on Bt-cotton–wheat cropping system and their residual effects on yield. Zn biofortification in direct effect of 5 kg Zn/ha to wheat exhibited higher NR and BCR in CWCS. Significant enhancement in system productivity, PE, EE, and water productivity were obtained with a successive increase in Zn-levels up to 5 kg Zn/ha. The residual effects of 5 and 7.5 kg Zn/ha applied to Bt-cotton had an influence on productivity, profitability, and Zn biofortification of succeeding wheat in a semi-arid IGPR.

6.3.5.7 MINERAL NUTRITION

6.3.5.7.1 Nitrogen

Khan et al. (2017) made a review on nitrogen nutrition in cotton. They have also discussed the control approaches for greenhouse gas emissions. It is known through several studies in a cotton crop that among all the nutrients, nitrogen has high effects on yield, maturity, and lint quality. Excessive use of N fertilizers is of much environmental concerns as it leads to the occurrence of more amounts in ground water, and also leads to the production of an extremely potent greenhouse gas (GHG), namely, nitrous oxide. Essential nutrients for successful crop production are the nitrate and ammonium nitrogen. In plant N metabolism ammonia is a central intermediate which is assimilated in cotton by the mediation of glutamine synthetase, glutamine (z-) oxoglutarate amino-transferase enzyme systems in two steps, namely, in the first step NH3 is added to glutamate with the requirement of adenosine triphosphate (ATP) and forms the glutamine (Gln). Subsequently in the second step the NH3 from glutamine (Gln) is transferred to α-ketoglutarate to produce two glutamates. Additional amino acids are formed with the transfer of NH3 combined into glutamate by various transaminases to other carbon skeletons. For the production of low-molecular-weight organic N compounds (LMWONCs), for instance, amides, amino acids, ureides, amines, and peptides the glutamate and glutamine that are formed can be used rapidly. These low-molecular-weight organic N compounds further synthesized into high-molecular-weight organic N compounds (HMWONCs) like proteins and nucleic acids.

GuoJuan et al. (2017) studied the soil nitrogen availability and root biomass of cotton in arid region to understand the soil nitrogen transformation process, influenced by the stubbles that are retained in the soil. The results have shown that stubble returning to soil and fertilization has increased the ammonium, nitrate, and inorganic contents. At full-bloom stage and full-boll stage there was higher content of inorganic N, than at harvesting stage. However, it did not exhibit any significant effect on root biomass and fine/coarse root biomass ratio. Thus, they concluded that stubble returning to soil would increase the soil net mineralization, net nitrification, gross nitrification, and denitrification. It would also bring about an increase in the nitrate nitrogen, ammonium nitrogen, and absorbable inorganic nitrogen content and also an increase in the root biomass. Thus, in arid areas in cotton fields the soil nitrogen availability can be promoted by the adoption of stubble returning to soil and by different fertilization measures. This in combination with NPK fertilizer and chicken manure applications would help in the acceleration of the transformation of soil nutrients, improves the efficiency of fertilizers, the effective content of nutrient would increase. It would also result in the promotion of the root growth and also improve the carbon assimilation ability of crop.

6.3.5.7.2 *Effective Nitrogen Doses*

Wajid et al. (2017) reported the nitrogen dosage has positive correlation with biomass accumulation (growth), yield, and yield components, namely, cotton growth, seed cotton yield, and traits increases significantly with nitrogen increments. To prove this under arid climate of Multan an experiment was conducted in cotton–wheat cropping system of Punjab for evaluating the CIM-496 and NIAB-111 cotton cultivars adaptability and response to 50, 100, 150, and 200 kg N ha^{-1} nitrogen doses.

6.3.5.7.3 *Method of Nitrogen (N) Fertilization*

In China in the Yellow River Valley area the conventional method of nitrogen (N) fertilization is applications of urea at 1:1 ratio before planting and at the initial flowering stage. Li et al. (2017) conducted a pot culture experiment in a randomized complete block design using CCRI 79 and CCRI 60 modern cotton cultivars with the application of same total N amount 3.5 g N per 35 kg soil per pot at all the growth stages in four fertilization timing/ratios, that is the conventional fertilization method (1:1), application of urea before planting and at the initial flowering stage at ratios of 1:2 and the application of urea all at the initial flowering stage and all at the budding stage. The results showed that compared with conventional fertilization method the 1:2 ratios of urea before planting and at the initial flowering stage increases the seed cotton yields, 15N recovery and decreases the FNL of cotton plant.

6.3.5.7.4 Adaptability and Response to Different Nitrogen (N) Doses

Wajid et al. (2017) carried out a study under arid climate of Multan in cotton–wheat cropping system of Punjab using CIM-496 and NIAB-111 cotton cultivars for evaluating their adaptability and response to different nitrogen (N) doses (50, 100, 150, and 200 kg N ha^{-1}). They found that N increments have positive correlation with biomass accumulation (growth), yield and yield associated components, namely, with nitrogen increments the cotton growth, seed yield, and its components increases considerably.

6.3.5.7.5 Nitrogen use efficiency

XiangBei et al. (2016) investigated wheat–cotton cropping systems for the influence of nitrogen use efficiency of cotton (*Gossypium hirsutum* L.). Crop N balance analysis signified that the surplus N in preceding wheat (*Triticum aestivum* L.) in wheat–cotton rotations contributed to higher N surpluses rather than from monoculture cotton. They mentioned that for the cotton by the reduction of a base fertilizer input and by an increase in the bloom application improvements in the N management can be obtained in wheat–cotton rotations.

Qi et al. (2016) studied cotton (*Gossypium hirsutum* L.) straw and its biochar application effects on NH_3 volatilization and N use efficiency in a drip-irrigated cotton field. They mentioned that an effective way to increase nitrogen (N) fertilizer use efficiency (NUE) is by decreasing the decrease of ammonia (NH_3) volatilization. In their study, they amended the soil once with either cotton (*Gossypium hirsutum* L.) straw (6 t ha^{-1}) or its biochar (3.7 t ha^{-1}) unfertilized (0 kg N ha^{-1}) or fertilized (450 kg N ha^{-1}). Later during the next two seasons they measured the soil inorganic N concentration and distribution, NH3 volatilization, cotton yield, and NUE. There was an increase in the inorganic N concentrations, cotton yield, cotton N uptake, and NUE during the first cropping season after application of straw amendment. This increase was not evident in the second season. In contrast, cotton N uptake and NUE during both the first and the second cropping seasons were increased by the application of biochar. These results indicated that cotton straw and cotton straw biochar can both effectively reduce the NH_3 volatilization, thereby they can also bring about an increase in the cotton yields, N uptake, and NUE.

Li et al. (2017) mentioned that in the Yellow River Valley, in Cotton (*Gossypium hirsutum* L.), conventional nitrogen (N) fertilization methods are the applications of urea before planting and at the initial flowering stage at ratios of 1:1. A pot culture study was taken up two contemporary cultivars (CCRI 79 and CCRI 60). Four fertilization timing/ratios with the same total N application amount 3.5 g N per 35 kg soil per pot over the entire growing stage, the conventional fertilization method (1:1), application of urea before planting and at the initial flowering stage at ratios of 1:2 and the application of urea at the initial flowering stage and at budding stage were taken up in these cultivars so as to determine the effects of increasing urea fertilization ratios at the initial flowering stage or fertilization once at the budding or the initial flowering stage on seed cotton yields. Entire application of urea at the budding stage and at the initial flowering stages resulted in increased 15N recovery and decreased the FNL. Furthermore, there was also an increase in the seed cotton yields with the application of urea before planting and at the initial flowering stage at ratios of 1:2. It did not increase the cotton plant 15N recovery and also led to a decrease in the FNL than the conventional fertilization method.

From the irrigated crop fields such as cotton (*Gossypium hirsutum* L.) the emissions of green-house gas can be reduced by improving the nitrogen (N) fertilizer use efficiency. Rochester (2012) reported that in high-yielding irrigated cotton the crop N use efficiency can be estimated by using seed nitrogen concentration. Over six cropping seasons (2004–2009), within two N fertilizer rate experiments lint yield, crop N uptake and cotton seed N concentrations were measured. The economic optimal N fertilizer rates and the internal crop N use efficiencies (iNUE) were determined. Higher cotton seed N concentrations exhibited correlation with high crop N uptake and above-optimal N fertilizer application. The cotton seed N concentrations were above optimal in 203 modules (45%), optimal in 201 modules (45%), and below optimal in 45 modules (10 for the 203 modules indicating high cotton seed N concentration (average 3.88% N), about 83 kg N/ha N had been applied in excess of the optimum. This work proved that to measure crop N use-efficiency seed N concentration of cotton can be used and specified where N fertilizer management can be improved.

6.3.5.7.6 PAM, Humic Acid and Compound Fertilizer Combination

Wan Tao et al. (2018) carried out a field experiment to observe the effects of five treatments: CK (no fertilizer), S (conventional compound fertilizer), P (PAM modified materials and compound fertilizer), H (cottonseed meal humicacid + composite fertilizer processing), and P + H (PAM modified materials + cottonseed meal + humate compound fertilizer) on soil nitrogen of cotton. The results revealed the application of PAM modified materials, meal humic acid and compound fertilizer showed improved nitrogen use efficiency, fertilizer nutrient release rate, increased nutrient absorption, and reduced nutrient leaching which in-turn increases cotton yield. Thus, it is determined for the nitrogen agronomic use PAM modified materials and cotton seed meal humic acid were enhanced by 19.4% and 27.8%, respectively and the PAM, humic acid, and compound fertilizer combination augments the soil alkaline nitrogen and improves nitrogen use efficiency.

6.3.5.7.7 Plant Density, Plant Pruning and N Fertilizer Rate

In the Yellow River Valley of China a usual combination of 52,500 plants ha^{-1}, intensive pruning, and 255 kg N ha^{-1}has been extensively utilized. JianLong et al. (2017) reported that the practice of high plant density, reduced N rate, and extensive pruning combination leads to 21.7% more net revenue by producing equivalent yield with less input compared with typical combination and use of this combination in the Yellow River Valley and other cotton-growing areas with related ecosystems ensures advantageous cotton production. Here the plant density (52,500 and 82,500 plants ha^{-1}), plant pruning (intensive and extensive), and N fertilizer rate (195 and 255 kg N ha^{-1}) individual and interaction effects on yield, plant biomass, and partitioning, N uptake and use efficiency in addition to input and output values were studied. The seed cotton yield was reduced by 6.9, 6.7, and 5.4% under low and high (195 kg N ha^{-1} and 255 kg N ha^{-1}) when pruned intensively and extensively and planted at moderate rate, while the equivalent yield to that of typical combination was produced by four combinations with high plant density (82,500 plants ha^{-1}), suggesting that regardless of pruning mode and N rate the high plant density gives steady yield.

6.3.5.7.8 Returning Stubble to Soil and Fertilization

To study the returning stubble to soil and fertilization effects on soil nitrogen availability and root biomass of cotton GuoJuan et al. (2017) undertook a field experiment in arid region. The results revealed at full-bloom stage and full-boll stage returning stubbles to soil and ammonium, nitrate and inorganic N increased fertilization resulted in higher inorganic N than at harvesting stage without any effect on root biomass and fine/coarse root biomass ratio and they determined that the combination of stubble returning, application of NPK and chicken manure encourages the root growth and carbon assimilation ability of crop and useful to hasten the soil nutrient transformation, to increase the fertilizer efficiency and active nutrient content in arid region.

6.3.5.7.9 N Fertilization and GHG Emission

On the basis of data from literature it is found that among all the primary plant nutrient nitrogen is most important nutrient influencing the yield, maturity, and lint quality of a cotton crop but the nitrate in ground water and release of most influential greenhouse gas (GHG) nitrous oxide due to the use of excessive nitrogen is a major environmental concern. In this article, Khan et al. (2017) studied the N fertilization effect in cotton and GHG emission control strategies. For effective crop production nitrate and ammonium are vital nutrients. Assimilation of NH_3 occurs in steps, that is, first the NH_3 is added to glutamate forming glutamine (Gln) and in next step from glutamine (Gln) NH_3 is transferred to α-ketoglutarate forming two glutamates and then to carbon skeletons with transaminases to produce surplus amino acids. This process is mediated by the glutamine synthetase, glutamine (z-) oxoglutarate amino-transferase enzyme systems. In addition to this high-molecular-weight organic N compounds (HMWONCs) like proteins, nucleic acids and amides, amino acids, ureides, amines, peptides like low-molecular-weight organic N compounds (LMWONCs) can be produced from glutamate and glutamine.

6.3.5.7.10 Boron Nutrition

For growth, yield, and fiber quality of cotton crop balanced boron (B) application is necessary. Shah et al. (2017) studied the effects of deficient and adequate levels of boron on growth, yield, and fiber quality of 10 genotypes. The effects were observed with the application of boron, though there were variations among the genotypes with respect to the degree of these effects. There was an increase in plant height (8.4%), sympodia per plant (6.4%), bolls per plant (10.0%), boll weight (8.6%), seed-cotton yield (18.7%), leaf B content (14.1%), lint production (5.2%) with adequate levels of B application, though there was a decrease in the microniare value by 7.1% of all genotypes. These genotypes exhibited variations in the growth, yield and quality traits, except fiber length under both B regimes. They suggested that the suitable level of B nutrition exhibit significant effects on different growth and yield related traits rather than the traits related to fiber quality.

6.3.5.7.11 Cotton Genotypes and Boron Levels

Shah et al. (2017) conducted an experiment in split plot design with three replications growing 10 cotton genotypes in sub-plots and in main plots two treatments B deficient (control) and B adequate (2.0 kg B ha^{-1}) were organized. Different genotypes exhibited a varied degree of effects on B application. The plant height (8.4%), sympodia per plant (6.4%), bolls per plant (10.0%), boll weight (8.6%), seed-cotton yield (18.7%), leaf B content (14.1%), lint production (5.2%) were increased with the application of adequate B, while the microniare value of all the genotypes was reduced by 7.1%, indicating that the various growth and yield associated traits are significantly affected with adequate level of B nutrition while the fiber quality traits are not directly associated with B level in cotton.

6.3.5.7.12 Potassium Permanganate on Yield

LingLi (2017) reported potassium permanganate treatment influences the physiological characteristics, yield and quality of cotton and cotton plants sprayed with potassium permanganate exhibits purplish red stem branches, deep green leaves without disease, big bolls and good boll opening at the late growth stage leading to improved yield and quality. Four potassium permanganate sprayings were tested on Zhongmian 50 cotton cultivar. From July 1st to September 20th 0.1% potassium permanganate was sprayed one time for every 10 days amount to nine times (A). Amount to six times (B) from July 1st to September 15th 0.1% potassium permanganate was sprayed one time for every 15 days, for every 20 days 0.1% potassium permanganate was sprayed one time from July 1st to September 20th, which amounts to five times (C) and no potassium permanganate applied (CK). Increased chlorophyll content, sucrose invertase (INV) activity, photosynthetic rate, superoxide dismutase (SOD) and peroxidase (preduced OD) activities, and reduced malondialdeyde (MDA) accumulation in leaves were exhibited by plants treated with potassium permanganate compared with CK. This resulted in improved lipid peroxidation of leaves cells membranes in late stage, delayed leaf senescence, improved physiological functions of leaves leading to higher boll weight, seed cotton yield, lint percent and seed index, improved fiber quality. The treatment A was observed as best treatment for improving cotton yield and fiber quality as it exhibited a higher fiber strength, quality and increased boll weight of 28.9%, 8.2%, and 19.7% compared with CK, treatments B and C, respectively.

6.3.5.7.13 Cotton Toppers

The traditional cotton toppers causes' high percentage of missed tops and overcutting damages to fruit branches because their topping height is not adjustable corresponding to the height of individual cotton plants. To overcome this problem a fine-tuned height-adjustable cotton topper was developed by ZhaoYang and Lei (2017) and evaluated in this study which helped to avoid the spindling of 83% of the plants.

6.3.5.7.14 Use of Delinted Seeds

At present the use of delinted cotton seeds in place of fuzzy cotton seeds has spread in cotton planting. In this study, Zeybek (2010) proposed cotton seeds fuzzy coating as a substitute to delintation. For the pneumatic spacing planter coating makes fuzzy cotton seeds more appropriate. Also, unlike delintation, sulphuric acid is not used for coating and this excludes the problems of seed loss, pollution, and threats to human health. The outcomes revealed that concerning the agronomic and technological characteristics of cotton seed the coated cotton seeds cultivation has no disadvantage; seed characteristics may get improved by coating and with pneumatic spacing drills the coated seeds can be planted easily and with the seed coating method organic seed procurement would be provided.

6.3.5.7.15 Photoprotection

YaLi et al. (2010) studied mechanism for photoprotection of leaves in *Gossypium barbadense* pima cotton and *G. hirsutum* upland cotton at the bolling stage under field conditions. They measured diurnal leaf movement, incident photon flux density (PFD) on leaves, leaf temperature, highest photochemical efficiency of PSII (Fv/Fm), PSII photochemical efficiency (ΦPSII), electron transport rate (ETR), photochemical quenching coefficient (qP), non-photochemical quenching (NPQ), net photosynthetic rate (Pn), stomatal conductance (Gs), and photorespiration (Pr) in pima and upland cotton. It was observed that: upland cotton had higher Pn and Gs than pima cotton. The ratio of photorespiration rate to gross photosynthetic rate was higher in pima cotton. Both these possessed identical degrees of photoinhibition. On the other hand, pima cotton had generally higher thermal energy dissipation. All results revealed that differences in leaf diaheliotropic movement and Gs in pima cotton and upland cotton caused the differences in both incident PFD on leaves and leaf temperature. Both these also preferentially used the electron transport flux and thermal energy dissipation, for light energy dissipation toward photoprotection to against photoinhibition. Nevertheless, more electron transport flux was distributed into the photorespiration pathway to prevent over-reduction of the photosynthetic electron transport chain and photoinhibition in pima cotton.

6.3.5.7.16 Super High-Yielding Cotton

XinGuo et al. (2010) studied relations of mutual growth on organs and cotton bolls' spatial distribution of super high-yielding cotton. It is observed that the relation of together growth of fruit branches' first cotton bud and the main stem leaves is n+1.5-n-1.2. The relation of together growth of fruit branches' first flower and the main stem leaves is n+6.2-n+8.8 (n for the bit of main stem leaf with the same frui-bit). They concluded that the proportion of the central-lower and interior part of bolls of super high-yielding reached 90%. Cotton is the main part of super high-yielding cotton bolls. The top part of bolls is about 10%, the top and external bolls is ultra-high yield potential; the proportion of the external bolls is nearly 40%, the top and external bolls is the key of super high-yielding cotton.

RenSong et al. (2010) analyzed super high-yielding cotton of yield component factor, LAI, SPAD value of leaf, the number and ratio of bolls, and sink capacity load of unit leaf area. The result showed that the harvestable bolls were most contribution rate (0.997**), effect of the rate of pre-reproductive cotton LAI growth on the number and ratio of bolls, LAI of super high-yielding cotton were investigated. LAI indicated that source-sink characters of super high-yielding cotton have grown faster than that of high yield cotton, mid-low yield cotton. It is concluded that leaf source were increased faster in early growth duration of super high-yielding cotton. It reached a peak value at middle and later stage, increasing source and enlarging sink were main channel of super high-yield cotton.

6.3.5.7.17 Perennial Cultivation of Cotton

The perennial cultivation of annual cotton fixes heterosis and reserve superior germplasm, hence it is very important for conserving genetic male sterile lines, producing hybrid seeds. GuoPing et al. (2010) studied the yield and quality characters of perennial cultivation of annual upland cotton to provide a theoretic proof of perennial cultivation. There was no difference in the fiber quality of the perennial cultivation with that of the annual cotton, however, the perennial cultivation yield was higher than the annual cotton. For production with finer prospect the perennial cultivation may be directly used and for perennial cultivation the insect-resistant hybrid cotton cultivars with high yield should be selected.

6.4 CONCLUSION

The productivity of a crop is a function of various physiological and biochemical functions, mineral nutrition, biotic, abiotic factors, apart from generic and agronomic practices employed. This chapter brings together research advances of various physiological functions contributing to cotton productivity. Different factors affect different stages of crop growth starting from germination, seedling establishment, vegetative growth and flowering and fruiting stage and yield. Besides it discusses various manipulation techniques used to determine their effects on cotton growth and productivity.

KEYWORDS

- **cotton physiology**
- **germination**
- **vegetative**
- **flowering**
- **fruiting**
- **mineral nutrition**
- **abiotic stresses**
- **productivity**

REFERENCES

Abdelraheem, A.; Fang, D. D.; Zhang JinFa. Quantitative Trait Locus Mapping of Drought and Salt Tolerance in an Introgressed Recombinant Inbred Line Population of Upland Cotton Under the Greenhouse and Field Conditions. *Euphytica* **2018,** *214*, 8. (http://rd.springer.com/journal/10681).

Afzalinia, S.; Ziaee, A. R.; Al-Mulla, Y. A.; Ahmed, M.; Jayasuriya, H. Effect of Conservation Tillage and Irrigation Methods on the Cotton Yield and Water Use Efficiency. *Acta Horticulturae* **2014,** *1054*, 119–126. (http://www.actahort.org/books/1054/1054_13.htm).

Ali, I.; Khan, N. U.; Mohammad, F.; Iqbal, M. A.; Abbas, A.; Farhatullah; Bibi, Z.; Ali, S.; Khalil, I. A.; Ahmad, S.; Mehboob-ur-Rahman. Genotype by Environment and gge-Biplot Analyses for Seed Cotton Yield in Upland Cotton. *Pak. J. Bot.* **2017,** *49*, 2273–2283. (http://www.pakbs.org/pjbot/papers/1512402598.pdf).

Arpat, A.; Waugh, M.; Sullivan, J. P. Functional Genomics of Cell Elongation in Developing Cotton Fibers. *Plant Mol. Biol.* **2004,** *54*, 911–929.

Aslam, M.; Haq, M. A.; Bandesha, A. A.; Haidar, S. NIAB-777: An Early Maturing, High Yielding and Better Quality Cotton Mutant Developed Through Pollen Irradiation Technique-Suitable for High Density Planting. *J. Animal Plant Sci.* **2018,** *28*, 636–646. (http://www.thejaps.org.pk/docs/v-28-02/36.pdf).

Bartimote, T.; Quigley, R.; Bennett, J. M.; Hall, J.; Brodrick, R.; Tan, D. K. Y. A Comparative Study of Conventional and Controlled Traffic in Irrigated Cotton: II. Economic and Physiological Analysis. *Soil Tillage Res.* **2017,** *168*, 133–142. (http://www.sciencedirect.com/science/journal/01671987).

Bennett, O. L.; Ashley, D. A.; Doss, B. D. Methods of Reducing Soil Crusting to Increase Cotton Seedling Emergence. *Agron. J.* **1964,** *56, 162–165.*

Beasley, C. A.; Birnbaum, E. H.; Dugger, W. M.; Ting, I. P.A Quantitative Procedure for Estimating Cotton Fiber Growth. *Stain Technol.* **1974,** *49*, 85–92.

Beasley, C.A.; Irwin, P. T. Effects of Plant Growth Substances on In Vitro Fiber Development from Unfertilized Cotton Ovules. *Am. J. Bot.* **1974,** *61* (2), 188–194.

Bednarz, C. W.; Roberts, P. M. Spatial Yield Distribution in Cotton Following Early-Season Floral Bud Removal. *Crop Sci.* **2001,** *41*, 1800–1808.

Bednarz, C. W.; Don Shurley, W.; Anthony, W. S.; Nichols, R. L. Yield, Quality and Profitability of Cotton Produced at Varying Plant Densities. *Agron. J.* **2005,** *97*, 235–240.

Bishnoi, U. R.; Delouche, J. C. Relationship of Vigour Tests and Seed Lots to Cotton Seedling Establishment. *Seed Sci. Technol.* **1980,** *8*, 341.

Bourland, F. M.; Oosterhuis, D. M.; Tugwell, N. P. Concept for Monitoring the Growth and Development of Cotton Plants Using Main-Stem Node Counts. *J. Prod. Agric.* **1992,** *5, 532–538.*

Bradow, J. M.; Davidonis, G. H. Quantitation of Fiber Quality and the Cotton Production–Processing Interface: A Physiologist's Perspective. *J. Cotton Sci.* **2000,** *4*, 34–64.

Bynum, J. B.; Cothren, J. T. Indicators of Last Effective Boll Population and Harvest Aid Timing in Cotton. *Agron. J.* **2007,** *100*, 1106–1111.

Carmi, A. Effects of Root Zone Volume and Plant Density on the Vegetative and Reproductive Development of Cotton. *Field Crops Res.* **1986,** *13*, 25–32.

Carmi, A.; Shalhevet, J. Root Effects on Cotton Growth and Yield. *Crop Sci.* **1983,** *23*, 875–878.

Cassman, K. G.; Kerby, T. A.; Roberts, B. A.; Bryant, D. C.; Higashi, S. L. Potassium Nutrition Effects on Lint Yield and Fiber Quality of Acala Cotton. *Crop Sci.* **1990,** *30*, 672–677.

Chun, F.; JianNing, Z.; Gang, L.; ZhiGuo, Y.; Hui, W.; DongMei, W.; Yu, H.; DianLin, Y. Effect of Transgenic Double-Gene Cotton on Enzyme Activity and Nutrient Content in Rhizosphere Soil. *Cotton Sci.* **2013,** *25*, 178–183 (http://journal.cricaas.com.cn/en/index0.htm).

Cook, D. R.; Kennedy, C. W. Early Flower Bud Loss and Mepiquat Chloride Effects on Cotton Yield Distribution. *Crop Sci.* **2000,** *40*, 1678–1684.

Cook, C. G.; Zik, K. M. Cotton Seedling and First-Bloom Plant Characteristics: Relationships with Drought-Influenced Boll Abscission and Lint Yield. *Crop Sci.* **1992,** *32*, *1464–1467.*

Dagdelen, N.; Başal, H.; Yılmaza, E.; Gürbüza, T.; Akçaya, S. Different Drip Irrigation Regimes Affect Cotton Yield, Water Use Efficiency and Fiber Quality in Western Turkey. *Agric. Water Manage.* **2009,** *96*, 111–120.

Deho, Z. A.; Laghari, S.; Abro, S. Impact of Irrigation Frequencies and Picking Timings on Fiber Quality and Seed Germination of Cotton Varieties. *Pak. J. Bot.* **2017,** *49* (6), 2347–2352. (http://www.pakbs.org/pjbot/papers/1512403656.pdf).

Farid, M. A.; Ijaz, M.; Hussain, S.; Hussain, M.; Farooq, O.; Sattar, A.; Sher, A.; Wajid, A.; Ullah, A.; Faiz, M. R. Growth and Yield Response of Cotton Cultivars at Different

Planting Dates. *Pak. J. Life Social Sci.* **2017,** *15*, 158–162. (http://www.pjlss.edu.pk/pdf_files/2017_3/158-162.pdf).

Faircloth, J. C.; Edmisten, K. L.; Wells, R.; Stewart, A. M. Timing Defoliation Applications for Maximum Yields and Optimum Quality in Cotton Containing a Fruiting Gap. *Crop Sci.* **2004,** *44*, 158–164.

Gao, W.; Zheng, Y.; Slussera, J. R.; Heisler, G. M. Impact of Enhanced Ultraviolet-B Irradiance on Cotton Growth, Development, Yield and Qualities Under Field Conditions. *Agric. Forest Meteorol.* **2003,** *120*, 241–248.

Gardner, B. R.; Tucker, T. C. Nitrogen Effects on Cotton: I. Vegetative and Fruiting Characteristics. *Soil Sci. Soc. Am. J.* **1967,** *31,780–785.*

Grimes, D. W.; Yamada, H. Relation of Cotton Growth and Yield to Minimum Leaf Water Potential. *Crop Sci.* **1982,** *22*, 134–139.

GuoPing, C.; Xin, Z.; RuiYang, Z.; HongTao, Z. A Study on the Changeable Law of the Yield and Quality Characters of Perennial Upland Cotton in Southern Guangxi. *Sci. Agric. Sinica* **2010,** *43*, 3106–3114. (http://www.ChinaAgriSci.com).

Gwathmey, C. O.; Bednarz, C. W.; Fromme, D. D.; Holman, E. M. Response to Defoliation Timing Based on Heat-Unit Accumulation in Diverse Field Environments. *J. Cotton* **2004,** *67*, 102–108.

Hafeez, F. Y.; Safdar, M. E.; Chaudhry, A. U.; Malik, K. A. Rhizobial Inoculation Improves Seedling Emergence, Nutrient Uptake and Growth of Cotton. *Australian J. Exp. Agri.* **2004,** *44*, 617–622.

Hicks, S. K.; Wendt, C. W.; Gannaway, J. R.; Baker, R. B. Allelopathic Effects of Wheat Straw on Cotton Germination, Emergence, and Yield Plant. *Crop Sci.* **1989,** *29* (4), 1057–1061.

Holman, E. M.; Oosterhuis, D. M. Cotton Photosynthesis and Carbon Partitioning in Response to Floral Bud Loss Due to Insect Damage. *Crop Sci.* **1999,** *39*, 1347–1351.

Jalota, S. K.; Sood, A.; Chahal, G. B. S.; Choudhury, B. U. Crop Water Productivity of Cotton (*Gossypium hirsutum* L.)–Wheat (*Triticum aestivum* L.) System as Influenced By Deficit Irrigation, Soil Texture and Precipitation. *Agric. Water Manage.* **2006,** *84*, 137–146.

Jat, R. D.; Nanwal, R. K.; Kumar, P.; Shivran, A. C. Productivity and Nutrient Uptake of Bt Cotton (*Gossypium hirsutum* L) Under Different Spacing and Nutrient Levels. *J. Cotton Res. Dev.* **2014,** *28*, 260–262. http://www.crdaindia.com

Ji, S. J.; Lu, Y. C.; Feng, J. X.; Wei, G.; Li, J.; Shi, Y. H.; Fu, Q.; Liu, D.; Luo, J. C.; Zhu, Y. X. Isolation and Analyses of Genes Preferentially Expressed During Early Cotton Fiber Development by Subtractive PCR and cDNA Array. *Nucleic Acids Res.* **2003,** *31*, 2534–2543.

JianLong, D.; WeiJiang, L.; DongMei, Z.; Wei, T.; ZhenHuai, L.; HeQuan, L.; XiangQiang, K.; Zhen, L.; ShiZhen, X.; ChengSong, X.; HeZhong, D. Competitive Yield and Economic Benefits of Cotton Achieved Through a Combination of Extensive Pruning and a Reduced Nitrogen Rate at High Plant Density. *Field Crops Res.* **2017,** *209*, 65–72. (http://www.sciencedirect.com/science/journal/03784290).

Jin Kim, H.; Barbara, A. T. Cotton Fiber Growth in Planta and In Vitro. Models for Plant Cell Elongation and Cell Wall Biogenesis. *Plant Physiol.* **2001,** *127*, 4.

Jost, P. H.; Cothren, J. T. Growth and Yield Comparisons of Cotton Planted in Conventional and Ultra-Narrow Row Spacing. *Crop Sci.* **2000,** *40*, 430–435.

Jones, J. W.; Hesketh, J. D.; Kamprath, E. J.; Bowen, H. D. Development of a Nitrogen Balance for Cotton Growth Models: A First Approximation. *Crop Sci.* **1974,** *14*, 541–546.

Jones, M. A.; Wells, R.; Guthrie, D. S. Cotton Response to Seasonal Patterns of Flower Removal: I. Yield and Fiber Quality. *Crop Sci.* **1996,** *36*, 633–638.

Kaware, D.; Bhoyar, S. M.; Deshmukh, P. W.; Khuspure, J.; Joshi, M. S. Study of Rhizosphere Microbial Activity in Organic Arboreum Cotton Grown on Vertisol. *Ann. Plant Physiol.* **2013,** *27*, 111–113.

Kent, L. M.; Lauchli, A. Germination and Seedling Growth of Cotton: Salinity–Calcium Interactions. *Plant Cell Environ.* **1985,** *8*, 155–159.

Khan, A; Tan, D. K. Y; Munsif, F; Afridi, M. Z; Shah, F; Wei, F; Fahad, S; Zhou, R. Nitrogen Nutrition in Cotton and Control Strategies for Green House Gas Emission. *Environ. Sci. Pollut. Res. Int.* **2017,** *24*, 23471–23487.

Khan, M. I.; Dasti, M. A.; Mahmood, Z.; Iqbal, M. S. Effects of Fiber Traits on Seed Cotton Yield of Cotton (*Gossypium hirsutum* L.). *J. Agric. Res. (Lahore)* **2014,** *52*, 159–166. (http://www.jar.com.pk/upload/1402782801_113_paper1.pdf).

Khan, N.; Ullah, N.; Ullah, I.; Shah, A. I. Yield and Yield Contributing Traits of Cotton Genotypes as Affected by Sowing Dates. *Sarhad J. Agric.* **2017,** *33*, 406–411. (http://researcherslinks.com/current-issues/Yield-and-Yield-Contributing-Traits-of-Cotton-Genotypes-as-Affected-by-Sowing-Dates/14/1/721/html).

Kwak, P. B.; Wang, Q. Q.; Chen, X. S.; Qiu, C. X.; Yang, Z. M. Enrichment of a Set of MicroRNAs During the Cotton Fiber Development. *BMC Genomics* **2009,** *10*, 457.

Larson, J. A.; Gwathmey, C. O.; Hayes, R. M. Effects of Defoliation Timing and Desiccation on Net Revenues from Ultra-Narrow-Row Cotton. *J. Cotton* **2005**.

Lee, J. J.; Woodward, A. W.; Chen, Z. J. Gene Expression Changes and Early Events in Cotton Fibre Development. *Ann. Bot.* **2007,** *100*, 1391–1401.

Ling, Y. R.; Llewellyn, D. J.; Furbank, R. T. Suppression of Sucrose Synthase Gene Expression Represses Cotton Fiber Cell Initiation, Elongation, and Seed Development. *Plant Cell* **2003,** 15.

Li XiaoGang; Ding Chang Feng; Wang XingXiang; Liu Biao. *Comparison of the Physiological Characteristics of Transgenic Insect-Resistant Cotton and Conventional Lines*. Nature Publishing Group: London, UK, Scientific Report, 2015, 5, p 8739. (http://www.nature.com/articles/srep08739.pdf).

Liang FengChao; Cheng HongXia; Hu LieQun; Li ShuAi; Ma LiYun.Method for Monitoring Cotton Growth During Growing Season Base on FY-3/MERSI Data. *Xinjiang Agric. Sci.* **2014,** *51*, 1381–1387. (http://xjnykx.periodicals.net.cn/default.html).

Li YanBin; Zhang Qin. Effects of Naturally and Microbially Decomposed Cotton Stalks on Cotton Seedling Growth. *Arch. Agron. Soil Sci.* **2016,** *62*, 1264–1270. (http://www.tandfonline.com/doi/full/10.1080/03650340.2015.1135327#abstract).

Li Wan Tao; Wang KaiYong. Effect of Two Kinds of New Materials and Compound Fertilizers Applying on Cotton Nitrogen Use Efficiency in Field. *Southwest China J. Agric. Sci.* **2018,** *31*, 513–518.

Li PeiLing; Zhang FuCang; Jia YunGang. Coupling Effect of Water and Nitrogen on Population Physiological Indices Under Alternative Furrow Irrigation. *Sci. Agric. Sinica* **2010,** 206–214. (http://www.ChinaAgriSci.com).

Li, P. C.; Liu, A. Z.; Liu, J. R.; Zheng, C. S.; Sun, M.; Wang, G. P.; Li, Y. B.; Zhao, X. H.; Dong, H. L. Pakistan Effect of Timing and Ratio of Urea Fertilization on 15N Recovery and Yield of Cotton. *J. Animal Plant Sci.* **2017,** *27*, 929–939. (http://www.thejaps.org.pk/docs/v-27-03/30.pdf).

Liu ShaoDong; Zhang SiPing; Zhang LiZhen. Above-Ground Dry Matter Accumulation of Cotton Genotypes with Application of Different Levels of Nitrogen. *Cotton Sci.* **2010,** *22*, 77–82.

Li, Y.; Allen, V. G.; Chen, J.; Hou, F.; Brown, C. P.; Green, P. Allelopathic Influence of a Wheat or Rye Cover Crop on Growth and Yield of No-Till Cotton. *Agron. J.* **2013,** *105*, 1581–1587. (https://www.crops.org/publications/aj/abstracts/105/6/1581).

Liu JingRan; Liu JiaJie; Meng YaLi; Wang YouHua; Chen BingLin; Zhang GuoWei; Zhou ZhiGuo. Effect of 6-BA and ABA Applications on Yield, Quality and Photosynthate Contents in the Subtending Leaf of Cotton with Different Planting Dates. *Acta Agron. Sinica* **2013,** *39*, 1078–1088. (http://www.chinacrops.org).

Li LingLi; Li Wen; Zhu Wei; Ma ZongBin; Yang Tie Gang. Effects of Spraying Potassium Permanganate on Physiological Characteristics, Yield and Quality of Cotton. *J. Henan Agric. Sci.* **2017,** *46*, 48–52. (http://www.hnagri.org.cn/hnnykx.htm).

Li XingXing; Wang YanTi; Yan QingQing; Wang LiHong; Zhang JuSong.Effect of Seed Coating Agent on Photosynthetic Characteristics of Cotton Seedlings Under Low Temperature. *Acta Botanica Boreali-Occidentalia Sinica* **2018,** 525–534.

Li Qi; Liao Na; Zhang Ni; Zhou GuangWei; Zhang Wen; Wei Xing; Ye Jun; Hou ZheNan. Effects of Cotton (*Gossypium hirsutum* L.) Straw and Its Biochar Application on NH3 Volatilization and N Use Efficiency in a Drip-Irrigated Cotton Field. *Soil Sci. Plant Nutr.* **2016,** *62*, 534–544. (http://www.tandfonline.com/loi/tssp20).

Luo QunYing; Bange, M.; Johnston, D.; Braunack, M. Cotton Crop Water Use and Water Use Efficiency in a Changing Climate. *Agric. Ecosyst. Environ.* **2015,** *202*, 126–134. (http://www.sciencedirect.com/science/journal/01678809).

Mai WenXuan; Xue XiangRong; Feng Gu; Yang Rong; Tian ChangYan. Can Optimization of Phosphorus Input Lead to High Productivity and High Phosphorus Use Efficiency of Cotton Through Maximization of Root/Mycorrhizal Efficiency in Phosphorus Acquisition. *Field Crops Res.* **2018,** *216*, 100–108. (http://www.sciencedirect.com/science/journal/03784290).

Mauney, J. R.; Fry, K. E.; Guinn, G. Relationship of Photosynthetic Rate to Growth and Fruiting of Cotton, Soybean, Sorghum, and Sunflower. *Crop Sci.* **1976,** *18*, 259–263.

Ma YunYan; Xu WanLi; Tang GuangMu; Gu MeiYing; Xue QuanHong. Effect of Cotton Stalk Biochar Application on Soil Microflora of Continuous Cotton Cropping Under Use of Antagonistic Actinomycetes. *Chinese J. of Eco-Agric.* **2017,** *25*, 400–409. (http://www.ecoagri.ac.cn/zgstnyen/ch/index.aspx).

Mauney, J. R.; Kimball, B. A.; Pinter, P. J.; Morte, R. L.; Lewin, K. F.; Nagy, J.; Hendrey, G. R. Growth and Yield of Cotton in Response to a Free-Air Carbon Dioxide Enrichment (FACE) Environment. *Agric. Forest Meteorol.* **1994,** *70*, 49–67.

Ma XiaoYan; Peng Jun; Ma Yan; Jiang WeiLi; Ma YaJie. Glyphosate Tolerance of Transgenic Glyphosate/Insect-Resistant Cotton During Different Growth Period. *Acta Phytophylacica Sinica*, **2013,** *40* (5), 457–462. (http://www.wanfangdata.com.cn).

Meinert, M. C.; Delmer, D. P. Changes in Biochemical Composition of the Cell Wall of the Cotton Fiber During Development. *Plant Physiol.* **1977,** *59*.

Mina, U.; Chaudhary, A.; Kamra, A. Effect of Bt Cotton on Enzymes Activity and Microorganisms in Rhizosphere. *J. Agric. Sci. (Toronto)* **2011,** 96–104. (http://ccsenet.org/journal/index.php/jas/article/view/7076).

Ming, X. H.; Chengyi, Z.; Jirka, Š.; Gary, F. 2015.Evaluating the Impact of Groundwater on Cotton Growth and Root Zone Water Balance Using Hydrus-1D Coupled with a Crop Growth Model. *Agric. Water Manage.* **2015,** *160*, 64–75.

Nalayini, P.; Raj, S. P.; Sankaranarayanan, K. Growth and Yield Performance of Cotton (*Gossypium hirsutum*) Expressing *Bacillus thuringiensis var. Kurstaki* as Influenced by Polyethylene Mulching and Planting Techniques. *Ind. J. Agric. Sci.* **2011,** *81*, 55–59.

Nichols, S. P.; Snipes, C. E.; Jones, M. A. Cotton Growth, Lint Yield, and Fiber Quality As Affected By Row Spacing and Cultivar. *J. Cotton Sci.* **2004**.

Nyakatawa, E. Z.; Chandra, K. R. Tillage Cover Cropping and Poultry Litter Effects on Cotton: I. Germination and Seedling Growth. *Agron. J. Waste Manage.* **2000,** *92*, 992–999.

Onanuga, A. O.; Jiang Ping'an; Adl, S. Residual Level of Phosphorus and Potassium Nutrients in Hydroponically Grown Cotton (*Gossypium hirsutum*). *J. Agric. Sci. (Toronto)* **2012,** *4*, 149–160. (http://ccsenet.org/journal/index.php/jas/article/view/13664/10918).

Paul, T.; Rana, D. S.; Choudhary, A. K.; Das, T. K.; Rajpoot, S. K. Crop Establishment Methods and Zn Nutrition in Bt-cotton: Direct Effects on System Productivity, Economic-Efficiency and Water-Productivity in Bt-Cotton-Wheat Cropping System and Their Residual Effects on Yield and Zn Biofortification in Wheat. *ICAR—Ind. J. Agric. Sci.* **2016,** *86*, 1406–1412. (http://epubs.icar.org.in/ejournal/index.php/IJAgS/article/view/62892/25645).

Pettigrew, W. T.; Bruns, H. A.; Reddy, K. N. Growth and Agronomic Performance of Cotton When Grown in Rotation With Soybean. *J. Cotton Sci.* **2016,** *20*, 299–308. (http://www.cotton.org/journal/2016-20/4/upload/JCS20-299.pdf).

Pettigrew, W. T. M. Deficit Effects on Cotton Lint Yield, Yield Components, and Boll Distribution. *Agron. J.* **2004,** *96, 377–383.*

Pettigrew, W. T.; Heitholt, J. J.; Meredith, W. R. Early Season Floral Bud Removal and Cotton Growth, Yield, and Fiber Quality. *Agron. J.* **1992,** *84*, 209–214.

Pettigrew, W. T. The Effect of Higher Temperatures on Cotton Lint Yield Production and Fiber Quality. *Crop Sci.* **2008,** *48*, 278–285.

Premdeep; Ram Niwas; Khichar, M. L.; Abhilash; Sagar Kumar; Dhawan, A. K.; Chauhan, S. K.; Walia, S. S.; Mahdi, S. S. Quantification of Relationship of Weather Parameters With Cotton Crop Productivity. *Ind. J. Ecol.* **2017,** *44* (4), 286–297. (http://indianecologicalsociety.com/society/indian-ecology-journals).

Plant, R. E.; Munk, D. S.; Roberts, B. R.; Vargas, R. L.; Rains, D. W.; Travis, R. L.; Hutmacher, R. B. Relationship Between Remotely Sensed Reflectance Data and Cotton Growth and Yield. *Trans. ASAE* **2000,** *43*, 535–546.

Pu, L.; Li, Q.; Fan, X.; Yang, W.; Xue, Y. The R2R3 MYB Transcription Factor GhMYB109 Is Required for Cotton Fiber Development. *Genetics* **2008,** *180*, 811–820.

Qi, C.; ChunYan, H.; DengWei, W.; LiJuan, X. Correlation Between CWSI from Infrared Thermography and Photosynthetic Parameters During Flowering and Boll-Forming Stage of Cotton. *Xinjiang Agric. Sci.* **2012,** *49*, 999–1006 (http://www.xjnykx.periodicals.com.cn).

RenSong, G.; HongGuo, W.; YanRong, F.; JuSong, Z.; LiWen, T.; Tao, L. Study on Dry Matter Accumulate or Distribution and Nutrient Absorption or Transfer of Super High-Yield Cotton in South Xinjiang. *Xinjiang Agric. Sci.* **2011,** *48*, 410–418.

Qi Hong; Duan LiuSheng; Wang ShuLin; Wang Yan; Zhang Qian; Feng GuoYi; Du HaiYing; Liang QingLong; Lin YongZeng. Effect of Enhanced UV-B Radiation on Cotton Growth and Photosynthesis. *Chinese J. Eco-Agric.* **2017,** *25*, 708–719. (http://www.ecoagri.ac.cn/zgstnyen/ch/index.aspx).

Qin YaPing. In *Effects of Different Groundwater Levels on Growth and Development of Cotton*, Proceedings of the Annual Meeting of China Society of Cotton Sci-Tech in 2016, Jiangsu, China, August 8–9, 2016, pp. 119–124.

Rahman, H.; Malik, S. A.; Saleem, M. Heat Tolerance of Upland Cotton During the Fruiting Stage Evaluated Using Cellular Membrane Thermostability. *Field Crops Res.* **2004,** *85*, 149–158.

Rajpoot, S. K.; Rana, D. S.; Choudhary, A. K. Influence of Diverse Crop Management practices on Weed Suppression, Crop and Water Productivity and Nutrient Dynamics in Bt-cotton (*Gossypium hirsutum*) Based Intercropping Systems in a Semi-Arid Indo-Gangetic Plains Region. *Indian J. Agric. Sci.*, **2016,** *86*, 637–641. (http://epubs.icar.org.in/ejournal/index.php/IJAgS/issue/view/1998).

Read, J. J.; Reddy, K. R.; Jenkins, J. N. Yield and Fiber Quality of Upland Cotton as Influenced By Nitrogen and Potassium Nutrition. *Eur. J. Agron.* **2006,** *24*, 282–290.

Reddy, K. R.; Koti, S.; Kakani, V. G.; Zhao, M. D.; Gao, W. In *Genotypic Variation of Soybean and Cotton Crops in Their Response to UV-B Radiation for Vegetative Growth and Physiology*, Proceedings of Ultraviolet Ground- and Space-based Measurements, Models, and Effects, 2005.

Reddy, K. R.; Hodges, H. F.; Mc Kinion, J. M.A Comparison of Scenarios for the Effect of Global Climate Change on Cotton Growth and Yield. *Australian J. Plant Physiol.* **1997,** *24*, 707–713.

Reema Vistro; Chachar, Q. I.; Chachar, S. D.; Chachar, N. A.; Aziz Laghari; Sarang Vistro; Irum Kumbhar. Impact of Plant Growth Regulators on the Growth and Yield of Cotton. *Int. J. Agric. Technol.* **2017,** *13*, 353–362. (http://www.ijat-aatsea.com/pdf/v13_n3_%20 2017_May/5_IJAT_13(3)_2017_Reema%20Vistro_Biotechnology.pdf).

RenSong, G.; Bin, L.; Gang, X.; YunGuang, G.; Qiang, Z.; JuSong, Z. A Study on Source-Sink Characters of Super High-Yielding Cotton in South Xinjiang. *Xinjiang Agric. Sci.* **2010,** *47*, 649.

Rikin, A.; Chalutz, E.; Anderson, J. D. Rhythmicity in Ethylene Production in Cotton Seedlings. *Plant Physiol.* **1984,** *75*.

Rochester, I. J. Using Seed Nitrogen Concentration to Estimate Crop N Use-Efficiency in High-Yielding Irrigated Cotton. *Field Crops Res.* **2012,** *127*, 140–145. (http://www.sciencedirect.com/science/journal/03784290).

RuiZhi, C.; TianXing, C.; Landivar, J. A.; ChengHai, Y.; Maeda, M. M. Monitoring Cotton (*Gossypium hirsutum* L.) Germination Using Ultra High-Resolution UAS Images. *Precision Agric.* **2018,** *19*, 161–177. (http://rd.springer.com/journal/11119).

Saad, M.; Muhammad, W.; Aqueel, A. Effect of Climate Change on Cotton Vegetative: A Reproductive Growth, Influence of Temperatyre on the Seasonal Abundance of Predatory Mites Euseius Scutalis in Few Cotton Cultivars. *Int. J. Agric. Appl. Sci.* **2014,** *6*.

Sadras, V. O. Compensatory Growth in Cotton After Loss of Reproductive Organs. *Field Crops Res.* **1995,** *40*, 1–18.

Sanaullah Yasin; Hafiz Naeem Asghar; Fiaz Ahmad; Zahir. Impact of Bt-Cotton on Soil Microbiological and Biochemical Attributes. *Plant Prod. Sci.* **2016,** *19*, 1–10.

Sawan, Z. M.; Fahmy, A. H.; Yousef, S. E. Direct and Residual Effects of Nitrogen Fertilization, Foliar Application of Potassium and Plant Growth Retardant on Egyptian Cotton Growth, Seed Yield, Seed Viability and Seedling Vigor. *Acta Ecol. Sinica* **2009,** *29*, 116–123.

Sawan, Z. M. Response of Flower and Boll Development to Climatic Factors in Egyptian Cotton (*Gossypium barbadense*). *Climatic Change* **2009,** *97*, 553–591. (http://springerlink.metapress.com/openurl.asp?genre=journal&issn=0165-0009).

Schubert, A. M.; Benedict, C. R.; Berlin, J. D.; Kohel, R. J. Cotton Fiber Development-Kinetics of Cell Elongation and Secondary Wall Thickening.*Crop Sci.* **1994,** *13*, 704–709.

Seelanan, T.; Schnabel, A.; Wendel, J. F. Congruence and Consensus in the Cotton Tribe (Malvaceae). *Syst. Bot.* **1997,** *22*.

Seagull, R. W.; Oliveri, V.; Murphy, K.; Binder, A.; Kothari, S. Cotton Fiber Growth and Development 2.Changes in Cell Diameter and Wall Birefringence. *J. Cotton Sci.* **2000,** *4*, 97–104.

Seagull, R. W. Cytoskeletal Involvement in Cotton Fiber Growth and Development. *Micron* **1993,** *24*, 643–660.

Shah, M. A.; Farooq, M.; Shahzad, M.; Khan, M. B.; Hussain, M. Yield and Phenological Responses of Bt Cotton to Different Sowing Dates in Semi-Arid Climate. *Pak. J. Agric. Sci.* **2017,** *54*, 233–239. (http://pakjas.com.pk/papers/2695.pdf).

Shopan, J.; Azad, A. K.; Rahman, H.; Mondol, M. S.; Hasan, M. K. Evaluation of Cotton Based Inter Cropping for Northern Region of Bangladesh. *Int. J. Agr. Agri. R.* **2012,** *43*, 43–49.

Shah, J. A.; Zia-ul-Hassan; Sial, M. A.; Abbas, M.; Sindh. Disparity in Growth, Yield and Fiber Quality of Cotton Genotypes Grown Under Deficient and Adequate Levels of Boron Agriculture University, Tandojam, Pakistan. *Pak. J. Agr. Agri. Eng. Vet. Sci.* **2017,** *33*, 163–176. (http://pjaaevs.sau.edu.pk/index.php/ojs/article/view/205).

Siebert, J. D.; Stewart, A. M. Influence of Plant Density on Cotton Response to Mepiquat Chloride Application. *Agron. J.* **2006,** *98*, 1634–1639.

Singh, Y.; Rao, S. S.; Regar, P. L. Deficit Irrigation and Nitrogen Effects on Seed Cotton Yield, Water Productivity and Yield Response Factor in Shallow Soils of Semi-Arid Environment. *Agric. Water Manage.* **2010,** *97*, 965–970. (http://www.sciencedirect.com/science?_ob=ArticleURL&_udi=B6T3X-4YKFMSB-1).

Snipes, C. E.; Byrd, J. D. The Influence of Fluometuron and MSMA on Cotton Yield and Fruiting Characteristics. *Weed Sci.* **1994,** *42*, 210–215.

Stewart, J. D.; Lee, C. H. In-Ovulo Embryo Culture and Seedling Development of Cotton (*Gossypium hirsutum* L.) *Planta* **1977,** *137*, 113–117.

Steiner, J. J.; Jacobsen, T. A. Time of Planting and Diurnal Soil Temperature Effects on Cotton Seedling Field Emergence and Rate of Development. *Crop Sci.* **1992,** *32*, 238–244.

Stephenson, D. O.; Barber, L. T.; Bourland, F. M. Effect of Twin-Row Planting Pattern and Plant Density on Cotton Growth, Yield, and Fiber Quality. *J. Cotton Sci.* **2011,** *15*, 243–250. (http://journal.cotton.org/journal/2011-15/3/upload/JCS15-243.pdf).

Sun Hao; Li MingSi; Li JinShan; Han QiBiao; Feng JunJie; Jia YanHui. Influence of Dripper Discharge on Cotton Root Distribution and Water Consumption Under Pot Cultivation. *J. Drainage Irrigation Machinery Eng.* **2014,** *32*, 906–913. (http://zzs.ujs.edu.cn).

Kumar, S.; Kumar, D.; Sekhon, K. S.; Choudhary, O. P. Influence of Levels and Methods of Boron Application on the Yield and Uptake of Boron By Cotton in a Calcareous Soil of Punjab. *Comm. Soil Sci. Plant Anal.* **2018,** *49*, 499–514. (http://www.tandfonline.com/loi/lcss20).

Sui, R. X.; Byler, R. K.; Fisher, D. K.; Barnes, E. M.; Delhom, C. D. Effect of Supplemental Irrigation and Graded Levels of Nitrogen on Cotton Yield and Quality. *J. Agric. Sci. (Toronto)* **2014,** *6*, 119–131. (http://ccsenet.org/journal/index.php/jas/article/view/32601/19272).

Tewolde, H.; Adeli, A.; Sistani, K. R.; Rowe, D. E. Potassium and Magnesium Nutrition of Cotton Fertilized with Broiler Litter. *J. Cotton Sci.* **2010,** 1–12. (http://www.cotton.org/journal/2010-14/1/upload/JCS14-1.pdf).

Terry, I. L. Effect of Early Season Insecticide Use and Square Removal on Fruiting Patterns and Fiber Quality of Cotton. *J. Econ. Entomol.* **1992,** *85*, 1402–1412.

Tian YouSheng; Ye ChunXiu; Zhang GuoLi; Dong YongMei; Li QuanSheng; Wang ZhiJun; Zhao ZengQiang; Ma PanPan; Sun GuoQing; Xie Zong Ming. Drought Resistance Evaluation and Indexes Screening of Cotton During Bud Stage to Flowering Stage. *J. Agric. Sci.* **2018,** *31*, 306–312.

Transactions of the Chinese Society of Agricultural Engineering, **2011,** *27*, 271–275. (http://www.tcsae.org).

Turner, J. H.; Ramey, H. H.; Worley, S. Influence of Environment on Seed Quality of Four Cotton Cultivars. *Crop Sci.* **1976,** *16*, 407–409.

Wajid, A.; Ahmad, A.; Awais, M.; Habib-ur-Rahman, M.; Raza, M. A. S.; Bashir, U.; Arshad, M. N.; Ullah, S.; Irfan, M.; Gull, U. Nitrogen Requirements of Promising Cotton Cultivars in Arid Climate of Multan. *Sarhad J. Agric.* **2017,** *33*, 397–405. (http://researcherslinks.com/current-issues/Nitrogen-Requirements-of-Promising-Cotton-Cultivars-in-Arid-Climate-of-Multan/14/1/717/html).

Walford, S. A.; Wu, Y.; Llewellyn, D. J.; Dennis, E.S. GhMYB25-like: A Key Factor in Early Cotton Fibre Development. *Plant J.* **2011,** *65*, 785–797.

Wall, G. W.; Amthor, J. S.; Kimball, B. A. COTCO2: A Cotton Growth Simulation Model for Global Change. *Agric. Forest Meteorol.* **1994,** *70*, 289–342.

Wang Jin; Zhang Jing; Fan XinYan; Wang ShaoMing; Tian LiPing; Jin LuSheng. Effect of Enhanced UV-B Radiation on Physiological Indices, Quality and Yield of Cotton. *Cotton Sci.* **2010,** *22*, 125–131.

Wang XinXin; Liu ShengLin; Zhang ShaoMin; Li HongBo; Maimaitiaili, B.; Feng Gu; Rengel, Z. Localized Ammonium and Phosphorus Fertilization Can Improve Cotton Lint Yield By Decreasing Rhizosphere Soil pH and Salinity. *Field Crops Res.* **2018,** *217*, 75–81. (http://www.sciencedirect.com/science/journal/03784290).

Wang XueJiao; Pan XueBiao; Wang Sen; Hu LiTing; Guo YanYun; Li XinJian. Dynamic Prediction Method for Cotton Yield Based on COSIM Model in Xinjiang, *Trans. Chinese Soc. Agric. Eng.* **2017,** *33*, 160–165. (http://www.tcsae.org/nygcxben/ch/index.aspx).

Wang ShuLin; Qi Hong; Wang Yan; Zhang Qian; Feng GuoYi; Lin YongZeng. Effects of Restructuring Tilth Layers on Soil Physical and Chemical Properties and Cotton

Development in Continuous Cropping Cotton Fields. *Acta Agron. Sinica* **2017,** *43*, 741–753. (http://www.chinacrops.org).
Wang HaiDong; Wu LiFeng; Cheng MingHui; Fan JunLiang; Zhang FuCang; Zou YuFeng; Chau Wai [Chau, W. H.]; Gao ZhiJian; Wang XiuKang. Coupling Effects of Water and Fertilizer on Yield, Water and Fertilizer Use Efficiency of Drip-Fertigated Cotton in Northern Xinjiang, China. *Field Crops Res.* **2018,** *219*, 169–179. (http://www.sciencedirect.com/science/journal/03784290).
Wang, Y.; Gutierrez, A. P.; Oster, G.; Daxl, R. A Population Model for Plant Growth and Development Coupling Cotton–Herbivore Interaction. *Can. Entomol.* **1977,** *109*, 1359–1374.
Wang ZiSheng; Xu Min; Liu RuiXian; Wu XiaoDong; Zhu He; Chen BingLin; Zhou ZhiGuo. Effects of Nitrogen Rates on Biomass and Nitrogen Accumulation of Cotton with Different Varieties in Growth Duration. *Cotton Sci.* **2011,** *23*, 537–544. (http://journal.cricaas.com.cn/en/index0.htm).
Wells, R. Leaf Pigment and Canopy Photosynthetic Response to Early Flower Removal in Cotton. *Crop Sci.* **2000,** *41*, 1522–1529.
Wells, R.; William, R.Comparative Growth of Obsolete and Modern Cotton Cultivars. II. Reproductive Dry Matter Partitioning. *Crop Sci.* **1984,** *24*, 863–868.
Wullschleger, S. D.; Oosterhuis, D. M. Photosynthetic Carbon Production and Use by Developing Cotton Leaves and Bolls. *Crop Sci.* **1990,** *30*, 1259–1264.
XiangBei, D.; BingLin, C.; YuXiao, Z.; WenQing, Z.; TianYao, S.; ZhiGuo, Z.; Yali, M. Nitrogen Use Efficiency of Cotton (*Gossypium hirsutum* L.) as Influenced By Wheat-Cotton Cropping Systems. *Eur. J. Agron.* **2016,** *75*, 72–79. (http://www.sciencedirect.com/science/journal/11610301).
Xiaoyu Li X.; Liu, L.; Yang, H.; Li, Y. Relationships Between Carbon Fluxes and Environmental Factors in a Drip-Irrigated, Film-Mulched Cotton Field in Arid Region. *PLOS One* **2018,** *24*, 156–162.
Xin ChengSong; Dong HeZhong; Luo Zhen; Tang Wei; Zhang DongMei; Li WeiJiang; Kong XiangQiang. Effects of N, P and K Fertilizer Application on Cotton Grown in Saline Soil in Yellow River Delta. *Acta Agron. Sinica* **2010,** *36*, 1698–1706. (http://www.chinacrops.org).
Xu DaBing; Wang QiuJun; Shen QiRong; Huang QiWei. Effects of Pig Manure Compost Extracts on the Growth and Nutrient Utilization of Cotton Plant. *J. Agro-Environ. Sci.* **2010,** *29*, 1239–1246. (http://cnki50.csis.com.tw/kns50/Navi/Bridge.aspx?LinkType=BaseLink&DBCode=cjfd&TableName=cjfdbaseinfo&Field=BaselD&Value=NHBH).
Xu WenXiu; Luo Ming; Li YinPing; Han Jian; Wang Jiao; Shu ChunXia; Yu Hong. Effects of Crop Stubbles on Cotton Yield and Soil Environment in Continuously Cropped Cotton Field. *Trans. Chinese Soc. Agric. Eng*. **2011,** *27* (3), 271–275.
Yang, S. S.; Cheung, F.; Lee, J. J. Accumulation of Genome-Specific Transcripts, Transcription Factors and Phytohormonal Regulators During Early Stages of Fiber Cell Development in Allotetraploid Cotton. *Plant J.* **2006,** *47*, 761–775.
Yang Rong; Su Yong Zhong. Responses of Net Photosynthetic Rate at the Flowering and Boll- Forming Stages and Cotton Yield to Irrigation and Nitrogen Fertilizer Application. *Plant Nutr. Fertilizer Sci.* **2011,** *17*, 404–410.

Yan YinFa; Han ShouQiang; Zhou ShengXiang; Song ZhanHua; Li FaDe; Zhang ChunQing; Zhang XiaoHui; Wang Jing. Parameter Optimization on Improving Aged Cotton Seeds Vigor by Extremely Low Frequency High Voltage Pulsed Electric Field. *Chinese Soc. Agric. Eng.* **2017,** *33*, 301–307. (http://www.tcsae.org/nygcxben/ch/index.aspx).

Ye, Z. H.; Zhu, J.; Chuan, Y. Genetic Analysis on Flowering and Boll Setting in Upland Cotton (*Gossypium hirsutum* L.). III. Genetic Behavior at Different Developing Stages. *Acta Genet. Sinica* **2000,** *27*, 800–809.

Yin Tie Song; Ge Feng. In *Effects of Bt Cotton and Incubation of Arbuscular Mycorrhizal Fungi (Amf) on Soil Nutrients Content and Soil Enzyme Activity in Plant Rhizospere*, Proceedings of the 5th Symposium of the NianFu; Yang Bing; Li National Agricultural Environmental Sciences Sponsored by the Chinese Society of Agro-Ecological Environment Protection in 2013, Jiangsu, China, April 19–22 April, 2013, pp. 351–358.

YiZhen, C.; XiangQiang, K.; Zhong, D. H. Removal of Early Fruiting Branches Impacts Leaf Senescence and Yield by Altering the Sink/Source Ratio of Field-Grown Cotton. *Field Crops Res.* **2018,** *216*, 10–21. (http://www.sciencedirect.com/science/journal/03784290).

Yong, H. S.; Sheng, W. Z.; Xi, Z. M.; Jian, X. F.; Yong, M. Q.; Zhang, L.; Cheng, J.; Wei, L. P.; Wang, Z. Y.; Zhu, Y. X. Transcriptome Profiling, Molecular Biological, and Physiological Studies Reveal a Major Role for Ethylene in Cotton Fiber Cell Elongation. *Plant Cell* **2006,** *18*.

YuLian; Tan LanLan; An MengJie; Li WanTao; Wang KaiYong; Guangxi. Effects of Polymer Compounds on Soil Aggregate Composition and Cotton Yield in Salted Cotton Field. *J. Southern Agric.* **2017,** *48*, 1989–1993. (http://www.nfnyxb.com/EN/Default.aspx).

Zeybek, A.; Dogan, T.; Ozkan, I. The Effects of Seed Coating Treatment on Yield and Yield Components in Some Cotton (*Gossypium hirsitum* L.) Varieties. *Afr. J. Biotechnol.* **2010,** *9*, 5523–5529. (http://www.academicjournals.org/AJB/PDF/pdf2010/23Aug/Dogan%20et%20al.pdf).

Zhang YaLi; Luo Yi; Yao HeSheng; Tian JingShan; Luo HongHai; Zhang WangFeng. Mechanism for Photoprotection of Leaves at the Bolling Stage Under Field Conditions in *Gossypium barbadense* and *G. hirsutum*. *J. Plant Ecol. (Chinese Version)* **2010,** *34*, 1204–1212.

Zhang Jun; Wei ChangZhou; Wang XiaoJuan; Min Wei; Li MeiNing. Study on Yield Effects of a Chelated B Fertilizer on Cotton Crop. *Xinjiang Agric. Sci.* **2012,** *49*, 400–404. (http://www.xjnykx.periodicals.com.cn).

Zhang HaiPeng; Zhou Qin; Zhou TaoHua.Comparative Studies on Physiological Characteristics of Main-Stem Functional Leaves of Coloured Cotton Varieties. *Cotton Sci.* **2012,** *24*, 325–330. (http://journal.cricaas.com.cn/en/index0.htm).

Zhang XiaoMeng; Liu SongJiang; Gong WenFang; Sun JunLing; Pang BaoYin; Du XiongMing. Effects of Plant Growth Regulators on Fiber Growth and Development in Colored Cotton Ovule Culture In Vitro. *Acta Agron. Sinica* **2017,** *43*, 763–776. (http://www.chinacrops.org).

Zhang GuoJuan; Pu XiaoZhen; Zhang PengPeng; Zhang WangFeng. Effects of Stubble Returning to Soil and Fertilization on Soil Nitrogen Availability and Root Biomass of Cotton in Arid Region. *Scientia Agric. Sinica* **2017,** *50*, 2624–2634. (http://www.chinaagrisci.com/EN/abstract/abstract19756.shtml).

Zhao XinHua; Qu Lei; Chen BingLin; Zhou ZhiGuo. Changes of Carbon and Nitrogen Contents in Subtending Leaf of Cotton Boll and Its Relationship to Biomass of Cotton Boll, Seed and Fibre. *Chinese Acad. Agric. Sci.* **2010,** *22*, 209–216.

Zhang DongMei; Li WeiJiang; Tang Wei; Dong HeZhong; Li ZhenHuai; Luo Zhen; Lu HeQuan. Interaction of Plant Density with Retention of Vegetative Branches on Yield and Earliness of Upland Cotton. *Cotton Sci.* **2010,** *22*, 224–230.

Zhang XinGuo; Chen Qian; Zhang JuSong; Chen GuanWen; Song JiHui.A Study on Relation of Together Growth of Organs and Cotton Bolls Spatial Distribution of Super High-Yielding Cotton. *Xinjiang Agric. Sci.* **2010,** *47*, 36–41.

Zhang YanLi; Dai XiJun; Lyu ShuangQing. Multivariate Statistical Analysis on Seed Cotton Weight and Main Characters of Mulched-Cotton by Drip Irrigation in the South of Xinjiang. *J. Southern Agric.* **2015,** *46* (8), 1373–1377. (http://www.nfnyxb.com/EN/Default.aspx).

Zhang DongLin; Li Wen; La GuiXiao; Li LingLi; Yang Tie Gang. Characteristics of Growth & Development and Yield of Cotton Using the Technique of Factory Mutual Aid Seedling Raising of Cotton with Wheat in Same Holes and Machinery Transplanting After Winter Wheat Harvest. *J. Henan Agric. Sci.* **2013,** *42*, 51–54.

Zhang Hao; Hao Liu; Chitao Sun; Yang Gao; Xuewen Gong; Jinsheng Sun; Wanning Wang. Root Development of Transplanted Cotton and Stimulation of Soilwater Movement Under Different Irrigation Methods. *Water* **2017,** *9*, 503.

ZhaoYang, C.; Lei, S. Development and Test of an Automatic Height-Adjusting Cotton Topper. *Int. J. Agric. Biol. Eng.* **2017,** *10*, 44–55. (http://www.ijabe.org).

ZhongLing, G.; XingAng, X.; JiHua, W.; HuaAn, J.; Hao, Y. Cotton Yield Estimation Based on Similarity Analysis of Time-Series NDVI. *Trans. Chinese Soc. Agric. Eng.* **2012,** *28*, 148–153 (http://www.tcsae.org).

Zhong, L. X.; Xu, M.; Gao, W.; Raja, R. K.; Kunkel, K.; Schmoldt, D. L.; Samel, A. N. A Distributed Cotton Growth Model Developed from GOSSYM and Its Parameter Determination. *Agron. J.—Biometry Modeling Statistics* **2011,** *104*, 661–674.

CHAPTER 7

Research Advances in Abiotic Stress Resistance in Cotton

ABSTRACT

Many abiotic stresses such as drought, high and cold temperature, salinity, flooding, etc., exert their influence on growth and productivity of cotton. Several concerted research studies were directed in studying their effects, developed techniques, and enabled in understanding physiological, biochemical, and molecular mechanisms to some of these stress factors. The research advances that occurred are briefly discussed in the following sections.

7.1 ABIOTIC STRESS

7.1.1 IDENTIFICATION OF GENE RESOURCES FOR STRESS TOLERANCE

Vinocur and Altman (2005) mentioned that crop losses occur largely across the world due to salinity and drought which form the main causes of yield decreases. They discussed about the advances that occurred in engineering of the genes for incorporation of tolerance to abiotic stress, their achievements, limitations, that arose in detail. They asserted that adaptation of plants to environmental stresses is to some extent dependent on a cascade of events of molecular events/origin that are activated. These might be those which are involved in the perception of stress, signal transductions, and also those which are involved in the specific expression of certain genes which are stress related or to some metabolites also. Stress tolerance in plants can be increased to some extent by the engineering of those genes that are involved in protection and maintenance of the functions and structures of the cellular components. A large gap exists in understanding of the metabolisms that are associated with stress and the existing knowledge. Under these conditions for attainment of success through molecular breeding for development of stress tolerant plants, it becomes a prerequisite to have a comprehensive profiling of the metabolites which were thought to be stress associated. Furthermore, there is also a need to identify and discover additional gene resources conferring stress tolerance from both wild, model plants and crops plants for future development of stress tolerance in the crop plants through molecular techniques.

7.1.2 ROLE OF GLYCINE BETAINE AND PROLINE IN OSMOTIC TOLERANCE

Ashraf and Foolad (2007) mentioned the importance of two osmolytes, viz., glycine betaine and proline in the improvement of abiotic stress (drought, salinity, temperature, etc.,) resistance in plants. They accumulate in a number of plant species in various parts as a response to the environmental stresses. Their key actual roles in imparting osmotolerance in plants yet remained controversial. However, it is ascertained that they exhibit positive effects on enzyme and membrane integrity. They do bring about the adaptive roles in plants by mediating osmotic adjustment under stress situations. Studies have indicated a positive correlative relationship in accumulation of glycine betaine and proline and plant stress tolerance. Few research studies have conferred that as a product of stress conditions their accumulation occurs but not as a mechanism of response of adaptation. They mentioned that genetically engineered plants are facing the limitation and are unable to produce sufficient amounts of glycine betaine and proline in spite of the presence of transgenes within them for production of above compounds for amelioration of effects of stress. Application of GB or proline exogenously to plants in stress conditions may be an alternative approach that has gained some importance. In their review, they mentioned many examples of their applications in improving stress tolerance in plants.

7.1.3 CLIMATE/ENVIRONMENTS

Hebbar et al. (2013) simulated three climatic scenarios (A2, B2, and A1B). He used the simulation model Infocrop-cotton and analyzed the effect of this change on production of cotton in India. His model projections indicated a rise in mean temperatures by 3.95, 3.20, and 1.85°C in many regions of India where cotton is grown. Furthermore, it has shown a decline in cotton yields by 477 kg ha^{-1} for the A2 scenario. They predicted that yield declines would be more in the northern zone of India, with a marginal decline in productivity. Some adaptive measures are to be adopted for increasing cotton production.

7.1.3.1 EFFECT OF LIGHT DISTRIBUTION ON COTTON CROP

One of the climatic factor influencing the cotton growth and productivity is the capture and distribution of light into cotton crop when intercropped by component crops. Wang Shi Wei et al. stated that (2017) intercropping with the jujube trees in cotton affected the saturated PAR in space time window in canopy of cotton. There was a decrease in the duration with an increase in jujube tree age. This variation in saturated PAR space–time window size of the cotton canopy exhibited a direct effect on cotton production. They concluded jujube of 5 years old planted in less than 4 m is not suitable for intercropping system in South Xinjiang basin for cotton.

7.1.3.2 EFFECT OF BT COTTON ON ENVIRONMENT

Kerur et al. (2014) study revealed that Bt cotton is less harmful to environment. They worked out the environmental impact quotient (EIQ) field use ratings for Bt and non-Bt cotton which were in range of 19.0591 and 38.0036.

7.1.3.3 RADIATION AND THERMAL USE EFFICIENCY

Niwas et al. (2017) computed radiation and thermal use efficiency at various phenophases of cotton. Cotton sown in April 2nd week had taken higher indices of thermal and radiation indices. A better association of weather parameters was seen at 50% flowering phenophase than others. The maximum and minimum temperatures had a positive correlation in boll opening stage, while at vegetative, flowering, and boll opening stage the relative humidities of morning and evening had a negative correlation with seed cotton parameters. They influence cotton yield to more than 82%.

7.1.3.4 TEMPERATURE AND RAINFALL

Extreme temperature and rainfall affect the growth and development and also the yields of dryland cotton. Anwar et al. (2017) observed that in dry land cotton growing regions in Australia, there was a high variation in interseasonal yields. This variation is to the extent of 33–49%. Highest yields were obtained when this dry land cotton was grown in those soils which had a higher availability of soil water storage capacities along with more growing season and summer rains. There was a reduction in yields of cotton when there were extreme temperatures that exceeded more than ≥35°C.

7.1.4 DROUGHT

Several systematic and sequential studies have been undertaken by various authors on drought resistance of cotton. Various studies have been undertaken to investigate the effect of drought and its interaction on cotton growth and productivity (Fig. 7.1), few of which are cited herein.

7.1.4.1 RESPONSE OF PARENTAL STRAINS TO WATER STRESS

Rosenow et al. (1983) screened primitive photoperiodic-sensitive strains of cotton from the world collections. They made an effort to identify the parental strains. These parental strains exhibited a unique response to water stress conditions. They determined the reasons that were behind these responses. They also undertook evaluations of physiological traits to identify and define some of the traits of potential use. They later incorporated some of them into those germplasm lines which were found to be desirable agronomically and screened that under different water regimes and locations and observed their performance at different growth stages.

7.1.4.2 ROOT GROWTH UNDER DIFFERENT LEVELS OF SOIL WATER CONTENT

Hu XiaoTang et al. (2009) mentioned that cotton root growth exhibited a logistic growth curve in three water regimes of soil of water contents of

FIGURE 7.1 Applying drought cycles in polyhouse (ranging from 20 to 30 days): Clear variation among cotton hybrids in the aspect of plant expression under drought condition.

90%, 75%, and 60% θf (θf is field water capacity). A faster rate of root growth with increased dry weight was observed in 75% θf. As the water content was more than the suitable conditions for root growth a drastic decline or slow root growth was observed at 90% θf. There was a decrease in root weight with an increase in soil moisture as well as soil depth. There was more distribution of cotton roots in narrow and wide row soils that were mulched by plastic film under 75% θf, with highest root weight also. There was also a decrease in the root to shoot ratio in cotton as there was an increase in the soil water content indicating that increased soil moisture levels have adverse effect on root growth of cotton (Fig. 7.2).

7.1.4.3 IMPROVEMENT IN THE TOLERANCE OF HEAT AND DROUGHT STRESS

Dabbert and Gore (2014) discussed in cotton about the challenges, perspectives for bringing about the improvement in the tolerance of heat and drought stress. They emphasized that future production of cotton has to face various abiotic stresses which become more prevalent affecting the productivity under field conditions in many regions. Extreme, prolonged high temperatures and water deficits may show an increase, as a result there is an increasing need to observe these effects together in a more comprehensive

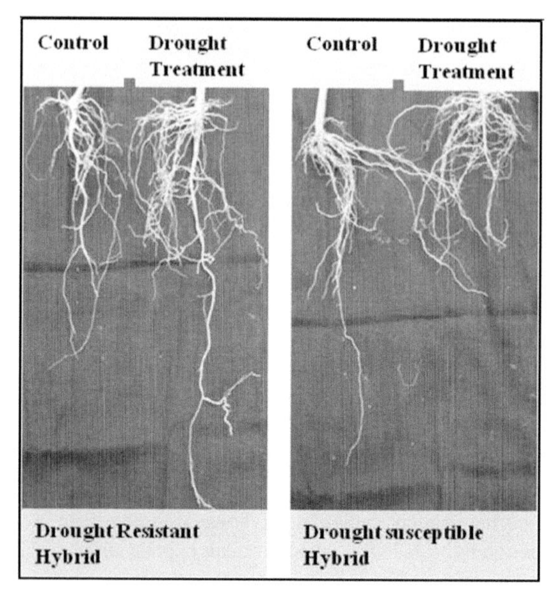

FIGURE 7.2 Clear variation among cotton hybrids in the aspect of root growth under drought condition: Drought-resistant hybrid showing increased root length and inclined lateral roots (robust root system) under drought condition.

manner. In their comprehensive review on the combined influences of these on cotton growth and on model plant species discussed about the importance of the genetic dissection of tolerant traits and their potentiality in cotton germplasm lines. They also discussed about the methods of phenotyping the cotton plants under these conditions (Fig. 7.3).

7.1.4.4 EFFECT OF DROUGHT STRESS ON LIGHT SATURATION POINT, CARBON DIOXIDE AND NITROGEN ASSIMILATION

Zhang YaLi et al. (2011) in his experiment on cotton observed that drought stress led to a decrease in the assimilation rate of carbon dioxide, light saturation point, and also the rater of photorespiration. There was an increase in leaf nitrogen content in cotton under drought stress. Furthermore, drought led to an increase in PSII quantum yield (ΦPSII) in leaves with a slight decrease in cyclic electron flow (CEF) around PSI. Cotton appeared to be more dependent on increased electron transport flux for utilization of light energy in drought situations in enhancing nitrogen assimilation of nitrogen within plants.

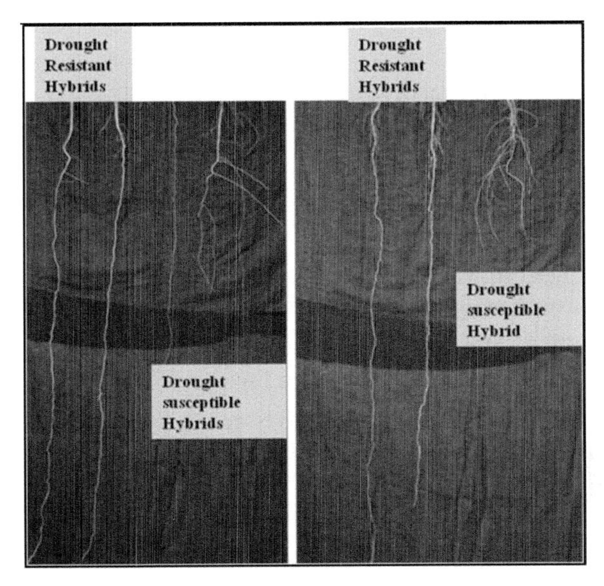

FIGURE 7.3 Brick chamber study for root growth: At flowering stage, noticed higher root growth in drought resistant cotton hybrids in normal condition.

7.1.4.5 COTTON CROP WATER STRESS INDEX DETERMINATION

Ünlü et al. (2011) in East Mediterranean region, Turkey, experimented in determination of the crop water stress index (CWSI) of cotton under drip irrigation in a heavy textured clay soil. They initiated irrigation at leaf water potentials of −15 bar for full (I100), −17 bar for DI70, and −20 bar for DI50 irrigation treatments and calculated the CWSI values and also measured the canopy temperatures air and vapor pressure deficits. Recording of measure of canopy temperature from full irrigated plots, they developed a non-water stressed baseline (lower baseline) equation for cotton as: Tc − Ta = −1.7543VPD + 1.56; $R^2 = 0.5327$ and by using canopy temperature data taken from continuous stress plots a non-transpiring baseline (upper baseline) equation as: Tc − Ta = −0.0217VPD + 3.2191 was built. CWSI and plant parameters exhibited a linear relationship. An increase in deficit of soil water led to a decline in dry matter and seed cotton yield. The relationship obtained SY = −2.3552CWSI + 3.5657; $R^2 = 0.499$ between seed cotton yield (SY) and seasonal mean CWSI values may be helpful in predicting the seed cotton yields and in scheduling of irrigations. Their results indicated that when CWSI approaches 0.36 cotton crop should be irrigated under the above climatic and soil conditions of that region.

7.1.4.6 EFFECT OF DROUGHT STRESS ON AGRONOMIC TRAITS, WUE AND FIBER QUALITY

Chen Yu Liang et al. (2013) in their study on effect of drought stress on various agronomic traits, fiber quality, and WUE in few upland cotton varieties (lines) of different color observed that there was a decline in all the agronomic and quality traits except lint yield in drought stress in these lines. They exhibited a significant decrease in yield with a reduction in quantity of irrigation by half. The decrease in seed cotton was more in brown cotton cultivar with a higher WUE, under drought stress than green cotton cultivar. They mentioned that WUE is related to leaf number, the boll number per plant, number of fruiting branches, number of fruit nodes, plant height, harvest index, weight per boll, length between fruit branches and main stem, seed index, the stem diameter, and lint percentage.

7.1.4.7 IMPACTS OF WATER SUPPLY RATE ON COTTON WUE AND YIELDS

Luo QunYing et al. (2015) used a stochastic weather generator, LARS-WG for the construction of models to suit the local climate scenarios of key cotton areas of production in eastern regions of Australia. The scenarios along with increased atmospheric carbon dioxide concentrations were linked to cotton model (CSIRO OZCOT) and quantified the potential impacts on cotton WUE and yields. Their model predicted an increase in WUE to the extent of 2–22% by 2030 and observed that in the future scenarios of climate change a water supply of 2 ML/ha may be sufficient for bringing positive effects in irrigated cotton. Furthermore, they mentioned for rain-fed cotton cultivation type of solid planting configuration would be more beneficial.

7.1.4.8 FLEA HOPPER DAMAGE UNDER WATER STRESS CONDITIONS

Brewer et al. (2016) in their study has observed that the damage by flea hopper in cotton is more under water stress conditions, bringing about a higher lint yield losses. Variability in damage was observed among different cotton cultivars, with variable irrigation schedules and plantings. They found that this pest has attained a threshold peak particularly at the beginning of the second stage of squaring.

7.1.4.9 INTERCROPPING

Tri and Dhiahul (2018) observed that at Asembagus Experimental Garden, Situbondo, cotton lines of the strain 03017/13 and 03008/24 have the highest consecutive acceptance of suitability to intercropping of IDR 17,860,681 and IDR 17,520,879 maize lines, to bring about an increase in income when intercropped with maize.

7.1.4.10 TECHNIQUE

Different techniques have been adopted to reduce the effect of drought and increase cotton productivity.

7.1.4.10.1 Mulching

In a field experiment, Nalayini et al. (2009) reported about the mulching with polyethylene (30, 50, 75, and 100 micron) in intercropping of cotton and maize. They mentioned that there was efficient moisture conservation and poly mulching has recorded highest net returns and also resulted in a highest benefit cost ratio.

7.1.4.11 CORRELATIONS AND PATH COEFFICIENT ANALYSIS UNDER DROUGHT STRESS CONDITIONS

Karademır et al. (2009) undertook correlations and path coefficient analysis in 20 genotypes of cotton (*Gossypium hirsutum* L.) between leaf chlorophyll content, yield, and yield components under drought stress conditions. The genotypes exhibited significant differences between leaf chlorophyll content, seed cotton weight per boll, 100 seed weight, plant height, and ginning out turn. Significant correlations existed between leaf chlorophyll content, seed cotton yield ($r = 0.231^{*}$) and ginning out turn ($r = 0.320^{**}$). Path analysis revealed that in cotton under drought stress conditions, the leaf chlorophyll content, plant height, number of monopodial branches, ginning out turn, and 100 seed weight have a direct effect on seed cotton yield. Their study indicated that improvement in the leaf chlorophyll content may act as an indicator for improvement in yield also particularly under drought stress conditions.

7.1.4.12 MECHANISMS OF DROUGHT RESISTANCE

Several studies have been undertaken to study physiological, biochemical, and molecular basis of mechanism of drought resistance in cotton.

7.1.4.12.1 Biochemical

A strong sensitivity of stomata to water stress conditions is seen with the nitrogen nutrition. Radin et al. (1981) in their study observed that there was closure of stomata at −9 bars of water potential in those plants which were well watered with a nutrient solution containing 0.31 millimolar N and were closed at −2 bars only when there was an increase in N level in nutrient solution. The N nutrition did not exhibit any influence on stomatal closure when there was an increase in the intercellular CO_2 concentrations, however, to this, CO_2 concentrations an exogenous application of ABA was found to bring about an increase in the stomatal sensitivity. They exhibited these responses when the water potentials were very high and especially seen in those plants which were N deficient. The study has concluded that some interaction occurs between the water stress and N nutrition, regulating the control of ABA accumulation within the plant.

7.1.4.12.2 Accumulation of Osmolytes

Kuznetsov et al. (2002) mentioned that during periods of water stress/ water deficiency, resistance to drought can be improved to suit a drought of 40 days, by imparting a preliminary heat–shock (HS) treatment (45°C for 1.5 h) even when the soil moisture content declines to 20% of field moisture capacity. Their study has indicated an accumulation of amino acids and amides, viz., arginine, proline, and asparagine in leaf cell sap induced by HS. These contributed to an increase in osmotic pressure of the cell sap of leaf. It not only increased resistance to deficiency of water but also aided in reduction of overheating of leaves. Cotton cv. INEBR-85 was found to be resistant.

7.1.4.12.3 Molecular

Yan et al. (2004) reported that overexpression of the *Arabidopsis* 14-3-3 Protein GF14λ in cotton produces a "stay-green" phenotype. It also has the ability to improve stress tolerance, particularly under moderate conditions of drought. They incorporated the *Arabidopsis* gene *GF14λ* into cotton plants. They explored its physiological roles. In transgenic cotton plants its expression exhibited variation. Some transgenic lines exhibited a "stay-green" phenotype. These lines also exhibited tolerance to water stress, had less wilting and higher photosynthesizing ability. They specified that stomatal conductance was regulated by this gene and it was also involved in the regulation of transporter proteins as H^+-ATPase through its interaction with 14-3-3 proteins. These proteins are thought to be involved in regulation of a number of metabolic processes and expression of this gene proves to be beneficial in cotton plants in water stress conditions.

Lian et al. (2009) transferrred H^+-PPase gene, *TsVP* from *Thellungiella halophila* into two cotton (*Gossypium hirsutum*) varieties (Lumianyan19 and Lumianyan 21). This foreign gene was found to be integrated well into cotton genome and its expression of higher activity was more in the vacuolar membrane vesicles. Its overexpression not only increased both shoot and root growth but also led to an increase in chlorophyll content, photosynthesis, relative water content, and increased resistance of these transgenic plants to osmotic/drought stress. The transgenic plants had improved root development and also a lower solute potential and these transgenic plants produced more yields than other plants and water stress conditions created for 21 days at flowering stage. There were 51% higher seed cotton yields.

Shi et al. (2011) reported that *GhMPK16*, a novel stress-responsive group D MAPK gene from cotton, is associated with disease resistance and drought sensitivity. They mentioned that mitogen-activated protein kinase (MAPK) cascades play vital part in mediation of biotic and abiotic stress responses. In plants, these are divided into four major groups (A–D) on the basis of their sequence homology and conserved phosphorylation motifs. Members of group A and B are extensively characterized, but very little information on the group D MAPKs was reported. They isolated and characterized *GhMPK16*, first group D MAPK gene detected in cotton. Their southern blot analysis suggested that it is a single copy in the cotton genome. Furthermore, RNA blot analysis revealed that its transcripts are accumulated to a large extent after the pathogen infection. It has a group of putative cis-acting elements that are related to stress responses in the promoter region. Subcellular localization analysis suggested that this gene is more functional in the nucleus.

Said et al. (2013) undertook a comprehensive meta-QTL analysis for fiber quality, yield, yield related and morphological traits, drought tolerance, and disease resistance in tetraploid cotton. The quantitative trait loci (QTL) studies in cotton (*Gossypium* spp.) were concentrated on those traits which are of agricultural value. Earlier research studies have identified a superfluity of QTLs related to fiber quality, disease and pest resistance, branch number, seed quality, and other related traits of yield and drought tolerance. Variations in the studies are due to the differences in the usage of different genetic populations, markers and marker densities, and testing environments. They mentioned that there is a requisite of updated comparative QTL analysis.

Wang HaiXia et al. (2014) studied drought resistance of cotton with *Escherichia coli* catalase gene. They investigated the T3 generation of KatE transgenic upland cotton for drought resistance. There was not much difference in the physiological indexes in the drought stress to squaring stage. In conditions of water stress to budding peak stage and also at the full-bloom stage, it was observed that this transgenic cotton leaves had higher relative water content, Fv/Fm, CAT activity, leaf net photosynthetic rate (Pn), stomatal conductance (Gs), etc. Drought stress to boll opening period, the transgenic cotton plant height, branch number, and boll number water stress affected the growth period of cotton.

7.1.4.12.4 Morphological

As a research advisor in a seed company in India, Dr. Maiti has identified few morpho-anatomical traits in Bt cotton which were found to be related to drought resistance in cotton (unpublished). Bt cotton hybrids showing drought resistance in field have small thick leaves, stout long petiole with thick collenchyma, profuse trichome on leaf surface, thick cuticle, compact one or two layered palisade cells which prevent excess loss of transpiration. Initially, they selected visually the 100 Bt cultivars for leaf traits mentioned and finally they selected only 16 Bt hybrids. These were grown in drought-prone areas in Maharashtra and were found to be well adapted and good yielders. These need to be confirmed by breeders in futures.

7.1.5 SALINITY

7.1.5.1 SALINITY DURING GERMINATION AND EARLY SEEDLING GROWTH

Ashraf (2002) reported that in three cotton cultivars (NIAB-Karishma, NIAB-86, and K-115) salinity during germination and early seedling growth has induced changes in α-amylase activity. An increase in NaCl concentrations led to a decrease in α-amylase activity and in all the cultivars it resulted in breakdown of starch into reducing and non-reducing sugars. This seems to be more pronounced in NIAB-86. K-115 recorded more germination.

7.1.5.2 SALINITY AT GERMINATION AND SEEDLING STAGE

Ahmad et al. (2002) mentioned that cotton is susceptible to salinity at stages of germination and seedling. Salt stress seriously affected the biomass production. This was attributed to a decline in leaf area, stem thickness, shoot and root weight. These ultimately lead to a decrease in seed cotton yield. They mentioned that 7.7 dS m^{-1} was the threshold salinity level where the initial yields of cotton exhibited a decline, while at 17.0 dS m^{-1} a 50% decline in yield was observed. There was also a decline in strength, length, and micronaire of fibers. However, ginning out turn has exhibited an increase in saline conditions in both *Gossypium hirsutum* and *Gossypium barbadense*. Leaf N contents and photosynthetic ability were affected with high saline levels. In cotton leaves there were increased concentrations of Na^+ and Cl^- but reduced K^+, Ca^{2+}, and Mg^{2+} under salinity. Several researches have reported a mild increase in K^+ and a modest accumulation of Na^+ with an increase in salinity, in some crops thus as a selection criteria of salinity tolerance the K^+/Na^+ ratio is used. Generally, in cotton it was observed that Na^+ exclusion was associated with salt tolerance. Salinity also led to a decline in uptake of N and P in cotton.

7.1.5.3 EFFECT OF SALINITY IMPOSED FROM 3 WEEKS TO MATURITY

Ashraf and Ahmad (2000) studied the influence of salinity imposed in cotton from 3 weeks to maturity. Their results reported a decline in root and shoot weight and also in cotton yield components in both sensitive and salt tolerant cultivars of cotton. However, three salt-tolerant (B-557, Culture 728-4, MNH-156) had a lower ginning out turn though they exhibited a higher staple length, strength, and fiber maturity. Salt sensitive lines [B-1580 (ne), culture 604-4, MNH-147] accumulated more Cl^- ions in leaves, while tolerant lines had more of K^+, Ca^{2+}, K/Na ratios and N in the leaves and seed oil content.

7.1.5.4 SALINITY TOLERANCE

7.1.5.4.1 Antioxidants in Salt Tolerance

Gossett et al. (1994) mentioned that in salt tolerant (cv. Acaia 1517-88 and Acala 1517-SR2) there were large constitutive levels of catalase (121–215%) and u-tocopherol (312–420%). A 38–72% increase in peroxidase activity, 55–101% increase in glutathione reductase activity was recorded in tolerant Acala cultivars. They also had a lower oxidized/reduced ascorbic acid ratio and a higher reduced/oxidized glutathione ratio. Their results have indicated protection from oxidative damage of antioxidant high levels and a very active ascorbate–glutathione cycle as in many plant species, here in cotton also it might be involved for development of salt tolerance.

7.1.5.4.2 Effect of Salinity on Metabolic Processes

Chaves et al. (2009) mentioned that during the life cycle plants often come across the water deficits and salinity. An understanding of their responses to above stress conditions enables to bring out the stabilization in yields. These might exhibit a direct or indirect effect on the plant metabolic processes. A direct effect on the photosynthesis is seen, where these stresses may bring about direct variations in the diffusion rates of stomata and mesophyll cells. Sometimes these stresses result in a secondary effect which arises from the multiple stresses in the form of oxidative stress damages. The recovery of a plant from the water or salt stress is more dependent on the carbon balance and a recovery in the photosynthetic ability (viz., velocity and degree of photosynthesis). The present knowledge on these aspects is limited. A comprehensive study conducted has indicated that a downregulation of a few photosynthetic genes may be brought about by both the drought and salt stress.

7.1.5.4.3 Transgenic Lines SNAC1 Gene Against Salinity and Drought

Liu et al. (2014) overexpressed *SNAC1* gene, in cotton cultivar YZ1 through *Agrobacterium tumefaciens*-mediated transformation. The transgenic cotton plants were found to be more vigorous in terms of root development and growth and had a higher content of proline under saline and drought treatments. They exhibited also an improved tolerance to these stresses. Though the transgenic lines exhibited a decrease in the transpiration rate they maintained a constant photosynthetic rate at flowering, indicating that overexpression of *SNAC1* I gene in cotton imparts high tolerance to drought and salt.

7.1.5.4.4 Reactive Oxygen Species (ROS) Production During Drought and Salinity Stress

Miller et al. (2018) reported about the reactive oxygen species (ROS) homeostasis and signaling that occurred during drought and salinity stresses. They mentioned that under high light intensity water deficit and salinity or in combination with other stresses, may affect largely the photosynthesis. It might also result in an increase in increase photorespiration and alter the normal homeostasis of cells. This alteration may lead to an increase in the production rates of reactive oxygen species (ROS) which may function as signal transduction molecules or act as toxic by-products.

7.1.5.4.5 Indexes of Salt Tolerance of Cotton

Zhang Yu et al. (2011) analyzed the indexes of salt tolerance of cotton in Akesu river irrigation district. They constructed linear subsected functions and the nonlinear S-shaped functions between the cotton relative yield and soil salt content in the salinized soil of this region at two soil depths, viz., 0–20 cm and 0–40 cm. A better response fitted relationship of above was observed with the nonlinear S-shaped function. They concluded that in that district, the cotton critical soil salt content was 0.302%, while the cotton threshold soil salt content was 1.119%. Similarly, salt content of 0.558% and 0.581% in 0–20 cm soil depth brought about a faster and 50% reduction in yield.

7.1.5.4.6 Saline Ice Water Irrigation Effect

Zhang XiuMei et al. (2012) in Xiaoshan, Haixing conducted an investigation of cotton emergence and yields under frozen saline water irrigation in winter. Their results had showed that during their period of investigation in the years of 2009, 2010, and 2011 soil salt and water contents at sowing was 0.32%, 0.29%, and 0.17% and 26.2%, 25.0%, and 24.2%, respectively, implying that irrigation of longer saline ice water irrigation exhibits a better soil desalination effect. It provided favorable soil water and salt conditions for growth, where there was 85% or higher germination percentage. There was also a decrease in seedling sodium contents by 57.6–64.5% under this irrigation, though there were higher contents in potassium and calcium. There was also an increase in seed cotton yields, thus their findings have indicated that frozen saline water irrigation in winter would bring about an improvement in cotton growth in coastal soils in north China.

7.1.5.4.7 Subsurface Pipe Drainage

In the arid area of China the soil salinization is a serious problem restricting the agricultural development. Liu YuGuo et al. (2014) studies indicated that an improvement in soil condition might be achieved by the shallow subsurface pipe drainage technology, which enabled in the control of soil salinity and also led to an enhancement in yields of cotton. They suggested that this technology in combination with the drip irrigation done under plastic mulch may prove a better suitable and effective method for the exploration as well as utilization of the saline soils in Xinjiang.

7.1.5.4.8 Combined Effects of Manganese (Mn) and Salinity (NaCl) on Cotton

The stress effects of salinity and few trace elements may be additive, antagonistic, or synergistic. Wang Xiao (2014) conducted a soilless culture experiment to have an understanding of the combined effects of manganese (Mn) and salinity (NaCl) on cotton growth and yield, in a greenhouse at Wuhan Botanical Garden, Chinese Academy of Sciences. The results showed that though there was an inhibition in cotton growth by salinity of NaCl increase from 15 to 45 mmol L^{-1}, it resulted in promotion of cotton yield. A good vegetative growth in cotton has resulted at NaCl concentration of 15–25 mmol L^{-1} in irrigation water. An increase in salinity led to a decrease in dry weight of cotton roots and shoots. Furthermore, Mn concentration at 36 μmol L^{-1} resulted in maximum growth and yield. The trace element Mn and salinity resulted in a significant impact on accumulation of Na and Mn in cotton leaf. Though salinity promoted an accumulation of Na accumulation, it resulted in an inhibition in the accumulation of Mn in cotton leaves. The Abbott equation that was used for analyzing the combination effects of Mn and salinity on cotton growth and yield exhibited a negative relationship between Mn and salinity. In other words, the research has indicated that the toxic effects on the growth and yield of cotton were less severe for combined Mn and salinity than that of the added individual effects.

7.1.5.4.9 Suitable Control Index of Soil Salinity

Feng Di et al. (2014) determined index controlling soil salinity for cotton field with salt irrigation. Expansion of secondary salinization of farmland increased over the recent years due to mismanagement of irrigation with saline water. They conducted a research study for exploring a favored suitable control index of soil salinity to lessen the impact on decline in yields. Their results indicated that there was an increasing trend in the micronaire of lint and fiber strength with increase in soil salinity. Their result has shown that the control index of soil salinity which was decided by net earnings was found to be lower than that was decided by the seed cotton yields. Thus, it proves to be an essential consideration of fiber quality index. They mentioned that consistent net earnings might be possible when the initial soil salinity in the soil and also when the mean soil salinity during the cotton growing period within a soil depth of 0–60 cm is less than 0.71 dS/m and 0.67 dS/m.

7.1.5.4.10 Optimal Time of Screening to Salinity Tolerance

Barrick et al. (2015) evaluated the efficacy of screening four introgressed upland genotypes from a backcross *Gossypium hirsutum* L. × *G. barbadense* L. inbred line population grown in two soils, that is, an organic farm loam soil or a conventional farm clay soil. They observed variations in genotypes in growth traits measured with response to NaCl at 3 and 6 weeks interval. They suggested that optimal time of screening to salinity tolerance is during early cotton establishment. Salinity did not exhibit any effect on chlorophyll content and fluorescence though it had a negative effect on all other growth traits. Their results indicated that soil type has very little influence on the screening of cotton genotypes for salt tolerance. An adequate stage of cotton for screening to salinity tolerance was second true leaf stage.

7.1.5.4.11 Insect Community Diversity in Transgenic Bt Cotton in Coastal Saline Alkaline Soils

In China, cotton is planted in coastal saline alkaline soils in and dry soils China's Yellow River and the Yangtze as it has the ability to tolerate high drought and salinity compared with other economic crops. Luo JunYu et al. (2016) studied insect community diversity in transgenic Bt cotton, in these soils in the Shandong Province and mildly saline and semi-dry soils of Zaoqiang County, Hebei Province for 2 years. Their results have shown that both in case of spray and no-spray, there was lower number of total number of individuals of insect communities and pest sub-communities in transgenic Bt cotton fields, though they exhibited variations. Though a higher insect diversity index and evenness index (of the insect community and pest sub-community) was observed in transgenic Bt cotton fields the dominant concentration index remained lower. Spraying decreased the total number of individuals within the insect community, pest sub-community and also the enemy sub-community, diversity index and evenness index in both Bt and non-Bt fields. These soils had a lower biodiversity and simpler ecosystem than non-saline-dry soils.

7.1.5.4.12 Biosafety Assessments of this Transgenic Cotton with Reference to the Rhizosphere Microbes and Salinity Interactions

In China, there was an increase in the area of cultivated transgenic Bt cotton, particularly in the saline alkaline soils. Therefore, for the biosafety assessments of this transgenic cotton with reference to the rhizosphere microbes and salinity interactions and the Bt protein in residues, much efforts were focused. Luo JunYu et al. (2017) studied these aspects for two years in 2013 and 2014 in a Bt cotton (variety GK19) and its parental non-transgenic cotton (Simian 3) which was cultivated at various salinity levels (1.15, 6.00, and 11.46 dS m^{-1}). Under soil salinity stress, the Bt cotton GK19 rhizosphere soil had trace amounts of Bt proteins though it exhibited an increase with growth of cotton and salinity levels. There was a decrease in the populations of slight halophilic, phosphate solubilizing, ammonifying nitrifying and denitrifying bacteria with an increase in soil salinity in the rhizosphere. However, in non-Bt in soil rhizosphere, an increase in salinity led to a decrease in the microbial biomass carbon. Microbial respiration and soil enzymatic activities, viz., catalase, urease, and alkaline phosphatase activity also exhibited a decline. Correlation analyses has shown an enhanced content in Bt protein in the rhizosphere soil of Bt cotton, was caused due to a slower decomposition of this protein by the soil microorganisms. This research finding has indicated that salinity alters soil microbial properties and population dynamics rather than the Bt protein in transgenic cotton.

7.1.5.4.13 Effects of Soil Salinity on Expression of Bt Toxin (Cry1ac) and Its Control Efficiency

Luo JunYu et al. (2017) analyzed the effects of soil salinity in field-grown transgenic cotton, on the expression of Bt toxin (Cry1Ac) and its control efficiency of *Helicoverpa armigera*. They observed that increasing soil salinity resulted in a decrease in the contents of Bt protein in the transgenic Bt cotton leaves. It led to a decline in the insecticidal activity of Bt cotton against cotton boll worm. Salinity levels and Bt protein contents were negatively correlated. The cotton boll worm populations seem to appear in higher number in the Bt cotton that was grown in soil with medium salinity. They attributed this to a change in plant nutritional qualities or other plant defensive characteristics by high salinity soil.

7.1.5.5 TECHNIQUES

7.1.5.5.1 Screening

As a research advisor in Seed Company in India Dr. Maiti developed a novel low-cost technique in screening cotton and other field and vegetable crops for screening in mass scale cultivars and finally selection of salt tolerant cultivars. The techniques involves in screening cotton cultivars in plastic pots in coconut peats in 1.5–2 mL NaCl solution and finally selecting salt-tolerant cultivars for high seedling growth and root density. It is observed that salt-tolerant cultivars produced profuse roots which might have facilitated in osmotic adjustment. To our wonders the salt-tolerant cultivars select in the laboratory test showed good performance in saline-prone areas in Gujarat and farmers had good demand for these cultivars.

Maiti et al. (2015) undertook a study on salt tolerance of Bt cotton hybrid subjected to NaCl stress at the seedling stage, at control, 0.1, 0.15, 0.20, and 0.25 M NaCl. With increasing salinity there was reduction in seedling emergence, root length, shoot and root dry weight. Few Bt Cotton hybrids were selected to be tolerant to salt stress, viz., Grace BG, hybrids 111m 114 (Figs. 7.4 and 7.5).

7.1.5.6 MECHANISM

7.1.5.6.1 Physiological-Biochemical and Molecular—Calcium-Dependent Protein Kinases (CPKs) Role

Gao Wei et al. (2018) mentioned that components of signaling which involve calcium (Ca^{2+}) or protein kinases which are dependent on calcium for downstream regulation play important roles in stress signals perception and transduction. They identified 98 predicted CPKs from upland cotton (*Gossypium hirsutum* L. "TM-1"). They classified them into four groups on the basis of phylogenetic analyses. Gene family distribution studies have revealed that the genome duplication events to the total number of GhCPKs have a substantial impact. Transcriptome analyses have shown that CPKs' expression in different organs is widely distributed. They selected 19 CPKs exhibiting a quick response at salt stress at the transcriptional level. Most of these were found to be induced by ethephon, which released ethylene. It suggested that there is a partial overlap in the responses of salinity and ethylene.

FIGURE 7.4 Variation among cotton hybrids for salinity tolerance. High seedling emergence % and seedling vigor under saline condition was noticed in salinity tolerant hybrid.

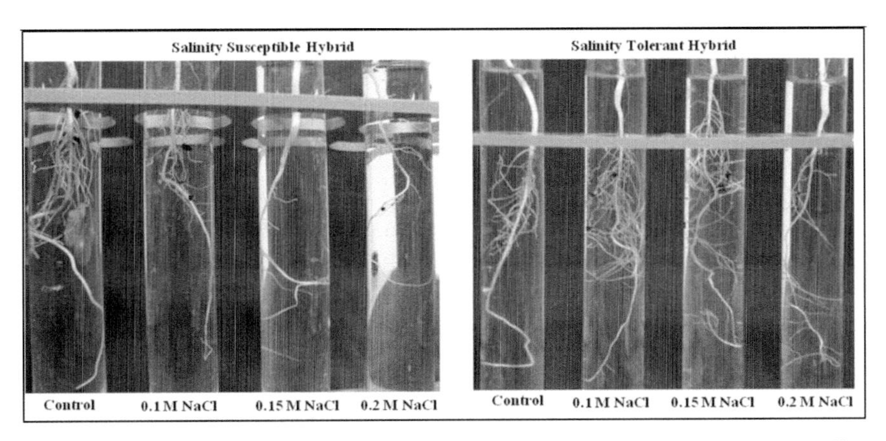

FIGURE 7.5 Increased taproot length and production of profuse lateral roots under saline condition (0.1 M NaCl) in salinity tolerant hybrid.

Silencing of four CPKs (GhCPK8, GhCPK38, GhCPK54, and GhCPK55) severely contributed to the initial/basal resistance to salt stress in cotton. It is concluded that in upland cotton CPKs are involved in multiple developmental responses, abiotic stresses. A cluster of the cotton CPKs were shown to exhibit their participation to some extent in the early signaling events.

7.1.5.6.2 Introduction of AtNHX1 and TsVP Gene

Cheng Cheng et al. (2018) introduced the AtNHX1 gene from *Arabidopsis thaliana* and the TsVP gene from *Thellungiella halophila* for coexpression in cotton (cv. GK35) for improving the salt tolerance. Cotton with overexpressed AtNHX1-TsVP genes had higher emergence rates and dry matter accumulation under salt stress. They stated that the growth of transgenic cotton with overexpression of the AtNHX1-TsVP genes under salt stress may be attributed to the accumulation of Na^+, K^+ and Ca^{2+} in leaves which enhanced the ability to maintain ion homeostasis and osmotic potential in plant cells under salt stress. This has led to the confirmation that the presence of higher relative water content in the cells results in maintenance of high carbon assimilation capacity. These results revealed that overexpression of AtNHX1-TsVP increased the tolerance of transgenic cotton.

7.1.5.7 ALLEVIATION

Few techniques are developed to alleviate salinity stress effect on cotton.

7.1.5.7.1 Nitrapyrin Soaking

Tao Rui et al. (2013) observed a marked inhibition of cotton seed germination under salt stress. Nitrapyrin soaking resulted in enhancement of seed germination rate, germination potential, and cotton seed vigor index, playing a role on alleviating the salt inhibition effect on cotton seed germination. It also led to an increase in the seed enzyme activities, viz., SOD, POD, and CAT by 57.2–282.7%, 8.3–139.3%, and 6.4–15.1%. Apart from these it also resulted in an increase in water-soluble organic substance content and promoted the transformation of complex organic substances to water-soluble organic substances during the germination process.

7.1.5.7.2 Bacteria from Noncultivated Plants

Irizarry et al. (2017) reported application of bacteria from non-cultivated plants (Malvaceae), Portia tree (*Thespesia populnea*), and wild cotton (*Gossypium hirsutum*) of Puerto Rico to promote growth, change root architecture, and alleviate salt stress of cotton. They inoculated *Bacillus amyloliquefaciens, Curtobacterium oceanosedimentum,* and *Pseudomonas oryzihabitans* onto acid delinted cotton seeds. The inoculation resulted in an increase in seed germination and radicle length of emerging seedlings. *B. amyloliquefaciens* inoculation resulted in a large percentage of seedlings with expanded cotyledons after 8 days. There was also an increase in primary and lateral root growth. They concluded that their research data has given a support to the hypothesis that non-cultivated plants in the Malvaceae growing in stressful environments possess bacteria that promote growth, alter root architecture and alleviate salt stress of cotton and okra seedlings.

7.1.5.7.3 Effect of Exogenous Salicylic Acid

Wang LiHong et al. (2017) studied effect of exogenous salicylic acid on the physiological characteristics and growth of cotton seedlings under 0.60% NaCl stress in "Zhongmiansuo 41" (tolerant varieties) and "Zhongmiansuo 49" (medium tolerant varieties). The results revealed that there was inhibition of cotton seedling growth by NaCL stress. NaCl stress with SA soaking + spraying treatment resulted in an increase in cotton seedling height, leaf area, dry mass, root activity, root/shoot ratio. There was an increase in malondialdehyde (MDA), soluble sugar (SS), soluble protein (SP), and proline (Pro) contents in leaves when there was exogenous application of SA, though there was a decrease in superoxide dismutase (SOD), peroxidase (POD), and catalase (CAT) activities in leaves and roots. The activities of SOD, POD, and CAT were less in roots of cotton seedlings than of leave under the SA treatment. They observed that SA soaking + spraying composite treatment on 0.60% NaCl mitigation effect was better and the 0.05 mmol L^{-1} soaking + 0.2 mmol L^{-1} spraying treatment was the best. In Zhongmiansuo 49' SA alleviation of stress intensity is more, with an improvement in root activity by 15.08–80.48%. The study revealed that exogenous SA could alleviate cell membrane damage and increase salt tolerance by regulating osmotic adjustment and antioxidant capacity of cotton seedlings.

7.1.6 HEAT AND COLD TEMPERATURE

7.1.6.1 HIGH TEMPERATURE

7.1.6.1.1 Effect of Planting Dates and Thermal Temperatures

Many uncontrollable environmental conditions exhibit their influence on growth and development in cotton. Temperature variabilities in the field are observed when planting is done at different dates. Ullah et al. (2016)

evaluated the effect of planting dates and thermal temperatures (growing degree days) on yield of four cotton genotypes, viz., CIM-598, CIM-599, CIM-602, and Ali Akbar-703. CIM-599 produced the highest seed cotton yield of 2062 kg ha^{-1}, due to a maximum boll number and boll weight. The highest seed cotton yield resulted when planting dates were from 15th April to 1st May. Early and delayed planting led to a reduction in yield, attributed to meager heat unit accumulation. Regression analysis has revealed that an increase of one unit (15 days) from early to optimum date (15th March to 15th April) resulted in an increase in yield by 93.58 kg ha^{-1}. Furthermore, a delay in planting also resulted in a similar ratio of decrease in seed cotton yield. Thus, they concluded that sowing of cotton between 15th April and 1st May may be beneficial to improve productivity.

7.1.6.1.2 Temperature-Sensitivity of Cotton Fiber Strength

Shu Hong Mei et al. (2009) studied the physiological mechanisms of changes in sensitivities to temperature of cotton fiber strength formation in different cotton cultivars (Kemian 1, a temperature-insensitive cultivar and Sumian 15, a temperature-sensitive cultivar). Results revealed that in normal sowing date, during fiber development period, when the mean daily minimum temperature was 24.0°C and 25.4°C, strength of fiber, sucrose synthase activity sucrose transformation rate, and cellulose accumulation in cotton fiber reached highest levels, while there was low activity of β-1,3-glucanase. On the other hand, when mean daily minimum temperature was lower than 21.1°C during late sowing dates, there was a decrease in enzyme activity and matter content related to cotton fiber development. There was a decrease in sucrose transformation rate, β-1,3-glucan and cellulose accumulation. Though sucrose synthase activity has shown a decline, there was an increase in β-1,3-glucanase activity. This might have resulted in lower fiber strength. Sensitivity of cultivars to low temperatures varied. In Sumian 15 change variations in sucrose transformation, cellulose accumulation, and enzyme activity were larger.

7.1.6.1.3 Effect of Elevated [Co2] and Warmer Temperatures

Broughton et al. (2017) reported that elevated [CO_2] and warmer temperatures affect the growth and physiology of an older and modern cotton genotype differentially. They quantified the physiological and growth capacity of two cotton genotypes older (DP16) and modern (Sicot 71BRF) grown in ambient and elevated atmospheric [CO_2] (CA: 400 μL L^{-1}, and CE: 640 μL L^{-1}) and two temperature (TA: 28/17°C and TE: 32/21°C, day/night) treatments under well-watered conditions. CE resulted in an increase in biomass and photosynthetic rates than those grown at CA, and similarly TE increased biomass compared with TA. The modern genotype Sicot 71BRF exhibited more photosynthetic rates, lower biomass, and leaf area. Modern genotypes with smaller, more compact growth habits, and higher photosynthetic rates may gain more carbon with less requirement of water and may confer advantage to grower's in the future climatic conditions, in the Australian production systems.

7.1.6.2 TECHNIQUES

7.1.6.2.1 Calculation of Heat Indexes With Grey Forecasting Model

Guo JianPing et al. (2010) used Grey forecasting model of heat index of cotton in Xinjiang. Xinjiang is base area of importance for production of commercial cotton in China. The main constraint to cotton's yield and quality in the area is the occurrence of unstable inter-annual thermal condition. One of the direct indexes of characterizing the thermal state of the environment is the heat index. The grey model (GM) is used to predict the trend of environmental elements time series across many areas as an effective and efficient method. They calculated the heat indexes by using mean temperature data from 1961 to 2005 (the mean temperature was calculated at a 10-day step), combined with the heat demand conditions of cotton at different growth stages and obtained monthly heat indexes and constructed GM (1,1). The results showed that the average regression-calculating accuracy of every model was more than 90%. In summary, the heat index forecast model for different cotton-growing regions in Xinjiang which was established by GM (1,1) model, could be used to predict thermal conditions of growth and development stages of cotton. Consequently, it laid the foundation for the forecasting cold damage of cotton. Thus, the prediction results of GM (1,1) regression model could be applied to guide agricultural production.

7.1.6.2.2 Screening of Breeding Lines for Thermotolerance

Kheir et al. (2012) developed a screening protocol based on the principle of "acquired tolerance," where the cotton seedlings are exposed to a sublethal level of specific stress and screened species and varieties of cotton for thermotolerance. Better thermotolerance was seen in Old World cotton species rather than New World cotton species. Variations in acquired thermotolerance were noticed among the 36 diverse *Gossypium hirsutum* germplasm lines. They identified a thermotolerant genotype *G. hirsutum* (H-28) by the TIR technique. This genotype exhibited an increase in cell viability, synthesizing capacity of proteins, during alleviation from high temperature stress. Their findings have shown that TIR can be used for effective screening of breeding lines or germplasms for identification of thermotolerant lines.

7.1.6.2.3 Selection of Pollen by Heat Treatment

Rodriguez and Barrow (1988) reported that pollen from cotton (*Gossypium hirsutum* L.) cultivars, which exhibits heat tolerance in the field, have higher fertility rates after heat treatment. They selected a highly heat tolerant breeding line, 7456, of *G. barbadense* L. as the donor parent and "Paymaster 404," a heat-sensitive cultivar, as the recurrent parent and made crosses. The results demonstrated that increased heat tolerance, measured on the basis of fertile pollen after heat treatment, is seen in plants of all populations, in parents, F_1, F_2, and first backcross populations. Their study has indicated that selection of pollen by heat treatment enables to screen for large numbers of genetic combinations in pollen. It also accounts as a valuable method of breeding for heat tolerance.

7.1.6.2.4 Heat Tolerance of Upland Cotton

Excessive high temperatures that prevail during the reproductive stage have a drastic effect on the reduction of the yields of cotton. Rahman et al. (2004) evaluated heat tolerance of upland cotton during the fruiting stage. They used the cellular membrane thermostability (CMT) assay as a measure of assessing heat tolerance, which indirectly measures the integrity of cellular membranes through quantifying electrolyte leakage that occurs after heat treatment. Higher CMT values were related to heat tolerance and higher yields in many crop species though with respect to cotton its utility and relationship was limited. The temperature regimes modified the relative ranking of the upland cultivars and hybrids, heat-tolerant and susceptible groups remained quite stable. Cultivars FH-900, MNH-552, CRIS-19, and Karishma emerged as relatively heat tolerant (thermo stable) and cultivars FH-634, CIM-448, HR109-RT and CIM-443 as heat susceptible. Exposure to high temperature prior to the CMT test resulted in a better distinction between heat-tolerant and heat-susceptible cultivars and hybrids. Stronger relationship is seen between CMT and SCY. Higher SCY was indicated by regression analysis. It was attributed to more of CMT in the presence of heat stress. CMT's positive relation to SCY was seen not only under supra-optimum greenhouse conditions, but also appeared at early and late field regimes too. Under greenhouse conditions which were optimum (non-stressed), CMT was found to show a negative relationship to SCY. This indicates that in the absence of heat stress susceptible cultivars and hybrids gave higher yields. The differential ability of the cotton cultivars and hybrids to adjust themselves to cellular membrane thermostability exhibited their mode of physiological adaption in these upland cotton genotypes against heat stress.

7.1.6.3 LOW TEMPERATURE

7.1.6.3.1 *Effects of Low Temperature on PSI and PSII*

Xiao Fei et al. (2016) investigated on the effects of low temperature on PSI and PSII photo inhibition in cotton variety Xinluzao 45 at leaf at boll stage. They observed that under low temperature stress there was a decline in the light-adapted maximum quantum yield of PSII (Fv′/Fm′), photochemical quenching coefficient (qP), and effective quantum yield of PSII [Y(II)], while it led to an increase in non-photochemical quantum yield of PSI. There was also an increase in the yield of regulated energy dissipation [Y(NPQ)] and non-regulated energy dissipation of PSII [Y(NO)], including reversible photo inhibition. Furthermore, low temperature stress also led to a significant decrease in the acceptor side limitation of PSI [Y(NA)] and an increase in donor side limitation of PSI [Y(ND)]. There was no decrease in the effective PSI complex content (Pm). This has suggested that PSI in cotton leaf is insensitive to low temperature than PSII. There was an enhancement in the quantum yield of cyclic electron flow [Y(CEF)] and the ratio of [Y(CEF)] to the effective quantum yield of PSII [Y(CEF)/Y(II)] by low temperature stress. Their research findings suggested that in cotton stimulation of cyclic electron flow has a key role in protecting PSI and PSII from photoinhibition that is caused by low temperature stress. Furthermore, the non-photochemical quenching (NPQ) and regulated heat dissipation [Y(NPQ)] exhibited a positive correlation with the quantum yield of cyclic electron flow [Y(CEF)]. This has indicated that the strong excess excitation energy that resulted due to the over closure of PSII reaction center might be responsible to bring reversible photo inhibition of PSII under low temperature stress.

7.1.6.3.2 Cold Resistance Characteristics of Different Cotton Varieties

Li XingXing et al. (2017) studied the characteristics of cold resistance of different cotton varieties. Their results revealed that Xinluzao 50 and 57 cotton cultivars had higher seed vigor, germination rate, and ability of resistance to low temperature stress. There was a higher suppression of the underground growth by low temperature conditions, with a severe inhibition in plant height of cotton seedling. Similarly, there was also a decrease in total root length and surface area and diameter of roots in cotton seedlings. Through cluster analysis, six cotton varieties were divided into three types: non-cold-resistant cotton varieties, Xinluzao 45 and 48; middle cold-resistant varieties, Xinluzao 50, 51, 60; strong hardy variety, Xinluzao 57.

7.1.6.3.3 cDNA Clone Ghdreb1 Against Low Temperature Stress

Peng et al. (2007) reported that in cotton a cDNA clone GhDREB1 isolated from cotton (*Gossypium hirsutum*) by cDNA library screening increases plant tolerance to low temperature. This transcription factor is found to be negatively regulated by gibberellic acid. Low temperatures and salt stress induced mRNA accumulation of *GhDREB1*. Its transcripts were also found to be regulated negatively by gibberellic acid in cotton seedlings. Its promoter region had one low temperature and four gibberellin responsive elements. Green fluorescent protein (GFP) signal intensity or β-glucuronidase (GUS) activity driven by the *GhDREB1* promoter was clearly enhanced by low temperature but repressed by GA_3. These results suggest that GhDREB1 functions as a transcription factor and plays an important role in improving cold tolerance, and also affects plant growth and development via GA_3.

7.1.6.3.4 GhTIP1;1 Protein in Cell Freezing-Tolerance

Li et al. (2009) reported that plant cells often increase their tolerance to cold and freezing stress by expressing some cold-related genes. A cotton gene encoding tonoplast intrinsic protein (TIP) from a cotton seedling was isolated, cDNA library, and designated as *GhTIP1;1*. GFP fluorescent microscopy revealed that it is localized to the vacuolar membrane. Assay on *GhTIP1;1* expression in *Xenopus laevis* oocytes showed that it displayed water channel activity and facilitated water transport to the cells. At normal conditions, its transcripts were found to be accumulated to a large extent in roots and hypocotyls of cotton. They observed that there was upregulation of expression of *GhTIP1;1* in cotyledons, but downregulated in roots of cotton seedlings within a few hours after cold treatment.

7.1.6.3.5 Gene Mpafp149 for Cold Tolerance

Liang Na et al. (2011) explored a rapid and effective method for improving cold tolerance of cotton plant. This was achieved through the inoculation of the endophyte of cotton that was transformed by an insect antifreeze protein gene Mpafp149 from the desert beetle *Microdera punctipennis*. The plasmid pBE2-Mpafp149-gfp is transformed into *Bacillus subtilis* M17 successfully and the transformed endophytes exhibited an increase in freeze tolerance. The cotton seedlings carrying this transformed M17 exhibited a lower relative conductivity. This indicated that the antifreeze protein expressed by the transformed endophyte exerted certain level of cryoprotective effect, though there was not much difference in the fluorescent intensity between the inoculated and the control cotton.

7.1.6.3.6 Principal Component Analysis for Chilling Tolerance

Wu Hui et al. (2012) investigated the chilling tolerance of cotton seedlings. They classified the 12 individual physiological indicators into seven independent comprehensive components by principal component analysis. By using membership function method and hierarchical cluster analysis they clustered 15 cotton varieties (lines) into three categories. They established a mathematical model for the evaluation of chilling tolerance of cotton seedlings, viz., D = 0.275 − 0.244 Fo1 + 0.206 Fv/Fm1 + 0.326gs2 − 0.056 SS + 0.225MDA + 0.038REC (R2 = 0.995) whose evaluation accuracy was more than 94.25%. They also screened six identification indicators which were found to be related closely to chilling tolerance, viz., including Fo1, Fv/Fm1, gs2, SS, MDA, and REC. They concluded that seedling leaves of cotton varieties (lines) with high chilling tolerance were less damaged under low temperature stress. They could maintain relatively high photosynthetic electron transport capacity and high stomatal conductance even after recovery treatment. They mentioned that under the same stress condition the six indicators can be adopted as determinants of rapid identification and prediction of the chilling tolerance in cotton lines or varieties.

7.1.7 FLOODING/EXCESS MOISTURE

7.1.7.1 Interaction of Waterlogging and High Temperature

Wu QiXia et al. (2015) studied response of cotton pertaining to its interaction to waterlogging and high temperature that was prevailing during flowering and boll-forming stage. They determined drainage index of Hubei plain area which receives heavy precipitation and high temperature. These conditions are prevalent during the flowering and boll-forming stage of cotton. Often, there is frequent occurrence of waterlogging and higher temperatures during cotton growth in the fields. Their results indicated the dwarf growth of cotton plants under normal temperature or high temperature. In the cotton plant that was under a minimum of 3 days waterlogging had an increase in soluble sugar (SS) and soluble protein (SP) content in the fourth leaf from the top. In normal temperature, with a minimum of 6 days waterlogging, there was a decrease in the numbers of fruit branches, with a concomitant increase in malionaldehyde content MDA content in the forth leaf from the top, however even with a minimum of 3 days waterlogging in high temperatures its content exhibited an increase. There was a decrease in the content of Chl a in the fourth leaf from the top with a minimum of 12 days waterlogging in normal temperature, while at high temperature it decreased with a minimum of 15 days waterlogging. The interactive effects were seen as interactive effects > waterlogging > high temperature in bringing about yield reduction. They attributed this reduction in yield by severe waterlogging stress due to a decrease in boll numbers and weight per boll. Assuming that the seed cotton yield would exhibit a decrease by 20% as the drainage waterlogging standard, and also waterlogging with 4 days of high temperature that appears during flowering and boll-forming stage of cotton they mentioned that the maximum waterlogged duration must be 3.4 days. Much before this, there is a requirement to drain off the surface waters to obtain normal yields without any yield declines due to water logging and hot temperatures that prevailed in above stages. Furthermore, the groundwater table must also be to a lower level of 80 cm or below within 3 days after drainage.

7.1.7.2 INCREASED ETHYLENE BIOSYNTHESIS IN WATERLOGGED COTTON

Increased ethylene biosynthesis in cotton tissues is found to be associated with acceleration of fruit abscission and yield losses in cotton grown under waterlogged conditions. In a series of glasshouse and field experiments, Najeeb (2016) investigated the effect of various application rates (0, 50, 100, and 150 [active ingredient, ai ha^{-1}) and time (pre- and post-waterlogging) of an anti-ethylene agent, amino ethoxy vinylglycine (AVG) on the growth and yield of cotton. It was observed that AVG (100-150 g [ai] ha^{-1}) applied 24 h prior to occurrence waterlogging had an ability to bring an increase growth and retention of fruits in both cottons. They validated its positive effects for 2 years. Their validation studies conducted in the field revealed that AVG (125 g [ai] ha^{-1}) applied at an early reproductive phase of cotton can bring about a significant increase in cotton yield under WL (13 %) and NWL (9%). Yield increase was due to an increase in the number of bolls, boll weight, and FR. It was observed that with application of higher AVG concentration (150 g [ai] ha^{-1}) there was no improvement in cotton yield, which has indicated a saturation of AVG on ethylene inhibition.

7.1.8 HEAVY METAL

7.1.8.1 COTTON FOR REMEDIATING HEAVY METAL-POLLUTED SOILS

Ma XiongFeng et al. (2017) discussed about the potential use of cotton for remediating heavy metal-polluted soils in southern China. In phytoremediation method, generally plants are used to a large extent to green the environment. Though cotton is not a standard hyperaccumulator plant, it has ability to produce a relatively large biomass. It also exhibits a greater tolerance ability and enrichment ability to heavy metals. Their results revealed that during the planting of cotton in southern China region for remediation of heavy metal pollution, the Cd concentration was less in cotton fiber than in other parts and it could be able to bring a better remediation and repair effect with beneficial ecological effects under these heavy metal polluted soils.

7.2 CONCLUSION

Plants under field conditions confront different abiotic stresses; their impact on plant and research advances that have were undertaken for understanding the effects of these stress factors has been discussed. Abiotic stresses caused by temperature and water are being intensified as consequence of climate change. To overwhelm this problem and meet the worldwide demand for food, tolerant crops must be developed by means of suitable technologies. Physiological aspects, mechanisms of tolerance, and management strategies for better crop production must be comprehensively studied to effectually use such technologies.

KEYWORDS

- **cotton**
- **abiotic stress**
- **research advances**
- **temperature**
- **water**

REFERENCES

Ahmad, S.; Khan, N. I.; Zaffar, M. I.; Hussain, A.; Hassan, M. Salt Tolerance of Cotton (*Gossypium hirsutum* L.). *Asian J. Plant Sci.* **2002,** *719.*

Anwar, M. R.; Darbyshire, R.; Leary, G. J.; Armstrong, R. D.; Hafner, L. In *Effect of Climate Variability on Australian Dryland Cotton Yield*, Australian Agronomy Conference, Ballarat, Victoria, Australia, 2017; pp 1–4.

Ashraf, M.; Ahmad, S. Influence of Sodium Chloride on Ion Accumulation, Yield Components and Fiber Characteristics in Salt-Tolerant and Salt-Sensitive Lines of Cotton (*Gossypium hirsutum* L.). *Field Crops Res.* **2000,** *66*, 115–127.

Ashraf, M. Y.; Sarwar, G.; Ashraf, M.; Afaf, R.; Sattar, A. Salinity Induced Changes in α-Amylase Activity During Germination and Early Cotton Seedling Growth. *Biologia Plantarum* **2002,** *45*, 589–591.

Ashraf, M.; Foolad, M. R. Roles of Glycine Betaine and Proline in Improving Plant Abiotic Stress Resistance. *Environ. Exp. Bot.* **2007,** *56*, 206–216.

Barrick, B.; Steiner, R.; Picchioni, G.; Ulery, A.; Zhang, J. Salinity Responses of Selected Introgressed Cotton Lines Grown in Two Soils From Organic and Conventional Cotton Production. *J. Cotton Sci.* **2015,** *19* (2), 268–278. (http://journal.cotton.org/journal/2015-19/2/loader.cfm?csModule=security/getfile&pageid=).

Brewer, M. J.; Anderson, D. J.; Parajulee, M. N. Cotton Water-Deficit Stress, Age, and Cultivars as Moderating Factors of Cotton Flea Hopper Abundance and Yield Loss. *Crop Protect.* **2016,** *86*, 56–61. (http://www.sciencedirect.com/science/journal/02612194).

Broughton, K. J.; Bange, M. P.; Payton, P.; Tan, D. K. Y.; Tissue, D. T.; O'Leary, G. J.; Armstrong, R. D.; Hafner, L. *Elevated [CO_2] and Warmer Temperatures Differentially Affect the Growth and Physiology of an Older and Modern Cotton Genotype*, Proceedings of the 18th Australian Agronomy Conference, Ballarat, Victoria, Australia, 2017; pp 1–4.

Chaves, M. M.; Flexas, J.; Pinheiro, C. Photosynthesis Under Drought and Salt Stress: Regulation Mechanisms from Whole Plant to Cell. *Ann. Bot.* **2009,** *103*, 551–560.

Cheng, C.; Zhang, Y.; Chen, X. G.; Song, J. L.; Guo, Z. Q.; Li, K. P.; Zhang, K. W. Co-expression of AtNHX1 and TsVP Improves the Salt Tolerance of Transgenic Cotton and Increases Seed Cotton Yield in a Saline Field. *Mol. Breed.* **2018,** *38* (2), 19. (http://rd.springer.com/journal/11032).

Chen, Y. L.; Shi, Y. T.; Luo, J. J.; Li, Z. W.; Hou, Y. Q.; Wang, D. Effect of Drought Stress on Agronomic Traits, Quality, and WUE in Different Colored Upland Cotton Varieties (Lines). *Acta Agron. Sinica* **2013,** *39* (11), 2074–2082. (http://www.chinacrops.org).

Dabbert, T. A.; Gore, M. A. Challenges and Perspectives on Improving Heat and Drought Stress Resilience in Cotton. *J. Cotton Sci.* **2014,** *18* (3), 393–409. (http://journal.cotton.org/journal/2014. 8/3/loader.cfm?csModule=security/getfile&pageid=158981).

Feng, D.; Zhang, J. P.; Sun, C. T.; Cao, C. Y.; Dang, H. K.; Sun, J. S. Index Controlling Soil Salinity for Cotton Field with Salt Irrigation Subject to High Quality and Yield. *Trans. Chinese Soc. Agri. Eng.* **2014,** *30* (24), 87–94. (http://www.tcsae.org).

Gao, W.; Xu, F. C.; Guo, D. D.; Zhao, J. R.; Liu, J.; Guo, Y. W.; Singh, P. K.; Ma XiaoNan; Long Lu; Botella, J. R.; Song ChunPeng. Calcium-Dependent Protein Kinases in Cotton: Insights Into Early Plant Responses to Salt Stress. *BMC Plant Biol.* **2018,** *18*, 15.

Gossett, R.; Millhollon, E. P.; Cran, M. L. Antioxidant Response to NaCl Stress in Salt-Tolerant and Salt-Sensitive Cultivars of Cotton. *Crop Sci.* **1994,** *34*, 706–714.

Guo, J. P.; Chen, Y. Y.; Zhao, J. F. Grey Forecasting Model of Heat Index of Cotton in Xinjiang. *Arid Land Geography* **2010,** *33* (5), 710–715.

Hebbar, K. B.; Venugopalan, M. V.; Prakash, A. H.; Aggarwal, P. K. Simulating the Impacts of Climate Change on Cotton Production in India. *Climatic Change* **2013,** *118*, 701–713.

Hu, X. T.; Chen, H.; Wang, J.; Meng, X. B.; Chen, F. H. Effects of Soil Water Content on Cotton Root Growth and Distribution Under Mulched Drip Irrigation. *Agri. Sci. China* **2009,** *8* (6), 709–716. (http://www.sciencedirect.com/science/journal/16712927).

Irizarry, I.; White, J. F. Application of Bacteria from Non-Cultivated Plants to Promote Growth, Alter Root Architecture and Alleviate Salt Stress of Cotton. *J. Appl. Microbiol.* **2017,** *122* (4), 1110–1120. (http://onlinelibrary.wiley.com/journal/10.1111/(ISSN)1365-2672).

Karademır, C.; Karademır, E.; Ekıncı, R.; Gencer, O. Correlations and Path Coefficient Analysis Between Leaf Chlorophyll Content, Yield and Yield Components in Cotton (*Gossypium hirsutum* L.) Under Drought Stress Conditions. *Notulae Botanicae Horti Agrobotanici Cluj-Napoca* **2009,** *37* (2), 241–244. (http://www.notulaebotanicae.ro).

Kerur, M.; Chinnapp, B.; Kumar, K.; Patil, R. Impact of Bt Cotton Cultivation on Environment. *Environ. Ecol.* **2014,** *32* (2), 487–490.

Kheir, E. A.; Sheshshayee, M. S.; Prasad, T. G.; Udayakumar, M. Temperature Induction Response as a Screening Technique for Selecting High Temperature-Tolerant Cotton Lines. *J. Cotton Sci.* **2012,** *16* (3), 190–199. (http://journal.cotton.org/journal/2012-16/3/190.cfm).

Kuznetsov, V. V.; Yu, V.; Rakitin, V. Y.; Zholkevich, V. N. Effects of Preliminary Heat–Shock Treatment on Accumulation of Osmolytes and Drought Resistance in Cotton Plants During Water Deficiency. *Physiologia Plantarum* **2002,** *26*, 10–18.

Lian, S. L.; Lian, L. J.; Tao, P. L.; Li, Z. X.; Zhang, K. W.; Zhang, J. R. Overexpression of *Thellungiella halophila* H^+-PPase (*TsVP*) in Cotton Enhances Drought Stress Resistance of Plants. *Planta* **2009,** *229*, 899–910.

Li, X. X.; Yan, Q. Q.; Wang, L. H.; Wei, X.; Zhang, J. S. Growth Analysis and Identification of Cold Resistance of Different Varieties of Cotton. *J. Nanjing Agri. Univ.* **2017,** *40* (4), 584–591. (http://nauxb.njau.edu.cn/).

Li, D. D.; Tai, F. T.; Zhang, Z. T.; Li, Y.; Zheng, Y.; Wu, Y. F.; Li, X. B. A Cotton Gene Encodes a Tonoplast Aquaporin That Is Involved in Cell Tolerance to Cold Stress. *Gene* **2009,** *438*, 26–32.

Liang, N.; Zhao, J.; Zhao, J.; Ma, J. Transformation of Insect Antifreeze Protein Gene Mpafp149 to an Endophyte of Cotton and Cold Tolerance of Cotton Carrying the Transformed Endophyte. *Xinjiang Agri. Sci.* **2011,** *48* (10), 1765–1772.

Liu, Y. G.; Yang, H. C.; Wang, K. Y.; Lu, T.; Zhang, F. H. Shallow Subsurface Pipe Drainage in Xinjiang Lowers Soil Salinity and Improves Cotton Seed Yield. *Trans. Chinese Soc. Agri. Eng.* **2014,** *30* (16), 84–90. (http://www.tcsae.org).

Liu, G.; Li, X.; Jin, S.; Liu, X.; Zhu, L.; Nie, Y.; Zhang, X. Overexpression of Rice NAC Gene SNAC1 Improves Drought and Salt Tolerance by Enhancing Root Development and Reducing Transpiration Rate in Transgenic Cotton. *PLoS One*, **2014,** *9* (1), e86895.

Luo, J. Y.; Zhang, S.; Zhu, X. Z.; Wang, C. Y.; Lü, L. M.; Li, C. H; Cui, J. J. Insect Community Diversity in Transgenic Bt Cotton in Saline and Dry Soils. *Biodiversity Sci.* **2016,** *24* (3), 332–340. (http://www.biodiversity-science.net/article/2016/1005-0094-24-3-332.html).

Luo, J. Y.; Zhang, S.; Zhu, X. Z.; Lu, L. M.; Wang, C. Y.; Li, C. H.; Cui, J. J.; Zhou, Z. G. Effects of Soil Salinity on Rhizosphere Soil Microbes in Transgenic Bt Cotton Fields. *J. Integr. Agri.* **2017,** *16* (7), 1624–1633. (http://www.sciencedirect.com/science/article/pii/S2095311916614569#!).

Luo, J. Y.; Zhang, S.; Peng, J.; Zhu, X. Z.; Lv, L. M.; Wang, C. Y.; Li, C. H.; Zhou, Z. G.; Cui, J. J. Effects of Soil Salinity on the Expression of Bt toxin (Cry1Ac) and the Control Efficiency of *Helicoverpa armigera* in Field-Grown Transgenic Cotton. *PLoS One*, **2017,** *12* (1). (http://journals.plos.org/plosone/article?id=10.1371/journal.pone.0170379).

Luo, Q. Y.; Bange, M.; Johnston, D.; Braunack, M. Cotton Crop Water Use and Water Use Efficiency in A Changing Climate. *Agri. Ecosyst. Environ.* **2015,** *202*, 126–134. (http://www.sciencedirect.com/science/journal/01678809).

Maiti, R; Gonzalez, H. R.; Rajkumar, D.; Vidya, P. S.; Gomez, M. V. M. Salt Tolerance of Pipe Line Hybrids of Bt Cotton Subjected to NaCl Stress. *In. J. Agric. Environ. Biotechnol.* **2015,** *2*, 123–132.

Ma, X. F.; Zheng, C. S.; Li, W.; Ai, S. Y.; Zhang, Z. G.; Zhou, X. J.; Pang, C. Y.; Chen, H. D.; Zhou, K. H.; Tang, M. D.; Li, L. F.; Wang, Y. H.; Li, Y. C.; Guo, L. S.; Dong, H. L.; Yang. Potential Use of Cotton for Remediating Heavy Metal-Polluted Soils in Southern China. *J. Soils Sediments* **2017,** *17* (12), 2866–2872. (https://link.springer.com/article/10.1007/s11368-017-1697-1).

Miller, G.; Suzuki, N.; Ciftci, S. Y.; Mttler, R. Reactive Oxygen Species Homeostasis and Signalling During Drought and Salinity Stresses. *Plant Cell Environ.* **2018,** *41*.

Najeeb, U.; Tan, D. K. Y.; Bange, M. P. Inducing Waterlogging Tolerance in Cotton Via An Anti-Ethylene Agent Aminoethoxyvinylglycine Application. *Arch. Agron. Soil Sci.* **2016,** *62* (8), 1136–1146. (http://www.tandfonline.com/doi/full/10.1080/03650340.2015.1113403#abstract).

Nalayini, P.; Anandham, R.; Sankaranarayanan, K.; Rajendran, T. P. Polyethylene Mulching for Enhancing Crop Productivity and Water Use Efficiency in Cotton (*Gossypium hirsutum*) and Maize (*Zea mays*) Cropping System. *Ind. J. Agron.* **2009,** *54* (4), 409–414.

Niwas, R.; Khichar, M. L.; Abhilash; Kumar,S.; Dhawan, A. K.; Chauhan, S. K.; Walia, S. S.; Mahdi, S. S. Quantification of Relationship of Weather Parameters With Cotton

Crop Productivity. *Ind. J. Ecol.* **2017,** *44*, 286–297. (http://indianecologicalsociety.com/society/indian-ecology-journals/).

Peng, D. S.; Guang, J. H.; Tao, Y. Y.; Hui, Y; G.; Ai, C. W.; Dong, Y. G.; Gao, Z. Cotton GhDREB1 Increases Plant Tolerance to Low Temperature and Is Negatively Regulated by Gibberellic Acid. *New Phytol.* **2007,** *176*, 70–81.

Radin, J. W.; Ackerson, R. C. Water Relations of Cotton Plants under Nitrogen Deficiency III. Stomatal Conductance, Photosynthesis, and Abscisic Acid Accumulation During Drought. *Plant Physiol.* **1981,** *67*.

Rahman, H.; Malik, S. A.; Saleem, M. Heat Tolerance of Upland Cotton During the Fruiting Stage Evaluated Using Cellular Membrane Thermostability. *Field Crops Res.* **2004,** *85*, 149–158.

Rodriguez, B. G.; Barrow, J. R. Pollen Selection for Heat Tolerance in Cotton. *Crop Sci.* **1988,** *28*, 857–859.

Rosenow, D. T.; Quisenberry, J. E.; Wendt, C.W.; Clarck, L. E. Drought Tolerant Sorghum and Cotton Germplasm. *Agric. Water Manage.* **1983,** *12*, 207–222.

Said, J. I.; Lin, Z.; Zhang, X.; Song, M.; Zhang J. A Comprehensive Meta QTL Analysis for Fiber Quality, Yield, Yield Related and Morphological Traits, Drought Tolerance, and Disease Resistance in Tetraploid Cotton. *BMC Genomics* **2013,** *14*, 776.

Shi, J.; Zhang, L.; An, H.; Wu, C.; Guo, X. *GhMPK16*, a Novel Stress-Responsive Group D MAPK Gene from Cotton, Is Involved in Disease Resistance and Drought Sensitivity. *BMC Mol. Biol.* **2011,** *12*, 22.

Shu, H. M.; Zhao, X. H.; Zhou, Z. G.; Zheng, M.; Wang, Y. H. Physiological Mechanisms of Variation in Temperature-Sensitivity of Cotton Fiber Strength Formation Between Two Cotton Cultivars. *Scientia Agric. Sinica* **2009,** 42 (7), 2332–2341. (http://www.ChinaAgriSci.com).

Tao Rui; Liu Tao; Chu GuiXin. Effect of Nitrapyrin Soaking on Cotton Germination Rate and Its Salt Resistant Physiological Characteristics in Salt Stress Condition. *Chinese Acad. Agric. Sci.* **2013,** *25* (5), 426–431.

Tri, K. F.; Dhiahul, K. A. Test of Cotton Lines with Drought Tolerant Intercropped with Maize. *Russian J. Agric. Socio-Econ. Sci.* **2018,** *1* (73), 17–24. (https://rjoas.com/issue-2018-01/article_03.pdf).

Ullah, K.; Khan, N.; Usman, Z.; Ullah, R.; Saleem, F. Y.; Shah, S. A. I.; Salman, M. Impact of Temperature on Yield and Related Traits in Cotton Genotypes. *J. Integr. Agri.* **2016,** *15* (3), 678–683. (http://www.sciencedirect.com/science/article/pii/S2095311916616490).

Ünlü, M.; Kanber, R.; Kapur, B.; Tekİn, S.; Koç, D. L. The Crop Water Stress Index (CWSI) for Drip Irrigated Cotton in a Semi-Arid Region of Turkey. *Afr. J. Biotechnol.* **2011,** *10* (12), 2258–2273. (http://www.academicjournals.org/AJB/).

Vinocur, B.; Altman, A. Recent Advances in Engineering Plant Tolerance to Abiotic Stress: Achievements and Limitations. *Curr. Opin. Biotechnol.* **2005,** *16*, 123–132.

Wang, X.; Wang, Z. M.; Jin, M. G. Antagonistic Effect of Mn and NaCl in Irrigation Water on Cotton Growth and Yield. *Chinese J. Eco-Agric.* **2014,** *22* (5), 571–577. (http://www.ecoagri.ac.cn).

Wang, L. H.; Li, X. X.; Sun, Y. Y.; Maimaitiali, A.; Zhang, J. S. Effects of Exogenous Salicylic Acid on the Physiological Characteristics and Growth of Cotton Seedlings Under NaCl Stress. *Acta Botanica Boreali-Occidentalia Sinica* **2017,** *37* (1), 154–162.

Wang, H. X.; Yang, J.; Zhang, Y. M.; Hu, Y. L.; Wang, A. Y.; Zhu, J. B.; Shen, H. T. Drought Resistance of Cotton with Escherichia Coli Catalase Gene. *Acta Botanica Boreali-Occidentalia Sinica*, **2014,** *34* (10), 2034–2040. (http://xbzwxb.nwsuaf.edu.cn).

Wang, S. W.; Pan, C. D.; Zhang, C. F.; Li, X.; Guo, J. H. Study on Light Environment in Intercropping Alley of Jujube and Cotton Southwest China. *J. Agric. Sci.* **2017,** *30* (4), 728–733.

Wu, Q. X.; Zhu, J. Q.; Yang, W.; Cheng, L. G.; Yan, J. Response of Cotton to Interaction of Waterlogging and High Temperature During Flowering and Boll-Forming Stage and Determination of Drainage Index. *Trans. Chinese Soc. Agric. Eng.* **2015,** *31* (13), 98–104. (http://www.tcsae.org/nygcxben/ch/index.aspx).

Wu, H.; Hou, L. L.; Zhou, Y. F.; Fan, Z. C.; Shi, J. Y.; Rouzi, A.; Zhang, J. S. Analysis and Evaluation Indicator Selection of Chilling Tolerance of Different Cotton Genotypes. *Agric. Sci. Technol. Hunan* **2012,** *13* (11), 2338–2346. (http://oversea.cnki.net/Kns55/oldnavi/n_item.aspx?NaviID=48&Flg=local&BaseID=HNNT&NaviLink=Search%3aagricultural+science+%26+technology).

Xiao, F.; Yang, Y. L.; Wang, Y. T.; Ma, H.; Zhang, W. F. Temperature Effects of Low Temperature on Psi and Psii Photoinhibition in Cotton Leaf at Boll Stage. *Acta Agron. Sinica* **2017,** *43* (9), 1401–1409. (http://www.chinacrops.org).

Yan, J.; He, C.; Wang, J.; Mao, Z.; Holaday, S. A.; Allen, R. D.; Zhang, H. Overexpression of the *arabidopsis* 14-3-3 protein gf14λ in Cotton Leads to a "Stay-Green" Phenotype and Improves Stress Tolerance Under Moderate Drought Conditions. *Plant Cell Physiol.* **2004,** *45*, 1007–1014.

Zhang, Y. L.; Hu, Y. Y.; Luo, H. H.; Chow, W. S.; Zhang, W. F. Two Distinct Strategies of Cotton and Soybean Differing in Leaf Movement to Perform Photosynthesis Under Drought in the Field. *Funct. Plant Biol.* **2011,** *38* (7), 567–575. (http://www.publish.csiro.au/nid/102.htm).

Zhang, Y.; Wang, L. H.; Sun, S. M.; Chen, X. L.; Liang, Y. J.; Hu, S. J. Indexes of Salt Tolerance of Cotton in Akesu River Irrigation District. *Scientia Agric. Sinica* **2011,** *44* (10), 2051–2059. (http://www.ChinaAgriSci.com).

Zhang XiuMei; Guo Kai; Xie ZhiXia; Feng XiaoHui; Liu XiaoJing. Effect of Frozen Saline Water Irrigation in Winter on Soil Salt and Water Dynamics, Germination and Yield of Cotton in Coastal Soils. *Chinese J. Eco-Agric.* **2012,** *20* (10), 1310–1314. (http://www.ecoagri.ac.cn).

CHAPTER 8

Cotton Biotic Stress

ABSTRACT

The biotic stress because of insects and diseases exert their influence on growth and productivity of cotton. Several concerted research studies were directed in studying their effects, developed techniques, and enabled in understanding physiological, biochemical, and molecular mechanisms to these stress factors. The research advances that occurred are briefly discussed in the following sections.

8.1 INSECTS PEST

8.1.1 COTTON PLANT BUG

Cotton plant bug at different developmental stages is one of the harmful pests of cotton. It migrates in and out of the cotton field and overwinters in some of the host plants in the absence of the cotton plants. In transgenic insect-resistant cotton these were studied by Yang ZhaoGuang et al. (2012), at Jiangxi province. They also studied its population dynamics and its main overwintering host plants. *Lygocoris lucorum* emerged as the main prevailing pest up to the bud period right from the initial period of seed bed, the population dynamics of this pest recorded two peak curve, while the bug *Orthotylus flavosparsus* prevailed as overriding pest species from stage of flowering formation of boll, and also prevailed from stage of boll opening to the end of harvest of cotton. A multi-peak and multi-valley curve is seen by its population dynamics. With a single-peak curve of the population dynamics right from stage of flowering to boll-forming stage, plant-bug species, *Adelphocoris suturalis* was dominant. The mixed population dynamic curves of these species in the fields exhibited a big peak and a small peak curve. The plant bugs *Adelphocoris suturalis* and *Lygocoris lucorum* successively were found to migrate out of cotton field and overwintered on the weeds that were present surrounding the cotton field.

8.1.2 FALL ARMY WORM

The fall armyworm, *Spodoptera frugiperda*, can act as a prominent pest of cotton in some years of cotton production, though it is not a main target pest of Bt cotton. Hardke et al. (2015) reported about its larval survivorship and fruiting injury in transgenic cotton lines which were expressing Cry1Ac (Bollgard), Cry1Ac+Cry2Ab (Bollgard II), and Cry1Ac+Cry1F (WideStrike) Bt proteins. The third instars enclosed within a nylon mesh exclusion cage and when placed on flower buds (squares), white flowers, and bolls exhibited a lower rate of survivorship with a minimal injury on wide strike cotton lines than others. Furthermore, it was observed that the development and survivorship of this pest was reduced in Bollgard II and Wide Strike cotton lines. Their results have suggested that this pest exhibit variable differences in performance in different Bt lines.

8.1.3 NONDISRUPTION OF TOP-DOWN FORCES IN BT COTTON

The usage of arthropod pests of higher trophic levels particularly some like the coccinellids and spiders for the regulation of other insect pests is generally referred to as top-down force. Some like lacewings and a few species of Hemipteran bugs act as natural enemies in the crop fields. Yao Yong Sheng et al. (2016) reported that in central China these top-down forces could not regulate the aphid population in the transgenic Bt cotton, wherein it could not disrupt those forces. Though the presence of natural enemies were observed even in Bt plots the presence of parasitoid mummies was less in abundance in Bt cotton plots, and these were not able to exhibit a differential top-down control on the population of *Aphis gossypii* in these plots. Thus, their study has indicated that these top-down forces seem not to be disrupted in Bt cotton in regulation of aphids population.

8.1.4 ENZYMES IN INDUCTION OF RESISTANCE AGAINST PEST INJURY

A well-known common marker in plants, inducing resistance when the plants are subjected to damage by wounding of insect pests like aphids, borers, etc., is the phenylalanine ammonia lyase activity. Lv Min et al. (2017) reported that on wounding by aphids this enzyme activity was tremendously increased in cotton seedlings. There was also an increase in the gene expression of this enzyme on wounding damage. An expression of this enzyme even in undamaged seedlings of cotton has indicated that there exists inter-species communication in cotton.

8.1.5 BT COTTON AND PRODUCTIVITY

Bakhsh et al. (2017) in Punjab province of Pakistan in his study for 2 years observed through their production function that those farmers who had used seed of Bt cotton obtained more yields than non-users of Bt cotton seed. They observed that on the cotton yields, variables of fertilizers and irrigation have a positive relation, in yield. Apart from the normal traditional inputs and other factors of socioeconomics their study has given an indication that farmers can be benefitted to a large extent with the usage of Bt cotton seeds provided they have an access to better quality seeds of this Bt cotton in the markets in the country.

8.1.6 ANTIXENOSIS TO FLEA HOPPER DAMAGE

Antixenosis, in simple, is the non-preference of a particular pest toward its host. It is one of the mechanisms of resistance that plants exhibit. A piercing-sucking pest of cotton feeding specially on the developing squares is the cotton flea hopper *Pseudatomoscelis seriatus* (Reuter) (Hemiptera: Miridae). Its heavy infestation results in the abscission of squares and in drastic yield reduction. McLoud et al. (2016) investigated on the resistance of cotton to this pest, through their studies concentrated on the stylet penetration of this pest in to the reproductive tissues, viz., the squares in particular. In a cultigen of upland cotton, derived from crosses with Pilose, they observed that it exhibited a heritable trait, depth of ovary and this trait impeded the ability of the flea hoppers to penetrate their stylets into the squares. They specified that in cotton antixenosis is a heritable trait, and in particular the ovary depth has an important part in conferring of the resistance of cotton squares against the injuries caused by flea hoppers, which seem to be an exploitable trait to develop resistance cultivars through resistance breeding.

8.1.7 WATER STRESS AND FLEA HOPPER INCIDENCES

Several factors like age of cotton crop, the cultivars, and the water stress conditions act as moderating and interacting factors in governing the abundance or scarcity of the flea hopper *Pseudatomoscelis seriatus* damage in cotton. Brewer et al. (2016) in South Texas and the Texas high plains conducted a few field experiments to analyze the effect of water deficits on the abundance of this pest and its likely damage in cotton. Simulations of a range of conditions of plant ages, cultivars, water stress has shown that cultivars did not have an impact on the yield reductions in spite of the abundant populations of flea hopper on them. They observed higher densities of this pest on older cottons which were not under water stress. Similarly, they observed that the water stress did not have an influence on the densities of this pest in South Texas. There was a drastic yield reduction of 50% when there was a decline in the soil moisture conditions. Though they detected the synergistic relationship between the cotton flea hopper populations and water stress, these were found to be highly variable with the intensity of water stress and the areas. In high plains there was a drastic reduction in the lint yield loss caused due to the damage of flea hopper in high water stress conditions. The flea hopper populations at the beginning of the 2nd week of squaring they exceeded the regional economic threshold values across the cultivars of cotton evaluated.

8.1.8 RESISTANCES EVOLVED AGAINST PINK BOLL WORM

A vital pest of cotton damaging lint quality with yellow spots on fiber, leading to not only yield reductions but also in the reduction in the market value of cotton fiber is the pink boll worm *Pectinophora gossypiella*. Rajput et al. (2017) made a study to compare the larval resistance of this pest in Bt and non-Bt cotton varieties. The study revealed that in IR-901 there was a maximum infestation of (1.30 ± 0.18) larval attack while it was minimal (1.15 ± 0.18) in CIM-602 Bt cotton varieties. But, its population was still higher (1.42 ± 0.19) in non-Bt cotton. Their study thus concluded that though this pest exhibited an attack on all the varieties of cotton, it did have some less preference of infestation and resistance evolved by pest on a few varieties which need to be further investigated.

8.1.9 CLCUVD AND WHITEFLY INFESTATION ON BT COTTON HYBRIDS

Arora et al. (2017) conducted to study for assessing the severity of damage caused due to infestation of CLCuVD and whitefly in Bt cotton hybrids (41 released + preleased) on yield reductions and on fiber quality. Their results revealed that all these Bt cotton hybrids were prone to the attack of CLCuVD and whitefly, except two Bt cotton hybrid, that is, NCS 855 and Bio 6539-2 which were found to exhibit moderate resistance against these two pests. There was no direct correlation of the fiber quality with the PDI of CLCuVD and the infestation of whitefly.

8.1.10 PROTEINS INDUCED BY JASMONIC ACID AGAINST LEAF CURL DISEASE

Several researches indicated that jasmonic acid confers resistance in plants against the incidences of pests and diseases through the synthesis of some proteins involved in the resistance or by some gene alterations. Raj et al. (2017) reported that in American cotton cultivars, viz., Ankur 3028 BGII and a desi cotton variety LD 694, treatment of 150 μM jasmonic acid at four to six leaf stage has induced some proteins which incurred some resistance against cotton leaf curl disease and white flies *Bemesia tabaci*. Proteins of 15–45 kDa along few other proteins as well were induced. These proteins affected the incidence as well as the severity of CLCuD and viruliferous whiteflies. Disease index of 40% was recorded in Ankur 3028 BGII. Their study has indicated that the application of jasmonic acid provided protection against the disease through the induction of some resistant proteins.

8.1.11 POLYPHENOL OXIDASE IN INSECT OR DISEASE RESISTANCE

A copper containing oxido reductase is polyphenol oxidase (PPO), an enzyme thought to confer resistance to some of the phytophagous insects or plant pathogens. The gene involved in this enzyme synthesis was GhPPOl. Cheng Hui et al. (2017) studied about its resistance to insects in cotton. They fed the cotton plants to 2nd instar larvae of cotton bollworm (*Helicoverpa armigera*), and obtained the expression of GhPPO1 gene in cotton plants, after GhPPO1 gene silencing. The results showed that tobacco rattle virus (TRV) system inhibits its expression and the expression of GhPPO1 gene was only 20%. This gene GhPPO1 in cotton had a significant silence effect. There was an increase in the weight of a boll worm by 86% after 72 h, when fed by a cotton leaf where this particular gene was silenced rather than the one with normal cotton feed. This has indicated that GhPPO1 was a type of defense mechanism gene in cotton against insect pest. The expression of GhPPO1 in the directly induced leaves with cotton bollworm was higher. The direct defense ability of cotton plants were weakened when this gene was silenced and resulted in an increase in the weight of the boll worm that was fed with these cotton leaves, thus indicating that GhPPO1 gene might be speculated to be involved in cotton direct defense responses.

8.1.12 E. BIPLAGA LARVAE RESPONSE TO BT COTTON

The Bt cotton Bollgard have Cry1Ac protein expression, while Bollgard ll is a stacked transgenic cotton variety which has the expression of both the Cry1Ac and Cry2Ab2 proteins. Fourie et al. (2017) evaluated the *E. biplaga* larvae response to these Bt cottons in South Africa for two growing seasons in 2013 and 2014. Three different bioassays involving squares (flower buds), cotton boll slices, or cotton bolls were carried out for evaluation of larval susceptibility to these two Bt cottons and a non-genetically modified cultivar which is isogenic to both Bt cottons. It was observed that after 10 days of initial feeding the larvae of *E. biplaga* were completely controlled on squares, cotton boll slices, and cotton bolls of both Bollgard and Bollgard II. The results confirmed that the larvae are susceptible to Bt Cry proteins. They were of the opinion that this pest may not that easily develop resistance to Bt cotton as it is highly susceptible to the Cry proteins development of resistance to Bt cotton, for this pest is highly unlikely due to the very high susceptibility of larvae to Cry proteins expressed by Bt cotton.

8.1.13 HOST EXPANSION OF COTTON SPECIALIZED APHIDS

Most of the insect pests have a wide variety of hosts, specifically though some are the preferable hosts. In the absence of their preferred natural host they expand to other plant species. The *Aphis gossypii* (Glover) host plant expansion which has a host specialization was studied in the laboratories by several researchers; however, as it remained meager in the field, Hu DaoWu et al. (2017) assessed its host expansion under field and lab conditions. They saw that these cotton-specialized aphids had the abilities to exhibit expansion on to new host plant (cucumber) where their number had exponentially increased on cucumber. A bioassay experiment that was conducted has shown that aphids preferred their natural host as cotton and cucumber; however, the clones from zucchini were found to exhibit a much higher preference for cucumber than cotton plant or the zucchini. Aphid individuals collected from mixed fields of cotton and cucurbit and also from individual fields of cotton and cucumber have shown that individuals from the cotton field were more of the cucurbit-specialized biotypes which were found to occur on cucumber. Similarly, the cotton-specialized biotypes were found to occur more on cotton and zucchini. A majority of aphids (>97.0%) collected from both the field cage and cotton farmland were found to be cotton-specialized individuals. The research study has indicated that if the intermediate host plants are eliminated, the outbreaks of aphids can be effectively suppressed, as very frequently, both cotton and cucumber are grown together by many farmers.

8.1.14 MONITORING OF APHID POPULATION

Zhang GuoLong et al. (2017) monitored the relationship between canopy spectral index, SPAD value, and cotton aphids, *Aphis gossypii*, in Shihezi, China. They observed that aphid amounts were negatively related to NDVI and NIRrefc and SPAD values, though the VI-2 exhibited a positive relationship to the aphid amount, as independent variables they exhibited variances in the aphid amount. With the wingless aphid amount, the nonlinear logarithm model expressed it as independent variable and was found to be the best for expression of association between these aphids amount and values of SPAD, with a variance of 85.48% of SPAD value. They demonstrated that their model is effective in providing a theoretical basis and an indirect technique for rapid monitoring of aphid population and its management.

8.1.15 PYRAMIDING OF CRY PROTEINS AGAINST RESISTANCE OF HELICOVERPA ARMIGERA

The boll worm *Helicoverpa armigera* has evolved *Bacillus thuringiensis* (Bt) cotton resistance that produces the Cry1Ac. This has progressed in many cotton varieties in China and to counter such resistance it was felt that there is a necessity to pyramid Cry proteins in Bt cotton and replace the Cry 1 Ac Bt cotton with these cotton types. Liu LaiPan et al. (2017) investigated the basic achievement of the refuge strategy for delaying resistance to Cry1Ac + Cry2Ab cotton, a pyramid, which was used widely outside China against this pest. They observed through their laboratory bioassays that the resistance was 130-fold to Cry2Ab. This resistance was found to be dominant in nature and was also inherited autosomally. It was also found to be under the control of many loci. Similarly, they observed 81-folds of strong cross-resistance occurring between Cry2Ab and Cry2Aa, while there was only a weaker cross-resistance of 18-fold between both the Cry2Ab and Cry1A toxins. They observed an increased An2Ab survival compared with An on cotton cultivars producing the fusion protein Cry1Ac/Cry1Ab or Cry1Ac. Survival on Cry1Ac + Cry2Ab cotton was also significantly higher in An2Ab than in An. It has shown that redundant killing on that pyramid was much incomplete. These results have indicated that switching over to three-toxin pyramided cotton would be of more value and of high durability in China.

Ni Mi et al. (2017) reported about the development as well as the first pyramids testing of cotton, to combine protection from a Bt toxin and RNA interference (RNAi). Two types of transgenic cotton plants those were able to produce double-stranded RNA (dsRNA) were developed. These were designed so as to interfere with the metabolism of juvenile hormone of *Helicoverpa armigera*, a global lepidopteran pest. They suppressed JH acid methyl transferase (JHAMT), which was vital for JH synthesis, and on JH-binding protein (JHBP), which was responsible for the transportation of juvenile hormone to organs. They observed that both types of RNAi cotton were more successful against Bt-resistant insects, though both the Bt cotton and RNAi have acted independently against the susceptible strains observed through simulation models. They observed that there was a delay in resistance of *H. armigera* in pyramided cotton that had a combination of both the Bt toxin and also the RNAi.

Transgenic cotton, that had the ability of expression of a single Bt toxin, was first utilized in the 1990s to bring about the control of *Helicoverpa armigera* and other pests of Lepidoptera. Baker and Tann (2017) reported that in Australia, New South Wales, Bt cotton crops Bollgard II was effective in replacing the Ingard which has only a single Bt toxin. The presence of two Bt toxins in Bollgard II led to a decrease in the moth populations of *H. armigera* and it exhibited a greater toxicity to the larvae and also caused natural enemies populations increase.

8.1.16 BEHAVIORAL AVOIDANCE OF BOLL WORM

The Bt cotton target pest is *Helicoverpa armigera.* Zhao DongXiao et al. (2017) observed that this pest exhibits behavioral response in BT and non-Bt cottons. They observed that it preferred much the non-Bt cottons for its oviposition and thus the number of eggs laid on Bt cotton were much less than on a non-Bt cotton. Similarly, its damage by 1st instar larvae in the form of consumed leaf area and holes in the Bt cotton was much less when they did not have an option of availability of non-Bt cotton crop. They concluded that this pest larva does exhibit behavioral avoidance to Bt cotton.

8.1.17 FEEDING BEHAVIORAL RESPONSE BOLL WORM

Zhao et al. (2016) provided insights underlying the resistance in Bt cottons which was found out during their examination of the *H. armigera* behavioral responses to Bt and non-Bt cottons. They observed that this pest exhibited avoidance to feeding as well as oviposition on Bt cotton.

8.1.18 GROWTH AND NUTRITIONAL INDICES OF BOLL WORM TO BT COTTON

Different larval instars have their preference of different plant parts. This also varies with the age of the cotton crop. Murthy et al. (2018) observed that there were variations between the different larval instars of 2nd to 5th of *Helicoverpa armigera* toward the preference of different plant parts of a 75-day-old Bt cotton crop. These larval instars irrespective of the cotton plant part that was fed exhibited a higher rate of consumption preference, digestibility, and more efficiency in conversion of ingested food from that of a non-Bt cotton. Their relative growth rate also was high when fed with different parts of a non-Bt cotton. The 3rd instar larvae recorded a maximum consumption index toward the boll feeding rather than the feed of leaves and squares of a Bt cotton. Similarly, the 5th instar larvae exhibited a lowest consumption index on squares of Bt cotton, while for these the 2nd instar larvae had a highest consumption index.

8.1.19 DIAMIDES IN BOLLWORM CONTROL

Diamides are supplementally applied for boll worm *Helicoverpa* spp. control. Its application in Bt cotton was evaluated by Little et al. (2017). In Bt cotton varieties Bollgard II and Widestrike 3, the mean larval count per 100 plants was much less compared with non-Bt cotton. For the haliothine control, the non-Bt cottons required an additional spray of diamide, whereas the toxin present in Bt cotton is found to be sufficient to bring down the larval population in Bt cotton without the additional spray of diamide.

8.1.20 BIOINOCULANTS AGAINST LEAF HOPPER

Manjula et al. (2017) reported that soil and foliar application of bioinoculant *Pseudomonas flurorescens* at 1 % decreased the leaf hopper *Amrasca devastans* populations and also resulted in high seed cotton yield in both Bt and non-Bt cottons.

8.1.21 SEASONAL ABUNDANCE OF COTTON THRIPS

Different thrips species, viz., *Thrips tabaci*, *Frankliniella occidentalis*, *F. schultzei* exhibit seasonal patterns in their abundance on the host plants. This relation was quantified with respect to their cotton crops invasion by Silva et al. (2018). *Thrips tabaci* was found to be the most abundant and dominant species present in numerous populations in nearly 31 types of plant species, while *F. occidentalis* was found in 35 plant species though its population was less. At very low densities the *F. schultzei* was present on 25 plant species. An analysis of mitochondrial CO1 gene sequences for identifying the genetic associations and identity of thrips has shown that *Thrips tabaci* and *F. occidentalis* which were collected from cotton were found to be genetically identical to conspecifics which were collected from weed plants. They move from the nearby weed source populations onto cotton plants. In their surveys though two members of *Frankliniella schultzei* complex species exist in Australia, only the black species was found to be present in large areas. *Thrips tabaci* was found to infest most of the early season sown cotton crop. Their study has indicated that the population ecology of thrips is mostly influenced by the presents of weeds, and with respect to any designing of pest management strategies, the role played by the weeds have to be taken into consideration. Furthermore, a clear understanding of the mechanisms of invasions of these can be known if there is a proper understanding about the role played by weeds.

8.1.22 SURVIVAL PATTERN OF COTTON BOLL WEEVIL

Pires et al. (2017) in their study observed that most of the adults (85%) of boll weevil remained the cotton reproductive structures, irrespective of the phenology of cotton plant. They survived in for more than 49 days after the fallow period. At the end of harvest, the boll weevils leave the cotton plant and look in for the alternate food sources. Thus, during the fallow period they did not take their shelter with in the cotton plant in Midwestern Brazil.

8.2 DISEASES

8.2.1 COTTON LEAF CURL DISEASE (CLCUD) RESISTANCE IN COTTON

Hassan et al. (2017) mentioned that since the early 1990s in Pakistan and India cotton leaf curl disease (CLCuD) caused by begomoviruses related with a particular satellite, viz., the cotton leaf curl Multan beta satellite (CLCuMB) is a major problem in cotton production and the cotton leaf curl Kokhran virus strain Burewala (CLCuKoV-Bur) recombinant begomovirus broke the (CLCuD) resistance introduced into cotton. In resistant cotton rarely this virus encodes only 35 of the common ~134 amino acids and lacks an intact transcriptional activator protein (TrAP) gene and in recent times in cotton breeding lines without earlier resistance CLCuKoV-Bur with a lengthier and still truncated, TrAP gene was detected, indicating that if the prior resistance is not conserved in current breeding programs further pathogenic viruses having full TrAP could reappear to produce CLCuD-resistant cotton varieties.

8.2.2 *P. BRACHYURUS* PATHOGENIC EFFECT

Machado et al. (2012) undertook two greenhouse trials to evaluate the *P. brachyurus* pathogenic effect on cotton cultivar Delta Opal using uninoculated, 12,000, 30,000, and 75,000 initial population densities of nematodes/400 cm^3 of soil. The damage plants were observed only at the two highest initial population densities and the cotton is found to be a suitable host for *P. brachyurus* from nematode reproduction data. Finally, they conclude that cotton is tolerant to *P. brachyurus* because the cotton plant did not show any severe symptoms and had insignificant effect on plant growth. This information can be utilized further in nematode management programs.

8.2.3 MIRNAS EFFECT ON CLCUMUV AND CLCUMB

In eukaryotes gene expression is regulated by micro (mi) RNAs and in plants Akmal et al. (2017) studied *Gossypium arboreum*-encoded miRNAs effect on CLCuMuV and CLCuMB. They selected two computationally predicted cotton-encoded miRNAs (miR398 and miR2950) having potential to bind multiple Open Reading Frames (ORFs; C1, C4, V1, and non-coding intergenic region) of CLCuMuV, and (βC1) of CLCuMB and in *G. hirsutum* var. HS6 by overexpression method miR398 and miR2950 functional validation was performed. They generated total 10 in vitro cotton plants from independent events and exposed to biological and molecular analyses. The transgenic cotton plants were inoculated with virus using viruliferous whitefly (*Bemisia tabaci*) insect vector to monitor the resistance. Only four transgenic lines remained symptom free and in healthy transgenic lines betasatellite was not detected in rolling circle amplification. In this study, the cotton leaf curl disease symptoms suppression in *Gossypium hirsutum* though miR398 and miR2950 overexpression is reported and experimentally demonstrated for the first time which could help in virus-resistant plants generation.

8.2.4 SCREENING OF COTTON AGAINST COTTON LEAF CURL VIRUS DISEASE

The cotton plants infected with cotton leaf curl virus disease (CLCuD) exhibits upward/downward curling of leaves, with the thickened veins prominent on the underside which later develops into leafy enations and plants become stunted, resulting in yield losses under severe conditions. Kumar et al. (2017) screened 100 cotton germplasm lines at Central Institute for Cotton Research Regional Station to study the cotton germplasm resistance against (CLCuD) and role of weeds in disease development. Among 100 germplasm lines, 20 lines showed resistance, 70 were moderately resistant, and 10 were moderately susceptible. Among naturally collected weeds, *Convulvu sarvensis* (Hirankhuri), Spinacea sp. (JungliPalak), *Solanum nigrum* (Blackberry nightshade), *Lantana camara* (Raimuniya), *Chenopodium album* (Bathua) were found to be associated with CLCuV by PCR amplification, suggesting the successful CLCuV transmission from cotton to weeds and vice versa.

8.2.5 CLCUD RESISTANCE/TOLERANCE IN COTTON

In Pakistan, Javed et al. (2017) analyzed CLCuD resistance/tolerance and other associated agronomical traits in 100 cotton genotypes using different statistical analytical tools such as correlation analysis, cluster analysis, and principal component analysis (PCA) to select the best genotypes. Among the variables based on significant amount of the variance, five principal components (PCs) were identified of which the first four components with greater than one Eigen values contributed more to the total variability. From the cluster analysis, one of the clusters genotypes showed better CLCuD tolerance and these genotypes can be used to develop varieties with increased CLCuD tolerance in future breeding programs.

8.2.6 ENDOPHYTIC FUNGI AGAINST VERTICILLIUM WILT

Yuan et al. (2017) isolated cotton roots to study the role of endophytic fungi against Verticillium wilt, caused by a defoliating *V. dahliae* strain Vd080. After inoculation in the greenhouse, decreased disease incidence and disease index were observed in all treatments at 25 days (d) with the control efficacy varying from 26% (CEF-642) to 67% (CEF-818). In heavily infected field at the first peak of the disease, the CEF-818 and CEF-714 gave better protection against Verticillium wilt with 46.9% and 56.6% in early July. Furthermore, the CEF-818 and CEF-714 treated plants increased cotton bolls by 13.1 and 12.2%. Their results showed that

V. dahliae can be controlled by cotton plants seed treatment with CEF-818 and CET-714 and the cotton yield can be improved with these endophytes which delays and reduces the wilt symptoms on cotton.

8.2.7 CLCUD-BEGOMOVIRUS CHARACTERIZATION

Godara et al. (2017) characterized five CLCuD-begomovirus and its associated satellite molecules on the basis of rolling circle amplification and complete genome sequencing. 82–99% nucleotide similarity was observed between them from sequence analysis. Based on phylogenetic analysis and identity matrix, three IARI-34, IARI-42, and IARI-50 CLCuMuV)-Rajasthan isolates members were determined out of five CLCuD-begomovirus isolates and named as CLCuMuV-Rajasthan-34, while other two IARI-30 and IARI-45 were members of CLCuKoV-Burewala isolates named as CLCuKoV-Burewala-45. High recombination events with the probability of ($P = 9.9 \times 10\text{-}10\text{-}3.2 \times 10 - 6$) were seen in IR, C1, and C4 genome regions of CLCuMuV-Rajasthan-34 recombinant isolate. Furthermore, three cotton leaf curl Burewala alpha satellite-related alpha satellites (1366–1396 nt) and *Gossypium darwinii* symptomless alpha satellite by 86% nt identity were found. Therefore, this study proved that three kinds of cotton leaf curl begomoviruses variants, two CLCuMuV strains, and one CLCuKoV strain by single or mixed infection with beta- and alpha-satellite molecules causes CLCuD.

8.2.8 COTTON TOMATO LEAF CURL BETASATELLITE AND LEAF CURL KOKHRAN VIRUS

After the cotton leaf curl disease second epidemic, Akram et al. (2017) found the tomato leaf curl betasatellite in cotton and parent strain of Cotton leaf curl Kokhran virus in the field for the first time by characterizing the begomovirus and is related satellites from cotton (*Gossypium hirsutum*) at molecular level. They amplified, cloned, and sequenced the begomovirus and associated molecules. The Cotton leaf curl Kokhran virus (HQ257374) isolate showed 98.2% sequence identity with begomovirus partial sequence, while the betasatellite partial sequence exhibited 96.4% sequence identity with tomato leaf curl betasatellites (EF068245, FR819710). The *Xanthium strumarium* alpha satellite (HF547408) had 98.2% sequence similarity with alpha satellite.

8.2.9 SMALL INTERFERING RNA (SIRNA)

In higher organisms and viruses, RNA interference (RNAi) is an established tool for gene expression knockdown. Ahmad et al. (2017) designed a small interfering RNA (siRNA) to target the CLCuKoV-Bu AC1 gene and the βC1 gene and the CLCuMB satellite conserved region. The cotton plants were transformed with the Vβ construct via agrobacterium-mediated embryo shoot apex cut method. Both transgenic and non-transgenic (susceptible) cotton plants were inoculated with CLCuKoV-Bu/CLCuMB complex using viruliferous whitefly vector and out of 11 T_1 plants only six plants were found with a Vβ transgene single copy from fluorescence in situ hybridization and karyotyping assays. The plant Vβ-6 did not show any symptoms and significantly reduced begomoviral-betasatellite amount with a single copy of the transgene on chromosome six, indicating the knockdown of CLCuKoV-Bu and CLCuMB expression successfully.

8.2.10 GARPL18 RIBOSOMAL PROTEIN AGAINST VERTICILLIUM WILT

In cotton (*Gossypium arboreum*) the salicylic acid-related GaRPL18 ribosomal protein confers resistance to Verticillium wilt caused by *Verticillium dahliae.* From a wilt-resistant cotton species (*Gossypium arboreum*), Gong Qian et al. (2017) cloned this L18 (GaRPL18) ribosomal protein gene and in cotton and *Arabidopsis thaliana* plants its function is characterized. GaRPL18 expression was induced by *V. dahliae* infections and in different cotton varieties with the level of disease resistance a stable GaRPL18 expression pattern was determined by virus-induced gene silencing technology. Furthermore, salicylic acid (SA) treatments upregulated the GaRPL18 expression, indicating its involvement in SA signal transduction pathway. Owing to decrease in immune-related molecules the wilt-resistant cotton species with silenced GaRPL18 became more susceptible to *V. dahliae* compared with control plants. Moreover, *A. thaliana* ecotype Columbia (Col-0) plants were also transformed with GaRPL18 using floral dip method and compared with the wild-type Col-0 plants, the plants overexpressing GaRPL18 were found to be more resistant to *V. dahliae* infections. Finally, this study suggested that the GaRPL18 are involved in SA-related signaling pathway mechanism and play a significant role in providing resistance against cotton Verticillium wilt.

8.2.11 VERTICILLIUM WILT RESISTANCE QUANTITATIVE TRAIT LOCUS MAPPING

Palanga et al. (2017) conducted quantitative trait locus mapping for Verticillium wilt resistance with SNP-based high density genetic map. From this QTL mapping, QTLs associated to VW resistance were detected. In cotton genome except chromosome 13 (c13) on 25 chromosome total 119 QTLs of disease index (DI) and disease incidence (DInc) were detected. Of which 59 QTLs for DInc elucidating 2.3–21.30% of the observed PV were detected on 19 chromosomes excluding c5, c8, c12-c13, c18-c19, and c26 while for DI seven QTLs with six sGK9708 alleles were identified. In these environments seven DI QTLs and 28 DInc QTLs were found to be stable, most of the stable QTLs were gathered into the clusters and on 13 chromosomes, viz., c1-c4, c6-c7, c10, c14, c17 c20-c22, and c24-c25 18 clusters of QTL comprising 40 QTLs were detected. In upland cotton, the identification of these QTLs and clusters provides the understanding of Verticillium wilt resistance complex genetic bases which could be applied in cloning of Verticillium wilt-resistant genes.

8.2.12 VERTICILLIUM WILT RESISTANCE IN COTTON

In cotton different primary defense mechanisms such as preformed defense structures comprising thick cuticle, phenolic compounds synthesis, and delaying or hindering the invader expansion via advanced measures for instance cell wall structure reinforcement, reactive oxygen species (ROS) accumulation, release of phytoalexins, the hypersensitive response, and the broad spectrum resistance development named as systemic acquired resistance (SAR) were observed. The long-lasting, cost–effective, and broad spectrum-resistant varieties can be developed in cotton using valuable information from these defense tactics to improve cotton breeding strategies and with this management method the usage of fungicides and other environmental risks can be reduced. In this chapter, Shaban et al. (2017) reviewed the mechanism of *V. dahliae* virulence and widely conferred the defense molecular mechanisms namely physiological, biochemical responses in cotton together with signaling pathways associated with resistance against Verticillium wilt.

8.2.13 CRY10AA PROTEIN EXPRESSION QUANTITATIVE AND QUALITATIVE ANALYSES

Ribeiro et al. (2017) performed Cry10Aa protein expression quantitative and qualitative analyses using T_0 GM cotton plants flower buds and leaves tissues and variable protein expression levels varying from 3.0 to 14.0 $\mu g\ g^{-1}$ fresh tissues were observed. Cotton boll worm adults and larvae were fed with leaves and flower buds of T_0 GM cotton and the significant entomotoxic effect and CBW mortality (up to 100%) were shown by CBW susceptibility bioassays further, molecular analysis showed that these effects to CBW were conserved. Finally they reported that Cry10Aa toxin gives high resistance to the cotton boll weevil in transgenic cotton and this is a great improvement in controlling the devastating CBW insect pest that can significantly affect cotton agribusiness.

8.2.14 COMPARATIVE TRANSCRIPTOME ANALYSIS AND CHROMOSOME SEGMENT SUBSTITUTION

Li PengTao et al. (2017) undertook cotton fiber development comparative transcriptome analysis of Upland cotton (*Gossypium hirsutum*) and *G. hirsutum* × *G. barbadense* Chromosome Segment Substitution Lines using Simple Sequence Repeat (SSR) markers two CSSLs, MBI9915 and MBI9749, along with the recurrent parent CCRI36 with tremendous fiber performance and comparatively clear chromosome substitution segments were selected during the fiber elongation development stages and secondary cell wall (SCW) synthesis (from 10DPA and 28DPA) and conducted a transcriptome sequencing to show the fiber development mechanism and the fiber development chromosome substitution segments of upland cotton contributed from Sea Island cotton. Total 705,433 million clean reads with average 45.13% of GC content and 90.26% of Q30 were generated by constructing and sequencing 15 RNA-seq libraries individually. Between these libraries, multiple comparisons were performed from which 1801 differentially expressed genes (DEGs) were detected and on the basis of GO annotation and KEGG enrichment analysis out of 1801 DEGs the 902 upregulated DEGs were found to be participated in organization of cell wall and response to oxidative stress and auxin, whereas the 898 downregulated DEGs involved in translation, transcription regulation, DNA-templated, and cytoplasmic translation. Afterward, DEGs temporal expression pattern was elucidated by performing STEM software and in the "oxidation–reduction process," two peroxidases and four genes of flavonoid pathway were detected which could be involved in development of fiber and quality formation. Furthermore, by 20 random genes quantitative real-time PCR the RNA-seq data reliability was confirmed. This study provided the understanding of fiber development molecular mechanism and new visions to investigate the likely role of *G. barbadense* substitution segments in fiber quality formation which would help in concurrent improvement of upland cotton fiber yield and quality via CSSLs.

8.2.15 MAJOR COTTON INFECTING VIRUSES

In Pakistan, single-stranded DNA viruses (Family, Geminiviridae) commonly infect the cotton crop. Saleem et al. (2017) found one-step guide PCR to screen the germplasm of cotton against most important cotton infecting viruses. To detect the three viruses, viz., CLCuMuV, CLCuKo-Bur, and CLCuKoV one-step PCR protocol was followed. Based on the symptoms and molecular diagnosis, 38 genotypes were screened out of which 13 were found to be positive for CLCuMuV, 24 were positive for Burewala strain, and one was positive for CLCuMuV and CLCuKoV-Bur strains while cotton leaf curl Kokhran virus was not found in the genotypes. The diagnostic assay established in this experiment will help in further advancement of the breeding programs.

8.2.16 COTTON BACTERIAL BLIGHT DISEASE

The bacterial blight disease commonly known as angular leaf spot, boll rot, and black leg is a major bacterial disease of cotton and caused by *Xanthomonas campestris pv. Malvacearum*. Under south Gujarat condition of India, Sandipan et al. (2018) studied the association between cotton bacterial leaf blight disease and different weather parameters. They recorded the bacterial blight disease incidence on G. Cot. Hy.12 (Non Bt) at weekly interval from 28 to 49th standard week, the incidence was observed with a maximum disease intensity in third week of September (23.5% PDI). The progress and development of bacterial blight disease was found to be nonsignificantly associated with abiotic factors.

8.2.17 PLANT–PARASITIC NEMATODE ASSOCIATED WITH COTTON

The important plant–parasitic nematode associated with cotton is reniform nematode (*Rotylenchulus reniformis Linford and Oliveira*); several research strategies varying from conventional breeding to triple species hybrids to marker-assisted selection have been applied in the United States cotton belt to introgress the resistant genes for reniform nematode into upland cotton from other wild species. Khanal et al. (2018) searched reniform nematode resistance in cotton. The resistant breeding lines developed from *G. longicalyx* exhibited stunting and a hypersensitive response to reniform nematode infection while breeding lines from *G. barbadense*, *G. aridum*, *G. armoreanum*, and other species exhibited a high resistance level. Therefore, in cotton, the reniform nematode-resistant cultivars can be developed using recent molecular techniques like CRISPER-Cas9 systems, specific double-stranded RNA to nematodes and by transferring suitable "cry" proteins.

8.2.18 SUITABLE SOWING DATES FOR LEAF CURL DISEASE (CLCUD) TOLERANCE

In Pakistan, CLCuD is a major menace to the sustainable cotton production which cause low crop stand the early sowing of the crop can avoid this problem but it decreases the seed germination percentage and coincides with wheat harvesting. Nazir et al. (2018) screened the cotton crop by sowing it at different dates, viz., early and late. March sown-transplanted cotton showed good escape to the disease and maximum yield was obtained from this crop while the May sown crop showed maximum incidence (22.2%) of the disease. Though extra labor cost incurred in transplanted cotton the income was 60% more than the May sowing owing to its improved crop stand, number of growing days, and escape from CLCuD. Thus, they recommended that in a cotton–wheat production cropping system the early planting of cotton by means of seedlings transplantation increases the cotton production by improving crop stand and escape from CLCuD.

8.3 CONCLUSION

Plants under field conditions confront biotic stress due to insects and diseases, their impact on plant and research advances that were undertaken for understanding the effects of these stress factors has been discussed. Biotic stresses caused by insects and diseases are being intensified as consequence of favorable conditions. To overwhelm this problem and meet the worldwide demand for food, tolerant crops must be developed by means of suitable technologies. Physiological aspects, mechanisms of tolerance, and management strategies for better crop production must be comprehensively studied to effectually use such technologies.

KEYWORDS

- **cotton**
- **biotic stress**
- **research advances**
- **insects**
- **diseases**

REFERENCES

Ahmad, A.; Zia-Ur-Rahman, M.; Hameed, U.; Rao. A. Q.; Ahad, A.; Yasmeen, A.; Akram, F; et al. Engineered Disease Resistance In Cotton Using RNA-Interference To Knock Down Cotton Leaf Curl Kokhran Virus-Burewala and Cotton Leaf Curl Multan Betasatellite Expression. *Viruses* **2017,** *9* (9), E257.

Akram, A.; Hussain, K.; Nahid, N.; Mahmood-ur-Rahman; Nasim, A.; Shaheen, S. Molecular Characterization of a Begomovirus and Associated Satellites from Cotton (*Gossypium hirsutum*) from Dera Ghazi Khan District of Pakistan, Lahore, Pakistan.

J. Animal Plant Sci. **2017,** *27* (4), 1245–1255. (http://www.thejaps.org.pk/docs/v-27-04/27.pdf).

Arora, R. K.; Kataria, S. K.; Paramjit Singh; Bhawana. Evaluation of Bt Cotton Hybrids Against Cotton Leaf Curl Virus Disease and Its Vector Bemisia tabaci. *Plant Dis. Res.* **2017,** *32* (2), 216–222.

Baker, G. H.; Tann, C. R. Broad-Scale Suppression of Cotton Bollworm, *Helicoverpa armigera* (Lepidoptera: Noctuidae), Associated with Bt Cotton Crops in Northern New South Wales, Australia. *Bull. Entomol. Res.* **2017,** *107*, 188–199. (https://www.cambridge.org/core/journals/bulletin-of-entomological-research).

Bakhsh, K.; Akram, W.; Jahanzeb, A.; Khan, M. Estimating Productivity of Bt Cotton and Its Impact on Pesticide Use in Punjab (Pakistan). *Pak. Econ. Soc. Rev.* **2016,** *54*, 15–24.

Brewer, M. J.; Anderson, D. J.; Parajulee, M. N. Cotton Water-Deficit Stress, Age, and Cultivars as Moderating Factors of Cotton Flea Hopper Abundance and Yield Loss. *Crop Protect.* **2016,** *86*, 56–61. (http://www.sciencedirect.com/science/journal/02612194).

Cheng Hui; Zhang Shuai; Luo JunYu; Rong Wei; Cui JinJie; Wang DengYuan. Study on Insect Resistance of GhPPO1 Gene in Cotton (*Gossypium arboretum*) by VIGS Technique. *J. Agric. Biotechnol.* **2017,** *25* (5), 722–728.

Fourie, D.; Berg, J. van den; Plessis, H. du. Evaluation of the Susceptibility Status of Spiny Bollworm *Earias biplaga* (Walker) (Lepidoptera: Noctuidae) to Bt Cotton in South Africa. *Afr. Entomol.* **2017,** *25*, 254–258. (http://www.bioone.org/loi/afen).

Godara, S.; Khurana, S. M. P.; Biswas, K. K. Three Variants of Cotton Leaf Curl Begomoviruses with Their Satellite Molecules are Associated with Cotton Leaf Curl Disease Aggravation in New Delhi. *J. Plant Biochem. Biotechnol.* **2017,** *26*, 97–105. (http://link.springer.com/article/10.1007/s13562-016-0370-x).

Gong Qian; Yang Zhaoen; Wang XiaoQian; Butt, H. I.; Chen ErYong; He ShouPu; Zhang ChaoJun; Zhang XueYan; Li FuGuang. Salicylic Acid-Related Cotton (*Gossypium arboreum*) Ribosomal Protein GaRPL18 Contributes to Resistance to *Verticillium dahliae*. *BMC Plant Biol.* **2017,** *17*, 59.

Hardke, J. T.; Jackson, R. E.; Leonard, B. R. Fall Armyworm (Lepidoptera: Noctuidae) Development, Survivorship, and Damage on Cotton Plants Expressing Insecticidal Plant-Incorporated Protectants. *J. Econ. Entomol.* **2015,** *108* (3), 1086–1093. (http://www.bioone.org/loi/ece).

Hassan, I.; Amin, I.; Mansoor, S.; Briddon, R. W. Further Changes in the Cotton Leaf Curl Disease Complex: An Indication of Things to Come. *Virus Genes* **2017,** *53* (6), 759–761. (https://link.springer.com/article/10.1007/s11262-017-1496-1).

Hu DaoWu; Zhang Shuai; Luo JunYu; LüLiMin; Cui JinJie; Zhang Xiao. An Example of Host Plant Expansion of Host-Specialized *Aphis gossypii* Glover in the Field. *PLoS One* **2017,** *12* (5), e0177981. (http://journals.plos.org/plosone/article?id=10.1371/journal.pone.0177981).

Javed, M.; Hussain, S. B.; Baber, M. Assessment of Genetic Diversity of Cotton Genotypes for Various Economic Traits Against Cotton Leaf Curl Disease (CLCuD). *Genet. Mol. Res.* **2017,** *16*, 16019446. (http://geneticsmr.com/sites/default/files/articles/year2017/vol16-1/pdf/gmr-16-01-gmr.16019446.pdf).

Khanal, C.; Mc Gawley, E. C.; Overstreet, C.; Stetina, S. R. The Elusive Search for Reniform Nematode Resistance in Cotton. *Phytopathology* **2018,** *108* (5), 532–541. (http://apsjournals.apsnet.org/loi/phyto).

Kumar, A.; Monga, D.; Kumhar, K. C. Screening of Cotton Germplasm Against Cotton Leaf Curl Virus Disease (Clcud) and Role of Weeds for Its Development. *J. Cotton Res. Dev.* **2017,** *31* (1), 87–96. (http://www.crdaindia.com/fileserve.php?FID=411).

Li PengTao; Wang Mi; Lu QuanWei; Ge Qun; Harun or Rashid, M.; Liu AiYing; Gong JuWu; Shang HaiHong; Gong WanKui; Li JunWen; Song WeiWu; Guo LiXue; Su Wei; Li ShaoQi; Guo XiaoPing; Shi YuZhen; Yuan YouLu. Comparative Transcriptome Analysis of Cotton Fiber Development of Upland Cotton (*Gossypium hirsutum*) and Chromosome Segment Substitution Lines from *G. hirsutum* × *G. barbadense. BMC Genomics* **2017,** *18*, 705. (https://bmcgenomics.biomedcentral.com/track/pdf/10.1186/s12864-017-4077-8).

Little, N. S.; Catchot, A. L.; Allen, K. C.; Gore, J.; Musser, F. R.; Cook, D. R.; Luttrell, R. G. Field Efficacy and Seasonal Expression Profiles for Terminal Leaves of Single and Double *Bacillus thuringiensis* Toxin Cotton Genotypes. *Southwestern Entomol.* **2017,** *42* (1), 15–26. (http://www.bioone.org/loi/swen).

Liu LaiPan; Gao MeiJing; Yang Song; Liu ShaoYan; Wu YiDong; Carrière, Y.; Yang YiHua. Resistance to *Bacillus thuringiensis* toxin Cry2Ab and Survival on Single-Toxin and Pyramided Cotton in Cotton Bollworm from China. *Evol. App.* **2017,** *10* (2), 170–179. (http://onlinelibrary.wiley.com/journal/10.1111/(ISSN)1752-4571).

Lv Min; Kong HaiLong; Liu Huaia; Lu YuRong; Zhang ChunMei; Liu JianFeng; Ji Chun-Ming; Zhu JinLei; Su JianKun; Gao XiWu. Induction of Phenylalanine Ammonia-Lyase (PAL) in Insect Damaged and Neighboring Undamaged Cotton and Maize Seedlings. *Int. J. Pest Manage.* **2017,** *63*, 166–171. (http://www.tandfonline.com/loi/ttpm20).

Manjula, T. R.; Kannan, G. S.; Sivasubramanian. Field Efficacy of *Pseudomonas fluorescens* Against the Cotton Leaf Hopper, *Amrasca devastans* Distant (Hemiptera: Aphididae) in Bt and Non-Bt Cotton. *Agric. Update* **2017,** *12* (4), 706–713. (http://www.researchjournal.co.in/online/AU.htminsect).

Machado, A. C. Z.; Ferraz, L. C. C. B.; Inomoto, M. M. Pathogenicity of *Pratylenchus brachyurus* on Cotton Plants. *J. Cotton Sci.* **2012,** *16* (4), 268–271.

McLoud, L. A.; Hague, S.; Knutson, A.; Smith, C. W.; Brewer, M. Cotton Square Morphology Offers New Insights Into Host Plant Resistance to Cotton Flea Hopper (Hemiptera: Miridae) in Upland Cotton. *J. Econ. Entomol.* **2016,** *109*, 392–398. (http://jee.oxfordjournals.org/content/109/1/392).

Mohd Akmal; Baig, M. S.; Khan, J. A. Suppression of Cotton Leaf Curl Disease Symptoms in *Gossypium hirsutum* Through Over Expression of Host-Encoded miRNA. *J. Biotechnol.* **2017,** *263*, 21–29. (http://www.sciencedirect.com/science/journal/01681656).

Murthy, K. D.; Manoharan, T.; Ravi, M.; Yadav, P.R. Impact of Bollgard II Bt Cotton on Nutritional Indices of American Bollworm, *Helicoverpa armigera* (Hübner) (Lepidoptera: Noctuidae). *J. Exp. Zool. India* **2018,** *21* (1), 507–510. (http://www.connectjournals.com/jez).

Nazir, M. S.; Khan, A. A.; Khan, R. S. A.; Cheema, H. M. N.; Amir Shakeel. Sustainable Cotton Production Under CLCuD Threat. *Pak. J. Agric. Sci.* **2018,** *55* (2), 279–285. (https://pakjas.com.pk/papers/2824.pdf).

Ni Mi; Ma Wei; Wang XiaoFang; Gao MeiJing; Dai Yan; Wei XiaoLi; Zhang Lei; Peng Yong-Gang; Chen ShuYuan; Ding LingYun; Tian Yue; Li Jie; Wang HaiPing; Wang XiaoLin; Xu GuoWang; Guo WangZhen; Yang YiHua; Wu YiDong; Heuberger, S.; Tabashnik, B. E.; Zhang TianZhen; Zhu Zhen. Next-Generation Transgenic Cotton: Pyramiding RNAi and Bt Counters Insect Resistance. *Plant Biotechnol. J.* **2017,** *15* (9), 1204–1213. (http://onlinelibrary.wiley.com/wol1/doi/10.1111/pbi.12709/full).

Palanga, K. K.; Jamshed, M.; Md. Harun-or-Rashid; Gong JuWu; Li JunWen; Iqbal, M. S.; Liu AiYing; Shang HaiHong; Shi YuZhen; Chen TingTing; Ge Qun; Zhang Zhen; Dilnur, T.; Li WeiJie; Li PengTao; Gong WanKui; Yuan YouLu. Quantitative Trait Locus Mapping for Verticillium Wilt Resistance in an Upland Cotton Recombinant Inbred Line Using SNP-Based High Density Genetic Map. *Front. Plant Sci.* **2017,** *8*, 382. (http://journal.frontiersin.org/article/10.3389/fpls.2017.00382/full).

Pires, C. S. S.; Pimenta, M.; Mata, R. A. da; Souza, L. M. de; Paula, D. P.; Sujii, E. R.; Fontes, E. M. G. Survival Pattern of the Boll Weevil During Cotton Fallow in Midwestern Brazil. *Pesquisa Agropecuária Brasileira* **2017,** *52* (3), 149–160. (http://seer.sct.embrapa.br/index.php/pab).

Raj, R.; Sekhon, P. S.; Sangha, M. K. Protection Against Cotton Leaf Curl Disease by Jasmonic Acid Induced Proteins. *J. Cotton Res. Dev.* **2016,** *30* (1), 84–89. (http://www.crdaindia.com/fileserve.php?FID=56).

Rajput, I. A.; Syed, T. S.; Khatri, I.; Lodhi, A. M. Comparative Resistance of Bt. and Non-Bt cotton Varieties Against Pink Bollworm, *Pectinophora gossypiella* Saund Larvae. *Pak. J. Agr. Agri. Eng. Vet. Sci.* **2017,** *33* (2) 220–226. (http://pjaaevs.sau.edu.pk/index.php/ojs/article/view/206).

Ribeiro, T. P.; Arraes, F. B. M.; Lourenço-Tessutti, I. T.; Silva, M. S.; Lisei-de-Sá, M. E.; Lucena, W. A.; Macedo, L. L. P.; Lima, J. N.; Amorim, R. M. S.; Artico, S.; Alves-Ferreira, M.; Silva, M. C. M.; Grossi-de-Sa, M. F. Transgenic Cotton Expressing CRY10aa Toxin Confers High Resistance to the Cotton Boll Weevil. *Plant Biotechnol. J.* **2017,** *15* (8), 997–1009. (http://onlinelibrary.wiley.com/wol1/doi/10.1111/pbi.12694/full).

Saleem, H.; Khan, A. A.; Azhar, M. T.; Nawaz-ul-Rehman, M. S. Molecular Screening of Cotton Germplasm for Cotton Leaf Curl Disease Caused by Viral Strains. *Int. J. Agric. Biol.* **2017,** *19* (1), 125–130. (http://www.fspublishers.org/published_papers/28221_pdf).

Sandipan, P. B.; Patel, R. K.; Faldu, G. O.; Patel, D. M.; Solanki, B. G. Relationship of Bacterial Leaf Blight Disease of Cotton with Different Weather Parameters Under South Gujarat Condition of India. *Cercetări Agronomice Moldova*, **2018,** *51* (1), 45–50. (http://www.univagro-iasi.ro/CERCET_AGROMOLD/index.php?lang=en&pagina=pagini/arhiva.html).

Shaban, M.; YuHuan, M.; Ullah, A.; Khan, A. Q.; Menghwar, H.; Khan, A. H.; Ahmed, M. M.; Tabassum, M. A.; Zhu LongFu. Physiological and Molecular Mechanism of Defense in Cotton Against Verticillium dahliae. *Plant Physiol. Biochem.* **2018,** *125*, 193–204. (http://wwwww.sciencedirect.com/science/journal/0981942).

Silva, R.; Hereward, J. P.; Walter, G. H.; Wilson, L. J.; Furlong, M. J. Seasonal Abundance of Cotton Thrips (Thysanoptera: Thripidae) Across Crop and Non-Crop Vegetation in an Australian Cotton Producing Region. *Agri. Ecosyst. Environ.* **2018,** *256*, 226–238. (http://www.sciencedirect.com/science/journal/01678809).

Yang ZhaoGuang; Zhang XingHua; Tian ShaoRen; Li Jie; Qiao YanYan; Tu QiJun. Population Dynamics of Cotton Plant-Bug in Transgenic Insect-Resistant Cotton. *Acta Agri. Jiangxi* **2012,** *24* (5), 28–33. (http://www.jxnyxb.com).

Yao YongSheng; Han Peng; Niu ChangYing; Dong YongCheng; Gao XiWu; Cui JinJie; Desneux, N. Transgenic Bt Cotton Does Not Disrupt the Top-Down Forces Regulating the Cotton Aphid in Central China; *PLoS One* **2016,** *11*. (http://journals.plos.org/plosone/article?id=10.1371/journal.pone.0166771).

Yuan Yuan; Feng HongJie; Wang LingFei; Li ZhiFang; Shi YongQiang; Zhao LiHong; Feng ZiLi; Zhu HeQin. Potential of Endophytic Fungi Isolated From Cotton Roots for Biological Control Against Verticillium Wilt Disease. *PLoS One* **2017,** *12*, e0170557. (http://journals.plos.org/plosone/article?id=10.1371/journal.pone.0170557).

Zhang GuoLong; Tao Xu; Zhang Ze; Du YinXi; LüXin. Monitoring of *Aphis gossypii* Using Green Seeker and SPAD meter. *J. Ind. Soc. Remote Sensing* **2017,** *45* (2), 361–367. (https://link.springer.com/article/10.1007/s12524-016-0585-2).

Zhao DongXiao; Liu Biao. Behavioral Responses of *Helicoverpa armigera* (Hübner) Larvae to Transgenic Bt Cotton. *J. Environ. Entomol.* **2017,** *39* (1), 152–159.

Zhao, D.; Zalucki, M. P.; Guo, R.; Fang, Z.; Shen, W.; Zhang, L.; Liu, B. Oviposition and Feeding Avoidance in *Helicoverpa armigera* (Hübner) Against Transgenic Bt cotton. *Appl. Entomol.* **2016,** *140* (9), 715–724. (http://onlinelibrary.wiley.com/journal/10.1111/(ISSN)1439-0418).

CHAPTER 9

Methods of Cultivation

A. I. ISSAKA[1], ABUL KALAM SAMSUL HUDA[1], SAMEENA BEGUM[2], ARUNA KUMARI[2*], and RATIKANTA MAITI[3]

[1]*Adjunct fellow, School of Science and Health, Western Sidney University, Australia*

[2]*College of Agriculture, Professor Jayashankar Telangana State Agricultural University, Hyderabad, India*

[3]*Forest Science Faculty, Universidad de Nuevo Leon, Mexico*

[*]*Corresponding author. E-mail: arunasujanagcjgl@gmail.com*

ABSTRACT

This chapter deals with the various methods involved in the cultivation of the cotton crop. Methods involved in cotton cultivation include the method of land preparation, the different types of sowing methods, how fertilizer is managed, methods of protection of the cotton crop, as well as research advances in cotton cultivation.

9.1 INTRODUCTION

Cotton cultivation, especially irrigated cotton, needs adequate and appropriate preparation of the land. In addition, an appreciable amount of time is required in getting the soil and the common lay of the land ready for cotton cultivation. Experts recommend that only certified cotton seeds should be used in sewing, since the seed is one of the key parameters for a good yield, and that seeds older than 2 years should not be used. Making the most of water (which is a valuable and limited resource) is critical. Creation of a whole farm water management plan and implementation of a water budget are important to optimizing water use efficiency (WUE) for the cotton crops. As with other crops, there is no universal scheme for fertilization of the cotton crop, as every field is different and has different needs. Insect pests may have a negative impact on the efficiency of cotton protection and on the cotton crop, and thus can hinder the hard work and investment of the farmer. Consequently, managing these pests is critical in the production of a healthy, plentiful, and sustainable harvest. This chapter presents the process of land preparation and sowing types for the cotton crop. It also discusses how water and fertilizer are managed during cotton cultivation, and also presents ways of protecting the crop. Research advances in cotton cultivation are also presented.

9.2 LAND PREPARATION

Cotton, especially irrigated cotton, land preparation. Cotton growing requires a bit of time to get the soil and the common lay of the land ready. Preparation of the soil normally begins immediately after the last cotton crop. Nowadays, to mulch the cotton stubbles back into the soil to add valuable nutrients and to help the soil preserve moisture by minimizing evaporation farmers commonly leave them standing in the field.

For irrigated cotton farming, to allow water flow in a controlled manner from the top of the field to the bottom the fields are levelled and graded in such a way that they have an exact slope or grade. If the field slopes are not adequately steep or if the sloping is not even, there is the tendency of water lying all around which waterlogs the soil. Moreover, if the fields are too steep, the water runs off fastly and does not seep into the soil profile. The slope of the field should be such that water can flow slowly down to ensure even watering of all plants. The excess run-off water can be recycled by constructing a tail drain at the end of the field.

Cotton grows best on lands where the surface soil is deep, with a high water holding capacity, adequate internal drainage, and a pH ranging between 5.5 and 7.0. Preparation of the land can be carried out with conventional or the minimum tillage method. With the conventional method, a typical upland field should be ploughed at least 15 cm deep. For a good soil tilth, adequate weed control, uniform seed germination, and adequate plant stand, the ploughed land should be harrowed two or three times. Furrows may then be made by passing an appropriate plough (e.g., animal-drawn or machine-drawn).

With the minimum tillage method, stubble should be cut close to the ground on a typical lowland field. Stubble cutting should be carried out soon after rice harvest, so that they will not affect the cotton seedlings. The field should then be cleaned by weeding by hand or by applying a suitable herbicide. In areas that are saline, furrow slices with one passing of a drawn plough (animal of machine) should be made at suggested row distances. This is to take away the highly saline soil surface which could unfavorably affect the germinating seedlings.

McGarry (1987) conducted a research to examine how aspects of soil physical condition and cotton growth are affected by the soil water content during land preparation (McGarry, 1987). He investigated the soil water content effects during presowing land preparation and cultivations intensity during the growing season on the physical conditions of the soil and furrow-irrigated cotton crop growth on cracking clay soil.

Constable et al. (1992) assessed different methods of preparing soil for cotton cultivation on a clay soil which is predisposed to compaction from traffic or tillage of wet soil. They compared a number of treatments, namely, crop rotation involving cotton and wheat, in which the good physical condition of the soil was maintained; continuous cotton with preparation of the soil by minimum tillage (min-till); continuous cotton with complete soil disturbance (max-till). Furthermore, because rates of nitrogen (N) fertilizer application and application approaches could vary depending on the soil preparation method, different N application methods, viz., anhydrous ammonia or urea applied before or after sowing, and placed under the crop row or between rows were evaluated with rates ranging between 0 and 225 kg N ha^{-1}. They repeated this experiment for 3 successive years.

Soil nitrate was reduced by max-till. Also, soil bulk density was improved by max-till; and root development was reduced by soil nitrate and in some years, water extraction reduction was more by soil nitrate than min-till and rotation cotton. In general, due to soil preparation under wet conditions the soil under the max-till treatment was found to be most compacted in the last two seasons.

On an average, 1.3% higher yield than for min-till and 7.1% higher yield than for max-till was obtained from cotton–wheat rotation. In the third year of the experiment the small difference between rotation and min-till indicated that min-till could take time to show production benefits. Optimum N fertilizer rates averaged for rotation 145 kg ha^{-1}, for min-till 189 kg ha^{-1}, and 210 kg ha^{-1} for max-till. In rotation cotton crop maturity somewhat was delayed and micronaire reduced to some extent.

The lower nitrogen uptake level was found in continuous cotton than cotton grown in rotation with wheat, which was consistent with soil nitrate tests taken before sowing. The application of anhydrous ammonia under the crop row 2–4 months before sowing was the best fertilizer application strategy. The similar response was observed with urea and the worst method of application was side-dressing of all N fertilizer after sowing mainly under max-till. This research concluded that for soil preparation min-till was a viable option on that soil type under wet conditions. However, the decision to change rotation and tillage practice could be affected with weed, insect, and disease factors.

Yilmaz et al. (2005) conducted a research to analyze use of energy and input costs for cotton production in Turkey. Their objective was to determine the cotton production direct input energy and indirect energy (in per hectare) and compare it with input costs along with analysis of farm size effect. With stratified random sampling technique the sample farms were selected and using face-to-face questionnaire data was collected from 65 farmers. The results unveiled the total energy consumption of cotton production was 49.73 GJ/ha of which diesel energy consumption (31.1%) was followed by fertilizer and machinery energy. For cotton the output–input energy ratio was 0.74 and energy productivity was 0.06 kg of cotton M/J. The net return per kilogram of seed cotton was found to be not sufficient to cover costs of production in the research area from cost analysis. The main cost items identified were labor, machinery costs, land rent, and pesticide costs. The large farms were more efficient in energy productivity and economic performance.

Murungu et al. (2003) recognized that the poor crop establishment is a major constraint to crop production for smallholder farmers in the semi-arid tropics which occurs because of poor land preparation methods and insufficient soil moisture. The researchers found that the solution to this problem is on-farm seed priming (soaking seed in water). But the interaction of this technology with soil conditions has not been well understood. To determine this interaction laboratory pot experiments on cotton and maize emergence and seedling growth were conducted with three treatments, viz., seed treatment (primed and non-primed), initial soil matric potential (−10, −50, −100, −200, and −1500 kPa), and aggregate size (<1, 1–2, 2–4.75, and 4.75–16 mm) of the soil used (a Chromic Cambisol) was collected from Save Valley Experiment Station in the southeastern Lowveld of Zimbabwe. After planting, the soil pots were allowed to dry out so as to simulate a deteriorating seedbed then emergence was monitored and after 8 days of planting plant growth was measured. In both crops the reduction in final percent emergence and seedling growth was observed with initial matric potential while final percentage emergence and seedling growth was increased with priming. The emergence and growth affected adversely with the overall, large aggregate sizes. The data obtained from this research were found to be reliable with the assumption that on-farm seed priming could partially compensate for the low soil matric potential and large aggregate sizes negative effects on crop establishment.

On a vertisol near Dalby, Queensland, Australia the two adjoining cotton crops differences in growth, yield, and root systems were explained by McGarry (1990) in terms of soil profile morphology and soil shrinkage indices. The platy structure in the 0.05–0.28 m layer and less air-filled specific pore volume in the 0.2–0.4 m layer were observed in soil beneath the strongly inferior crop. Furthermore, he observed that wet soil seedbed preparation before sowing cotton crop causes degradation of soil structure.

In agricultural systems, soil quality is regarded in terms of productive land which may keep up or increase farm profitability and maintain conservation of soil resources in order to create means of livelihoods for future farming generations. Tillage systems and crop rotations are among management practices which have the capacity to modify the quality of soil. A main percentage of Australian cotton is grown on Vertisols (approximately 75%), of which nearly 80% is irrigated. The clay contents of these soils are 40–80 g/100 g and strong shrink–swell capacities; however, they are frequently sodic at depth and predisposed to deterioration in the physical quality of soil if not correctly managed. As the extensive soil structure deterioration and declining fertility causes yield losses which are associated with tillage, trafficking, and picking under wet conditions, an important research program was started with the objective of developing systems of soil management which could increase cotton yields while simultaneously amending and upholding the soil structure and fertility. The result showed that the soil physical quality can be maintained with the cotton–winter crop sequences sown in a 1:1 rotation along with sustaining lint yields and minimizing fertility decline. As a result, in 2007, a large proportion (about 75%) of Australian cotton was grown in rotation with winter cereals such as wheat, legumes, or faba bean (Hulugalle and Scott, 2008).

In the early 1990s, a second phase of cotton rotations research in Vertisols was started, the key objective of which was to find sustainable cotton–rotation crop sequences, viz., crop sequences which can sustain and enhance soil quality, minimize disease incidence, facilitate soil organic carbon sequestration, and maximize economic returns and cotton WUE in the major commercial cotton-growing regions of Australia. The main findings of both these phases of Australian research are identification of soil quality and profitability and future areas for research. The soil quality indicators such as subsoil structure, salinity, and sodicity under irrigated and dryland conditions can be improved with wheat rotation crops under variety of cotton growing climates and irrigated and dry land conditions, whereas the available nitrogen may be increased by leguminous crops which fix atmospheric nitrogen and by decreasing N volatilization and leaching losses.

In spite of the fact that the rate of soil organic carbon decrease may be reduced by sowing crop sequences which return about 2 kg/m^2, there has been a decreased soil organic carbon in most locations over time. Except several of soil biodiversity beneficial effects on soil quality and few studies on soil macrofauna such as ants, the convincing field-based evidence to demonstrate this with regard to cotton rotations is not available.

Cotton monoculture gives lowest average lint yields per hectare and compared with cotton monoculture and other rotation crops higher average gross margins/ML irrigation water were obtained in the cotton–wheat systems. This shows that where irrigation water, not land, is the limiting resource, cotton–wheat systems would be more profitable.

Since the early to mid-2000s, the average yields and profitability of cotton were increased with the addition of vetch to the cotton–wheat system and profitability of cotton–wheat sequences differs with the cotton to wheat relative price. Owing to lower overall input costs, cotton–rotation crop sequence could be more resistant to fluctuations in the price of cotton lint, fuel, and nitrogen fertilizer than cotton monoculture (Hulugalle and Scott, 2008).

Regarding cotton–rotation crop sequences many issues were identified in the review where the knowledge and research was limited, in to "new" crop rotations, namely, comparative effects of soil quality on managing rotation crop stubble; machinery attachments for managing rotation crop stubble in situ in permanent bed systems; to increase soil organic carbon (SOC) levels from present values to the minimum amount of crop stubble which needs to be returned per cropping cycle; the comparative efficiency of C3 and C4 rotation crops related to carbon sequestration; the soil biodiversity, soil physical, and chemical quality indicators and cotton yields interactions; and the sowing rotation crops after cotton effects on farm and cotton industry economic indicators, for instance, the economic incentives for assuming new cotton rotations (Hulugalle and Scott, 2008).

Devkota et al. (2013) conducted a research on the tillage and nitrogen fertilization effects on yield and nitrogen use efficiency of irrigated cotton. In irrigated arid lands such as those of the Aral Sea basin they observed that management practices effect yield and nitrogen (N) use efficiency of irrigated cotton (*Gossypium hirsutum* L.). Under three different N application rates (0, 125, and 250 kg/ha) the effects of conservation tillage (permanent raised beds; PB) and conventional plough tillage (CT) on cotton growth, yield, N use efficiency, and N balance were compared. The results showed that with the increase, N fertilizer application rate from 0 to 125 kg/ha N yield and yield components of raw cotton increased significantly but these were not affected with tillage methods. However, much lower increment was observed with double N rate, that is, 250 kg/ha in both tillage methods. The applied N efficiency decreased when the N application rate was increased. Although tillage method had no effect on agronomic N use efficiency in both years, in the second year, apparent N recovery efficiency was higher in PB than in CT, whereas physiological N use efficiency with N-250 was 54% lower in PB than in CT. Furthermore, between tillage and N level a significant interaction was observed, viz., with N-0, CT had a 32% higher negative N balance than PB and with N-250, CT had a higher positive N balance (N loss), more than twice as high (66 kg/ha) as for PB. Therefore, they suggested that the long-run, cotton cultivation in PB with appropriate N application and the introduction of a winter cover crop would be a possible substitute to the unsustainable CT cotton mono-cropping system in irrigated arid lands.

A past study on decision processes on French farms and irrigated systems in Africa revealed that farmers plan their cyclical (recurrent) technical operations and that the planning process could be modeled. Taking cotton crop management in North Cameroon as a case study, Dounias et al. (2002) showed that with some changes, this type of modeling could also be carry out for rain-fed crop farming in Africa. The researchers noted that the adjustments were required to take in the social status differences among different fields on one farm and the consequences of the fact that farm work was chiefly manual. The authors found that these models can classify the farming practices into categories as per weather scenarios yield level as a function of weather scenario and provides a detailed understanding of the farming practices variability. It was shown that farms could be attributed to those types of model utilizing simple indicators about work organization. At the regional level an effective tool for organizing technical supervision of farmers could be produced with the analysis of North Cameroon farmers' decision processes for managing cotton crops. They stated that by quantifying some simple indicators advisers could work with those decision model types at farm level to foresee which types of model were appropriate without the heavy work of making individual decision models.

9.3 TYPES OF SOWING

There are two main types of sowing for cotton, namely drill sowing and hand dribbling. Drill sowing involves dropping the seeds in furrow lines in a continuous flow and covering them with soil. There may be one or more rows of planting. With this method of sowing, proper depth, spacing, and amount of seed to be sown in the field is achieved. Hand dribbling involves placing the seeds in holes (made at definite depth and at fixed spacing) made in seedbeds and covering them up. Prior to sowing, typical local practice cultivations were imposed on the three different states of soil that represented the three main treatments. These were: the "dry," "moist," and "wet" states.

According to this research, soil water content during the presowing land preparation influenced the physical condition of the soil along with the cotton plant growth. However, subsequent inter-rows cultivations did not have any influence on the soil physical condition and cotton plant growth. In the "dry" soil, after the first irrigation, a higher amount of water was taken out by the crop compared with the "moist" and "wet" treatments. There was a slight difference between the post-season soil profile morphology of the "dry" land and uncultivated pre-season profile; while there was a considerable alteration of the "wet" land, which is usually massive to a depth of 0.65 m. On most weeks of the season the "dry" land cotton crops grew tallest with the most bolls of any main treatment. Similar above-ground plant dry matter was produced on both "dry" and "moist" lands for most part of the season with higher amounts than the "wet" treatment. It was found that while plant population for each meter of row was equal on the 'dry" and "moist" treatments, it was lower in the "wet."

Yazgi and Degirmencioglu (2007) used response surface methodology (RSM) to optimize the seed spacing uniformity performance of a precision seeder and to confirm the variables optimum levels, viz., the vacuum on the seed plate, the diameter of seed holes, and the peripheral speed of the seed plate. The data acquired from the laboratory were divided into three diverse groups to get multiple index values, feed index quality, and miss index. They also suggested an additional performance criterion that is root-mean-square deviation from the theoretical seed spacing and utilized as a sowing performance indicator. Afterward the data was used to develop functions in polynomial form which permitted the optimum level calculation of each independent variable measured in their study. For precision seeding the optimum levels of vacuum pressure and the diameter of holes of cotton seeds were about 5.5 kPa and 3 mm. While for the seed plate peripheral speed optimum value was not obtained. Furthermore, they found the higher performance at the lower peripheral speed of the plate.

Karayel (2009) examined modified precision vacuum seeder performance for no-till sowing of maize (*Zea mays* L.) and soybean (*Glycine max* L.) following wheat (*Triticum aestivum*). The each unit of common precision vacuum seeder with a hoe opener on one row unit and a double disc-type opener on another row unit was mounted with a wavy-edged disc and side gauge wheels and at three forward speeds (1.0, 1.5, and 2.0 m/s) used for sowing.

Then together with row length, sowing depth uniformity, mean emergence time, and percent emergence, the multiple index, miss index, feed index quality, and seeds distribution precision were determined. With the increasing forward speed for both maize and soybean seeds, sowing depth uniformity, mean emergence time, and percent emergence were decreased while the seed distribution precision was increased along the length of the row and these were found to be better in the seeder equipped with the double disc-type opener than the seeder equipped with the hoe-type opener. For forward speeds of 1.0 and 1.5 m/s the seeds distribution precision along the length of the row was well below 29% and thus suitable for both maize and soybean seeds and based on the distribution of the seeds along the length of the row, sowing depth uniformity, and percent emergence the performance of the modified precision vacuum seeder was found to be best with the double disc-type furrow opener at the forward speed of 1.0 m/s.

A past research conducted by Ozmerzi et al. (2002) aimed to inspect the influence of different depths of sowing maize with reference to the precision sowing technique. At a fixed tractor forward speed of 6 km/h the maize seeds were sown at the nominal depths of 40, 60, and 80 mm. In field tests, two precision vacuum seeders were used and significant statistical differences were not found between horizontal distribution patterns sowing uniformity for the 40, 60, and 80 mm nominal sowing depths. For the first and the second precision vacuum seeder coefficients of variation of 429% and 493% were found at 60 mm nominal sowing depth, indicating that the 60 mm nominal sowing depth is most suitable sowing depth. Furthermore, at the nominal sowing depths of 40 and 60 mm, maximum emergence rate indices occurred while the lowest mean emergence time was 77 days for the nominal sowing depth of 40 mm. As a result of all treatments as per the sowing depth uniformity and emergence rate index, the 60 mm of nominal sowing depth of was found to be optimum.

9.4 WATER MANAGEMENT

Globally, water availability and local climate constitute the key contributing factors to the distribution and type of cotton grown. Generally, cotton may be grown as dry land (relying on rainfall) or as irrigated cotton (requiring supplementary water supply).

Requirements for dryland cotton include: full soil moisture profile in the beginning of the season, rainfall during the summer months, and extended heat periods with low humidity. For irrigated cotton, the requirements include a consistent supply of water, irrigation from surface and underground water sources as well as extended periods of heat and low humidity.

Cotton crops grow and develop faster when the average temperature and direct sunlight amount are higher throughout the growing season. The potential yield of the crops increases with the longevity and hotness of the growing season.

Due to the fact that farmers can control the moisture level in the soil, irrigated cotton is better suited to low rainfall environments. Under this condition, the effect of rainfall on the quality of the cotton is less when the bolls open as discoloration occurs when rain falls on open cotton bolls.

Depending upon the source of water and the economics of application, irrigation may be carried out by the furrow or hose method. It is recommended that the hose method should be used when there is scarcity of irrigation water. The hose is usually made of $1^1/_4$ plastic pipes. This method entails carrying the hose and guiding water discharge to the rows or hills.

The cotton field should be mulched (with suitable material such as rice straw, if it is cheap or cost-free and in abundance), in order to aid in minimizing the frequency and quantity of water application.

The cotton crops should be irrigated on the 6th, 8th, 11th, and 13th week after planting (the critical stages of the crop). If these recommended schedules are not appropriate, the crops may be irrigated on the basis of mid-day wilting and soil-feel method, because it relates to the soil moisture available.

Irrigation systems should be carefully designed to assure that water travels down the field at the appropriate speed without waterlogging so that all run-off water is collected and reused in the following irrigation, which leads to maximum water savings.

The production of more food from less water is the key challenge before the agricultural sector which could be realized by increasing crop water productivity (CWP). Zwart and Bastiaanssen (2004) found in all the earlier cases reported by FAO, the CWP ranges of cotton and other crops exceed. Worldwide for cotton seed and cotton lint the measured average CWP values per unit water depletion were found to be 0.65 and 0.23 kg/m^3, respectively.

The CWP range was very large, that is, 0.41–0.95 kg/m^3 for cotton seed and 0.14–0.33 kg/m^3 for cotton lint, therefore offering remarkable opportunities for maintaining or increasing agricultural production with 20–40% less water resources. They found that the CWP variability could be ascribed to climate, irrigation water management along with soil (nutrient) management, among others. The CWP showed an inverse relationship with vapor pressure deficit. With the latitude the decreased vapor pressure deficit was observed indicating the higher latitudes are favorable areas for irrigated agriculture. They concluded that the CWP could be significantly increased with the reduced irrigation and induced crop water deficit.

From being an innovation employed by researchers, to a known irrigation method of both perennial and annual crops the use of subsurface drip irrigation (SDI) has been progressed. Ayars et al. (1999) reviewed the SDI research conducted by scientists at the Water Management Research Laboratory over a period of 15 years. On tomato, cotton, sweet corn, alfalfa, and cantaloupe the irrigation and fertilization management data for both plot and field applications was presented. Significant increase in yield and WUE was found in all the crops. In high water table areas crops use of high frequency irrigation lead to reduced deep percolation and increased water use from shallow ground water. Moreover, uniformity studies revealed that if management procedures are followed to prevent root intrusion after 9 years of SDI operation, uniformity will be as good as at the time of installation.

In many arid and semi-arid regions, the population growth and high living standards results in competition for freshwater among different water-use sectors leading to reduced allocation to irrigation. To decrease the gap between freshwater availability and demand the other complementary non-conventional water resources, such as saline and/or sodic drainage and groundwater could be used. In the Indus Plains of Pakistan, Murtaza et al. (2006) evaluated different irrigation and soil management methods for using saline-sodic water to grow cotton (*Gossypium hirsutum* L.) and wheat (*Triticum aestivum* L.) on a sandy loam soil. Five treatments that were applied:(1) irrigation with freshwater (FW); (2) irrigation with saline-sodic water; (3) cyclic use of fresh and saline-sodic water through alternate irrigations (FW − SSW); (4) soil application of farm manure at 25 Mg/ha/year and irrigation with saline-sodic water (FM + SSW); and (5) soil application of gypsum equivalent to gypsum requirement of saline-sodic water and irrigation with the same water (G + SSW). By different treatments significant effects were not found on the seed yield of first cotton crop; however, subsequent wheat and cotton crops yields were lower in the SSW than other treatments indicating the negative effects of saline-sodic water used in the absence of a soil or irrigation management approach. Furthermore, the significant increase in soil ECe and SAR levels were observed in SSW treatment. Therefore, compared with SSW alone which increased the bulk density, the significantly higher infiltration rate was found in treatments where irrigation with saline-sodic water is combined with amendments. The maximum net benefit was from FW − SSW treatment after that from FW, G + SSW, FM + SSW, and SSW.

According to Rhoades (1984), despite the fact that expansion of irrigated agriculture has the potential to add significantly toward meeting the food and fiber needs in the world, there is the possibility of competition for ever more limited water supplies. However, the reassessment of water (and land) suitability criteria for irrigation can significantly expand the available supplies. Rhoades observed that water usually categorized as too saline for irrigation can be used successfully without harmful long-term consequences to crops or soils, even under conventional farming practices if the conservative standards used in the past were relaxed. Moreover, they found that the assumption of new crop and water management approaches could aid the saline waters use for irrigation expanding the irrigated agriculture.

The widely measured significant and sustainable production approach in dry regions is deficit irrigation (DI). The DI practice brings about water productivity maximization and stabilization instead of yield maximization by restrictive water applications to drought-sensitive growth stages. Geerts and Raes (2009) reviewed the DI literature around the world and summarized its advantages and disadvantages. The results of their reviews established that DI successfully increases the water productivity for many crops without affecting yield. This result anyhow needs guarantee of a certain minimum amount of seasonal moisture. They observed that as drought tolerance varies significantly by genotype and phenological stage the DI required detailed knowledge of crop response to drought stress. Finally, they mentioned that field research should be combined with CWP modeling in the development and optimization of DI strategies

WUE represents a certain level of biomass or grain yield per unit of water used by the cotton crop. As there have been increasing concerns on the water resources availability in both irrigated and rain-fed agriculture, there has been improved interest in an effort to understand how WUE could be enhanced and how farming systems could be altered to be more effective in water use. Hartfield et al. (2001) reviewed the literature to understand the role of soil management practices for WUE. They observed that soil management practices modify the available energy, soil profile available water, and the exchange rate between the soil and the atmosphere, affecting the evapotranspiration processes. Moreover, they found that through the plant physiological efficiency the plant management practices, such as the N and P addition, indirectly affects the water use. Across a range of climates, crops, and soil management practices, a large variation in measured WUE was revealed from the literature. They suggested that soil management practices involving tillage could increase WUE by 25–40% and nutrient management practices modification could increase WUE by 15–25%. Furthermore, they observed that more intensive cropping systems in semiarid environments and increased plant populations in more temperate and humid environments enhance the precipitation use efficiency which positively affects the crop yield.

9.5 FERTILIZER MANAGEMENT

Before planting, using the hill-drop method, basal fertilizer is applied along the furrows, which is then covered with thin soil. For the dibble planting method, just after planting or in 1 week after emergence, the fertilizer is dibbled a distance of between 5 and 8 cm away from the seeds. It is ensured that the fertilizer does not come directly in contact with the seeds. Soil is liberally applied and incorporated with the soil plant residues and animal manure or any low cost organic fertilizer available in the farm to maintain soil organic matter.

Mostly, cotton may be grown on cracking self-mulching clay soils found on flood plains nearby to rivers. Depending on the water content of the clay, these soil types expand and contract.

Before planting, cotton growers test the soil to check nutrient levels and access the quantity of fertilizer that may be required. Nitrogen, which is the key nutrient, required by cotton plants, is applied in the anhydrous ammonia form that is a liquid when applied directly to the soil, changes back into a gas and sticks to soil particles for the plants to use later. Three months before planting nitrogen may also be applied in granular form. In addition to this cotton crop needs some other nutrients like phosphorus, potassium, sulphur, and zinc.

Most cotton growers rotate crops on their fields. This is due to the fact that growing only one type of crop in a field causes nutrients deficiency and a build-up of soil pests and diseases. For instance, a grower might choose to plant cotton in a field for 3 years; then grows wheat the following year, followed by planting a legume crop or leaving the field crop-free (fallow), a practice referred to as crop rotation. Generally irrigated cotton growers rotate their cotton crop every 3 or 4 years.

Compared with non-legume rotation crops, most of the legume crops improves the nitrogen availability and reduces the soil strength. During the successive cotton crop growth the penetrometer resistance may increase in this order: faba bean, lablab, field pea, wheat, cotton, and soybean. The reduced soil strength enables the better root systems development which may lead to the improvement in lint yields.

Various researchers have applied different techniques of fertilizer application to cotton cultivation. For instance, Constable and Rochester (1988) studied the response of cotton to N fertilizer, either applied before sowing or after sowing, or both. In the soil with high pH anhydrous ammonia was applied and the soil nitrate N, N uptake, and lint yield were measured.

The crop at 120 days from sowing took up 30% of average applied N and with increased N fertilizer application a linear increase in N uptake was observed. The N recovery was higher in split application than single presowing application in two experiments, while it was reduced in further two experiments. When compaction, waterlogging, and a long growing season permitted better use of side dressed N the lint yield was increased in split application in only one experiment while in other experiments two N application methods gave similar yields. At 120 days from sowing when crop uptake was about 108 kg N/ha maximum yield was obtained. At 4 weeks before sowing soil nitrate N, sampled to a depth of 30 cm was found to be closely associated with subsequent plant N uptake in nil fertilizer treatments. So by applying moderate levels of N before sowing N fertilizer requirement can be precisely determined further to determine the requirement of side dressing the testing for soil N on nil fertilizer strips can be done. Therefore, they concluded that for this soil and climate side dressing is a viable practice and improvement of N fertilizer recovery increases the fertilizer use efficiency and productivity of cotton in this environment.

A past research (Freney et al., 1993) described field experiments in which the efficacy of several nitrification inhibitors in preventing fertilizer nitrogen (N) loss applied in cotton were evaluated. In north western New South Wales (NSW), Australia, the team of researchers evaluated the nitrapyrin, acetylene (provided by wax-coated calcium carbide), phenylacetylene, and 2-ethynylpyridine usefulness in preventing denitrification with the estimation of N recovery applied as ^{15}N labeled urea to a heavy clay soil in 1 m × 0.5 m microplots. In a next experiment, they determined the wax-coated calcium carbide effect on lint yield of cotton by supplying five levels of N on 12.5 m × 8 m plots at the same site.

The ^{15}N balance study indicated that when nitrification inhibitors were absent, in the plants and soil only 57% of the applied N was recovered at crop maturity. With the addition of phenylacetylene the recovery ($P <$ 0.05) was increased to 70%, while with nitrapyrin, coated calcium carbide and 2-ethynylpyridine addition it was increased to 74%, 78%, and 92%, respectively.

In the grey clay the addition of the wax-coated calcium carbide reduced the NH_4^+ oxidation rate for about 8 weeks. Excluding the highest N level addition the lint yield was increased with the addition of the inhibitor at all N levels. The indigenous N along with the applied N was conserved with the help of inhibitor. This study unveiled the urea fertilizer efficacy for cotton grown on the heavy clay soils which could be significantly improved using acetylenic compounds as nitrification inhibitors.

The study on effects of cropping systems and management practices on soil properties provides the information required to assess sustainability and environmental impact. In a past study (Ishaq, 2002), the effects of tillage and fertilizer rates on soil bulk density (BD) were estimated with minimum till (MT), conventional till (CT), and deep till (DT) treatments. The researchers applied low, medium, and high fertilizer rates to wheat and cotton. Bulk density was not altered by tillage, neither affected by fertilizer rates. Soil penetration resistance was found to be lower for DT than CT and MT. The concentration of soil P was affected by tillage methods, but N and K concentrations were not affected by tillage methods, while with the application of fertilizer the soil P and K concentrations were increased significantly. Compared with sub-soil N, P, K, and SOC concentrations were higher in the plough layer. The wheat grain yield showed a significantly negative association with penetration resistance and significantly positive association with soil P and K concentrations. The cotton yield exhibited a significantly negative association with soil BD. So the data provided in this study can be used to reevaluate the fertilizer rates and tillage methods recommendations for wheat and cotton production in Punjab by establishing long-term experiments to study agronomic and environmental effects of tillage methods, fertilizer rates, and cropping systems on productivity and quality of the environment.

To assess the effect of application of potassium fertilizer on the physiological parameters and yield of cotton cultivated on a soil that was potassium deficient, field test on a potassium fertilizer was conducted on a coarse silty loamy soil with low exchangeable K^+ in the Zhejiang Province (People's Republic of China) by a team of Chinese researchers (Xi et al., 1989). With the application of potassium fertilizer on this K^+-deficient soil the leaf area, concentration of chlorophyll, and the CO_2 assimilation of cotton were increased. The activity of the cytochrome oxidase and nitrate reductase, stomatal conductance were also increased with the application of potassium while the transpiration was decreased. The leaves with the well supplied K^+ were characterized by a full turgor and a well developed cuticle, whereas the leaf tissue of K^+-deficient plants was flaccid, the cuticle of leaves was poorly developed, and contrasted much. Therefore, from these physiological effects of K^+, it is found that potassium fertilizer application increases lint yield and improves the cotton fiber quality.

A past study in China (Cheng-Song et al., 2010) sought to guide fertilizer management for *Bt* cotton cultivar SCRC28. In the soils with high, middle, and low salinity under three fertilizer treatments, viz., (1) the control (no fertilizers); (2) NPK; (3) NP; and (4) NK the cultivar was planted. Then the different fertilization treatments effects on nutrient assimilation, Na^+ assimilation, net photosynthetic rate (P_n), dry matter accumulation, and lint yield of *Bt* cotton were investigated. The higher levels of nutrient uptake was found in plants under NP and NPK treatments in the three types of saline soils while lower levels of Na^+ uptake was found in the NPK treatment when compared with the control. Among the treatments the highest nutrient use efficiencies with highest levels of leaf area, chlorophyll content, and P_n were found in the NPK treatment cotton plants irrespective of the level of salinity. The highest biomass and lint yield were also obtained in the NPK treatment, the lint yields were improved by 2.53%, 28.67%, and 30.47% in the low, middle, and high salinity soils, respectively. Based on the soil salinity level, the NPK treatments were applied with N 165 kg ha^{-1} plus P_2O_5 38.57 kg ha^{-1} plus K_2O 111.5 kg ha^{-1} in the middle salinity soil and N 135 kg ha^{-1} plus P_2O_5 32.14 kg ha^{-1} plus K_2O 74.35 kg ha^{-1} in the high salinity soil. So in this study to alleviate nutrition obstacles and improve cotton nutrition which finally increases nutrient use efficiencies in agronomy and cotton yield the fertilization based on the soil salinity level is recommended.

In a past research between 1994 and 1997, the rotational effects of winter or summer legume crops grown either for grain or green manuring on following cotton (*Gossypium hirsutum* L.) (Rochester, 2001) were evaluated. Non-legume rotation crops, wheat (*Triticum* aestivum) and cotton were included for comparison and for the legume phase of each cropping sequence Net nitrogen (N) balances, along with N fixed with the nodulated roots, were calculated. The results showed that after seed harvest faba bean (*Vicia faba*-winter) fixed 135–244 kg N/ha and contributed up to 155 kg fixed N/ha, soybean (*Glycine max*-summer) fixed 453-488 kg N/ha contributing up to 280 kg fixed N/ha to the soil, whereas green-manured field pea (*Pisum sativum*-winter) and lablab (*Lablab purpureus*-summer) fixed 123–209 contributing 181–240 kg N/ha before the crops were cut and incorporated into the topsoil.

In a separate experiment, during the fallow between legume cropping and cotton sowing, viz., 5–6 months following summer crops and 9 months after winter crops the N loss from ^{15}N-labeled legume residues was 9–40%

of ^{15}N added; in comparison, applied ^{15}N fertilizer (urea) was loss from the non-legume plots was averaged 85% of ^{15}N added. During the growth of the subsequent cotton crop little legume-derived ^{15}N was lost from the system (Rochester et al., 2001).

The increased N fertility of the legume-based systems was emphasized by the enhanced N uptake and lint yield of cotton. From the fitted N response curve the economic optimum N fertilizer application rate between 0 and 200 kg N/ha (as anhydrous ammonia) was determined. The researchers found that cotton following non-legume rotation crops requires the 179 kg N/ha application, while following the grain-manured and green-manured legume systems required only 90 and 52 kg N/ha, respectively. Furthermore, they observed that following most legume crops than non-legume rotation crops lowers the soil strength along with improving N availability and the increased penetrometer resistance was found in the order faba bean, lablab, field pea, wheat, cotton, and soybean during the growth of the subsequent cotton. Therefore, they speculated that decreased soil strength lead to improvement in lint yields of the following cotton crops by aiding the better root systems development.

Decisions regarding fertilization optimum rates directly or indirectly entail fitting some type of model which could yield data collected by applying several fertilizer rates. Though to describe crop yield response to fertilizers, several different models are commonly used it is difficult to prefer one model over others. Therefore, Cerrato and Blackmer (1990) compared and evaluated several models such as linear-plus-plateau, quadratic-plus-plateau, quadratic, exponential, and square root which were normally utilized for describing the response of corn (*Zea mays* L.) to N fertilizer. Similar maximum yields were observed in all models and AH models fit the data equally well when evaluated by using the R^2 statistic. While predicting fertilization economic optimum rates marked discrepancies were found among models. Furthermore, the various models indicated that at a common fertilizer-to-corn price ratio the mean (across all site-years) fertilization economic optimum rates ranged from 128 kg N/ha to 379 kg N/ha. Statistical analyses showed that the most commonly used model, the quadratic model, did not provide an effective description of the yield responses and tended to indicate fertilization optimal rates that were too high. In this study, the yield responses were best described with the quadratic-plus-plateau model and the results clearly revealed that particularly among increasing concerns about the economic and environmental effects of overfertilization, the purpose of selecting one model over others requires more attention than it had in the past.

Blaise et al. (2005) evaluated the fertilizer with or without farmyard manure (FYM) application effects on cotton productivity and fiber quality. Using difference methods, viz., nutrient applied-crop removal and improved seed cotton yield with addition of FYM (5 Mg/ha), a partial nutrient balance was calculated. Compared with the control and without N plots (PK) the significantlly higher seed cotton yield was observed in plots applied with both N and P and the higher uniformity ratio and ginning outturn (GOT) were found in the FYM amended plots than the plots without FYM. Furthermore, in the fertilizer N and P applied plots, the N and P balance was positive while K balance was negative despite of applying K fertilizer which was positive only when FYM was applied. This study suggested that the application of FYM was beneficial since it enhanced fiber yield by means of improved GOT and constant positive nutrient balance.

In the arid deserts of Khorezm, Uzbekistan. at three sites nitrous oxide emissions was monitored for 2 years in irrigated cotton fields (Scheer, et al. 2008). The fields were applied with different fertilizer management approaches and irrigation water regimes. All through the vegetation season and between the sites N_2O emissions varied widely between years. Furthermore, the amount of irrigation water applied, the N fertilizer amount and type applied and temperatures of topsoil were found to have the greatest effect on the emissions.

In periods following N-fertilizer application in combination with irrigation events very high N_2O emissions of up to 3000 μg N_2O-N/m^2/h were observed. Additionally, the research revealed that between April and September these "emission pulses" accounted for 80–95% of the total N_2O emissions varying from 0.9 to 6.5 kg N_2O-N/ha. Corresponding to an average 1.48% of applied N fertilizer emission factors (EF) lost as N_2O-N, the uncorrected EF for background emission, ranged between 0.4% and 2.6% of total N applied which is consistent with the default global average value of 1.25% of applied N used in N_2O emissions calculations by the Intergovernmental Panel on Climate Change. The high soil moisture, high mineral N availability, and daily changes in topsoil temperature triggers the emission of pulses during which a clear diurnal N_2O emissions pattern was observed. For these periods, the best representatives of the mean daily N_2O flux rates were air sampling from 8:00 to 10:00 and from 18:00 to 20:00. The production of N_2O from NO_3^- fertilizers was promoted by the wet topsoil conditions but not from NH_4+ fertilizers signifying that the main process causing emissions of N_2O was denitrification. Therefore, the researchers said that there was more scope for reducing N_2O emission from irrigated cotton production, that is, through the limited use of NH_4^+ fertilizers. From this regionally dominant agro-ecosystem, the emission of N_2O could also be minimized with the advanced application and irrigation techniques namely subsurface fertilizer application, drip irrigation, and fertigation.

9.6 CROP PROTECTION

The cotton crop is often subject to attack by many insect species, which include destructive species such as the boll weevil, pink bollworm, cotton leaf worm, cotton flea hopper, cotton aphid, rapid plant bug, and conchuela. A certain degree of damage caused by insect pests may be controlled by following proper planting timing and other cultural practices, or carrying out selective breeding for varieties which are resistant to damage caused by insects.

Chemical insecticides seem to be the most effectual and efficient method of pests control. However, their use requires care and selectivity, due to ecological considerations. While conventional cotton production requires a greater quantity of insecticides than any other crop, organic cotton production, which depend on non-chemical insecticides, has been on the increase in many places of the world. In order to reduce the amount of pesticides needed, genetically modified "Bt cotton" was developed for the production of bacteria proteins which are toxic to herbivorous insects. Genetic engineering technology was used to develop glyphosate-resistant cotton, which has the capacity to tolerate the herbicide glyphosate.

It is well-established fact that to meet food and feed demands of human population that keeps rising as well as livestock production which keeps increasing, there must be a considerable increase in agricultural production in the foreseeable future. To safeguard the crop productivity from weeds, animal pests, pathogens, and viruses competition, crop protection is significant. A past study for 17 regions (Oerke and Dehne, 2004) has estimated the loss potential due to these pest groups and the actual losses for wheat, rice, maize, barley, potatoes, soybeans, sugar beet, and cotton for the period between 1996 and 1998, on a regional basis. Worldwide, among crops, variation of the pests loss potential ranged from less than 50% (on barley) to more than 80% (on sugar beet and cotton). For sugar beet, barley, soybean, wheat, and cotton, actual losses were valued between 26% and 30%, while for maize, potatoes, and rice estimated losses were 35%, 39%, and 40%. Generally, highest loss potential (32%) was observed for weeds followed by animal pests and pathogens—18% and 15%, respectively. Though in potatoes and sugar beets viruses cause serious problems in some areas, the global losses because of viruses ranged between 6% and 7% on these crops and for other crops it was less than 1–3%. The crop protection efficiency was low for food crops, viz., 43–50% and it was highest in cash crops 53–68%. Among regions, the variation coefficient of efficiency was highest in wheat with 28% and low in in cash crops (12–18%). Owing to the fact that using mechanical or chemical methods weeds can be controlled, the worldwide efficiency in weed control (68%) was significantly higher than the animal pests or diseases (39% and 32%, respectively) control, which depend heavily on pesticides. Without strengthening the pest control simultaneous to the crop production intensification it is not possible to meet the increasing demand through improved productivity per unit area.

Seven commercially marketed active ingredients present in neonicotinoid insecticides are imidacloprid, acetamiprid, nitenpyram, thiamethoxam, thiacloprid, clothianidin, and dinotefuran. The technical profiles of neonicotinoid insecticides and main differences between them, comprising their efficiency spectrum as being used for vector control, systemic properties, and versatile application forms, particularly seed treatment are described by Elbert et al. (2008) and to optimize the neonicotinoids bioavailability through enhanced rain fastness, better retention, and spreading of the spray deposit on the leaf surface, along with higher leaf penetration and new formulations were developed. Further to broaden the neonicotinoids insecticidal spectrum and to substitute the World Health Organisation (WHO) Class I products from older chemical classes researchers developed combined formulations with pyrethroids and other insecticides. In the next few years, neonicotinoids will turn into most important chemical class in crop protection because of these innovations for lifecycle management together with the introduction of generic products.

According to Viswanathan et al. (2001), plant growth that promotes rhizobacteria (PGPR) belonging to *Pseudomonas* spp. have been commercially used to induce systemic resistance against various pests and diseases for plant protection. Different PGPR strains mixtures increased the efficiency by inducing systemic resistance against several pathogens attacking the same crop. The PGPR seed treatment modifies the cell wall structure and causes biochemical/physiological changes resulting in the proteins and chemicals synthesis which are involved in plant defence mechanisms. The major determining factors of PGPR-mediated ISR are lipopolysaccharides, siderophores, and salicylic acid. Under field conditions, the PGPR performance was found to be successful against certain pathogens, insect, and nematode pests.

The research and technological advances in the field of remote sensing has been greatly enhanced the ability to detect and quantify physical and biological stresses affecting agricultural crop productivity. There is proof of the use of reflected light in the specific visible, near-infrared, and middle-infrared regions of the electromagnetic spectrum in detecting nutrient deficiencies, disease, and weed and insect infestations. In relation to measured or predicted climate variables multispectral vegetation indices resulting from crop canopy reflectance in relatively wide wavebands are capable of being used to monitor the plants growth response. Any deviation from the expected seasonal pattern indicates a potential problem and guarantees additional investigation by agricultural resource managers (Hatfield and Pinter, 1993). The thermal infrared developed from aircraft or satellite platforms has the capacity to identify areas which are susceptible to frost damage, quantify crop water stress, and make available some previsual disease detection abilities. Hartfield and Pinter suggested that continuous research is required to measure the association among crop stress and remotely sensed parameters and to implement surveillance over large areas and provide appropriate information to growers and consultants in time for preventative action to be taken and techniques need to be developed.

Wu et al. (2002) identified *Lygus lucorum* Meyer-Dür, *Adelphocoris fasciaticollis* Reuter, and *Adelphocoris lineolatus* (Goeze) (Hemiptera: Miridae) to be important secondary insect pests that may be found in cotton fields in northern China. These researchers compared the mixed populations seasonal dynamics on a transgenic variety with expressed insecticidal Bt protein *Cry1A* of these insect pests and a cotton line which express *Cry1A* and *CpTI* (cowpea trypsin inhibitor gene) proteins with seasonal dynamics on similar but non-transgenic varieties. Despite the fact that reduced number of insecticide sprays against *Helicoverpa armigera* results in significantly higher mirid damage on unsprayed transgenic cotton than the number of sprays in the normal cotton, the researchers did not find any significant differences between unsprayed normal cotton and unsprayed transgenic cotton regarding these bugs population densities. Findings of their research suggested that in transgenic cotton fields, the mirids had remained significant insect pests. Furthermore, their research suggested that if no further control measures were taken on, expansion of the area planted to transgenic cotton could lead to a further increase to damage to the cotton crop by these insect pests.

Laboratory work has led to impressive results (now in use) on the Bt-susceptible pests control and the first commercial Bt transgenic crops (Hilder and Boulter, 1999). The transgenics expressing various protease inhibitors, lectins, etc. laboratory trials have led to improved resistance to a wide spectrum of pests and some promising field trials have conducted. However, to merit serious efforts at commercialization, there has not been convincing evidence about the scale of effects produced by plant-derived insect control genes. There are limitations to both compounds classes. Limitations include: severe failures of Bt cotton in providing resistance to targeted pests; production of chronic rather than acute effects by most plant-derived resistance factors; and simply non-susceptibility of numerous serious pests to known resistance factors. Hilder and Boutler (1999) analyzed the characteristics necessary in an ideal transgenic technology. The characteristics included being environmentally gentle, comparatively cheap to develop, with a potentially broad spectrum of activity (although targetable at pests and not beneficials), generated by a flexible technology which allow any insect site to be targeted and readily adaptable in order for substitutes to be produced as requisite. Hilder and Boutler (1999) based on single-chain antibody genes expression made an effort to develop such a technology in crop plants which would be compatible with the possible trends in discovery of pesticide using biology-driven target-based methods. They emphasized the significance of a changed, more socially responsible attitude in this sector and the necessity of much improved benefits presentation and necessity of responsible utilization of genetically engineered crops.

Kranthi et al (2002) conducted a study in five major cotton insect pests from the key cotton-growing regions of India with stress on Andhra Pradesh and Maharashtra to determine insecticide resistance to generally used insecticide group's representatives, viz., pyrethroids-cypermethrin, organophosphates-chlorpyriphos, cyclodienes-endosulfan.

These researchers observed that the field strains of cotton bollworm *Helicoverpa armigera* (Hubner) recorded 23–8022-fold resistance to cypermethrin. Another observation was that in *H. armigera* the resistance to endosulfan and chlorpyriphos was low to moderate. Furthermore, they observed that the pink bollworm *Pectinophora gossypiella* (Saunders) exhibits low resistance to pyrethroids. However, the researchers observed that in some areas of Central India, 23–57-fold resistance levels were recorded to endosulfan and resistance to chlorpyriphos was high in the Medak, Bhatinda, and Sirsa strains from North India. In South India, most of the *Spodoptera litura* (Fab.) strains were found to exhibit of 61–148-fold resistance levels to cypermethrin. Additionally, only two strains from Bhatinda and Karimnagar in North India exhibited high level of resistance to endosulfan. A high level of resistance 45–129-fold to chlorpyriphos was found in *S. litura* strains from South India. In the Sirsa and Sriganganagar strains from North India, low to moderate insecticide resistance in *Earias vittella* (Fab.) was observed and though *Bemisia tabaci* (Genn.) exhibited moderately high resistance levels to cypermethrin, the resistance to endosulfan and chlorpyriphos was negligible in the field strains tested.

9.7 RESEARCH ADVANCES IN CULTIVATION

A 2002 preview of research advances in cotton cultivation revealed that there had not been any true advances in commercial eucheumatoid farming in more than 10 years (Ask and Azanza, 2002). The preview recommended that researches which are capable of replacing the "tie-tie" system which was being used (as at 2002) in the vegetative propagation of the crops as well as the possible application of spore/sporelings in production (as in other economic seaweeds) should be given priority. Furthermore, the preview recommended that multifactorial experiments which consider key features such as nutrients, salinity, and light should be carried out to deal with issues regarding seasonality in growth/production of crops.

Since 1996, Bt technology was made accessible to cotton farmers, resulting in higher yields, fewer trips with the sprayer and increase in the beneficial insects population on their farms.

"Bt" stands for *Bacillus thuringiensis*, a common soil bacterium, named because it was first isolated in the Thuringia region of Germany.

Bt produces a protein which paralyses the some harmful insects larvae that include the common plant pests, viz., cotton bollworm and the Asian and European corn borers, whose infestations result in devastating effects on important crops.

When the target insect larva ingest it, in the gut's alkaline condition, the Bt protein gets activated and punctures the mid-gut leaving the insect not capable to eat. The insect dies within a few days.

Much research is being done to utilize the agronomic value of organisms owing to its capability to produce the insecticidal protein. So far, more than 200 types of Bt proteins have been detected with variable degrees of toxicity to some insects.

Bt technology has led to three key changes in cotton cultivation: (1) Increased yield due to decreased plant stress: as the plant does not persistently deal with defending itself from repeat feeding insects, it can effort on making a fruit and making more of it. Bt provides a more stable protection which help cotton plants to reach their full yield potential; (2) Decreased trips across the field with insecticides. This saves time and money from fuel and insecticide; (3) Increase in the beneficial insects population. The reduction in sprays means farmers would have more insects, especially beneficial ones, in their fields. Keeping beneficial insects does not just make the farmer sound good to environmentalists; it also can help them control some of the more troublesome pests in their fields. Beneficial insects ensure that there is competition for new flights of moths, which decreases the likelihood of an epidemic, for instance.

The area under Bt cotton expressing insecticidal Cry proteins is increasing worldwide (Romeis et al., 2006) but the potential effect of Bt crops on non-target organisms including biological control organisms is a major problem with their adoption. From the laboratory and glasshouse research it was found that the Bt crops effects natural enemies only when *Bt*-susceptible, sub-lethally damaged herbivores were used as prey or host with no indication of direct toxic effects. So to have access to probable non-target effects of Bt crops, the research suggested that regulatory procedures should search for a step-wise (tiered) approach. Furthermore, a correspondence between the abundance and activity of parasitoids and predators in Bt and non-Bt crops has been shown from other studies. In contrast, conventional insecticides applications frequently give rise to negative effects on biological control organisms. There is a possibility that Bt crops contributes to integrated pest management systems with a strong biological control component owing to the fact that Bt-transgenic varieties can substantially reduce the use of insecticide some crops.

In genome research of cotton, massive advances have been made especially for basic and applied genetics, genomics, and breeding research (Zhang et al., 2008); many genomic resources and tools comprising different types of DNA markers, for example, restriction fragment length polymorphism (RFLP), randomly amplified polymorphic DNA (RAPD), amplified fragment length polymorphism (AFLP), resistance gene analogues (RGA), sequence-related amplified polymorphism (SRAP), simple sequence repeat (SSR) or microsatellites, DNA marker-based genetic linkage maps, QTLs and genes for the traits important to agriculture, expressed sequence tags (ESTs), arrayed large-insert bacterial artificial chromosome (BAC) and plant-transformation-competent binary BAC (BIBAC) libraries, and genome-wide, cDNA- or unigene EST-based microarrays have been developed. Attempts were made to sequence the important cotton species genomes and to develop the genome-wide, BAC/BIBAC-based integrated physical and genetic maps but owing to limited funds allotted to the species the cotton genome research lags behind compared with other major crops, such as rice, maize, and soybean.

9.8 CONCLUSION

This chapter presented the various aspects of cotton cultivation. These included land preparation, types of sowing cotton, water management, fertilizer management, and crop protection. The chapter also presented research advances in cotton cultivation.

KEYWORDS

- **cotton**
- **cultivation**
- **water**
- **fertilizer**
- **irrigation**

REFERENCES

Ask, E. I.; Azanza, R. V. Advances in Cultivation Technology of Commercial Eucheumatoid Species: A Review with Suggestions for Future Research. *Aquaculture* **2002,** *206*, 257–277.

Ayars, J.; Phene, C.; Hutmacher, R.; Davis, K.; Schoneman, R.; Vail, S.; Mead, R. Subsurface Drip Irrigation of Row Crops: A Review of 15 Years of Research at the Water Management Research Laboratory. *Agric. Water Manage.* **1999,** *42*, 1–27.

Blaise, D.; Singh, J.; Bonde, A.; Tekale, K.; Mayee, C. Effects of Farmyard Manure and Fertilizers on Yield, Fibre Quality and Nutrient Balance of Rainfed Cotton (*Gossypium hirsutum*). *Biores. Technol.* **2005,** *96*, 345–349.

Cerrato, M.; Blackmer, A. Comparison of Models for Describing; Corn Yield Response to Nitrogen Fertilizer. *Agron. J.* **1990,** *82*, 138–143.

Cheng-Song, X.; He-Zhong, D.; Zhen, L.; Wei, T.; Wei-Jiang, L.; Xiang-Qiang, K. Effects of N, P, and K Fertilizer Application on Cotton Growing in Saline Soil in Yellow River Delta. *Acta Agron. Sinica* **2010,** *36*, 1698–1706.

Constable, G.; Rochester, I. Nitrogen Application to Cotton on Clay Soil: Timing and Soil Testing. *Agron. J.* **1988,** *80*, 498–502.

Constable, G.; Rochester, I.; Daniells, I. Cotton Yield and Nitrogen Requirement is Modified By Crop Rotation and Tillage Method. *Soil Tillage Res.* **1992,** *23*, 41–59.

Devkota, M.; Martius, C.; Lamers, J.; Sayre, K.; Devkota, K.; Vlek, P. Tillage and Nitrogen Fertilization Effects on Yield and Nitrogen Use Efficiency of Irrigated Cotton. *Soil Tillage Res.* **2013,** *134*, 72–82.

Dounias, I.; Aubry, C.; Capillon, A. Decision-Making Processes for Crop Management on African Farms. Modelling from a Case Study of Cotton Crops in Northern Cameroon. *Agric. Syst.* **2002,** *73*, 233–260.

Elbert, A.; Haas, M.; Springer, B.; Thielert, W.; Nauen, R. Applied Aspects of Neonicotinoid Uses in Crop Protection. *Pest Manage. Sci.* **2008,** *64*, 1099–1105.

Freney, J.; Chen, D.; Mosier, A.; Rochester, I.; Constable, G.; Chalk, P. Use of Nitrification Inhibitors to Increase Fertilizer Nitrogen Recovery and Lint Yield in Irrigated Cotton. *Fertilizer Res.* **1993,** *34*, 37–44.

Geerts, S.; Raes, D. Deficit Irrigation as an On-Farm Strategy to Maximize Crop Water Productivity in Dry Areas. *Agric. Water Manage.* **2009,** *96*, 1275–1284.

Hatfield, J. L.; Sauer, T. J.; Prueger, J. H. Managing Soils to Achieve Greater Water Use Efficiency. *Agron. J.* **2001,** *93*, 271–280.

Hatfield, P.; Pinter Jr, P. Remote Sensing for Crop Protection. *Crop Protect.* **1993,** *12*, 403–413.

Hilder, V. A.; Boulter, D. Genetic Engineering of Crop Plants for Insect Resistance–A Critical Review. *Crop Protect.* **1999,** *18*, 177–191.

Hulugalle, N.; Scott, F. A Review of the Changes in Soil Quality and Profitability Accomplished by Sowing Rotation Crops After Cotton in Australian Vertosols from 1970 to 2006. *Soil Res.* **2008,** *46*, 173–190.

Ishaq, M.; Ibrahim, M.; Lal, R. Tillage Effects on Soil Properties at Different Levels of Fertilizer Application in Punjab, Pakistan. *Soil Tillage Res.* **2002,** *68* (2), 93–99.

Karayel, D. Performance of a Modified Precision Vacuum Seeder for No-Till Sowing of Maize and Soybean. *Soil Tillage Res.* **2009,** *104*, 121–125.

Kranthi, K.; Jadhav, D.; Kranthi, S.; Wanjari, R.; Ali, S.; Russell, D. Insecticide Resistance in Five Major Insect Pests of Cotton in India. *Crop Protect.* **2002,** *21*, 449–460.

McGarry, D. The Effect of Soil Water Content During Land Preparation on Aspects of Soil Physical Condition and Cotton Growth. *Soil Tillage Res.* **1987,** *9* (3), 287–300.

McGarry, D. Soil Compaction and Cotton Growth on a Vertisol. *Soil Res.* **1990,** *28*, 869–877.

Murtaza, G.; Ghafoor, A.; Qadir, M. Irrigation and Soil Management Strategies for Using Saline-Sodic Water in a Cotton–Wheat Rotation. *Agric. Water Manage.* **2006,** *81*, 98–114.

Murungu, F.; Nyamugafata, P.; Chiduza, C.; Clark, L.; Whalley, W. Effects of Seed Priming, Aggregate Size and Soil Matric Potential on Emergence of Cotton (*Gossypium hirsutum* L.) and Maize (*Zea mays* L.). *Soil Tillage Res.* **2003,** *74*, 161–168.

Oerke, E.C.; Dehne, H.W. Safeguarding Production—Losses in Major Crops and the Role of Crop Protection. *Crop Protect.* **2004,** *23*, 275–285.

Ozmerzi, A.; Karayel, D.; Topakci, M. Effect of Sowing Depth on Precision Seeder Uniformity. *Biosyst. Eng.* **2002,** *82*, 227–230.

Rhoades, J. Use of Saline Water for Irrigation. *California Agric.* **1984,** *38* (10), 42–43.

Rochester, I.; Peoples, M.; Hulugalle, N.; Gault, R.; Constable, G. Using Legumes to Enhance Nitrogen Fertility and Improve Soil Condition in Cotton Cropping Systems. *Field Crops Res.* **2001,** *70*, 27–41.

Romeis, J.; Meissle, M.; Bigler, F. Transgenic Crops Expressing Bacillus Thuringiensis Toxins and Biological Control. *Nat. Biotechnol.* **2006,** *24*, 63.

Scheer, C.; Wassmann, R.; Kienzler, K.; Ibragimov, N.; Eschanov, R. Nitrous Oxide Emissions from Fertilized, Irrigated Cotton (*Gossypium hirsutum* L.) in the Aral Sea Basin, Uzbekistan: Influence of Nitrogen Applications and Irrigation Practices. *Soil Biol. Biochem.* **2008,** *40*, 290–301.

Viswanathan, R.; Raguchander, T.; Prakasam, V.; Samiyappan, R. Induction of Systemic Resistance By Plant Growth Promoting Rhizobacteria in Crop Plants Against Pests and Diseases. *Crop Protect.* **2001,** *20*, 1–11.

Wu, K.; Li, W.; Feng, H.; Guo, Y. Seasonal Abundance of the Mirids, Lygus Lucorum and Adelphocoris spp.(Hemiptera: Miridae) on Bt Cotton in Northern China. *Crop Protect.* **2002,** *21*, 997–1002.

Xi, S.; Lihua, R.; Yongsong, Z.; Qizhao, Y.; Caixian, T.; Lianxiang, Q. Effect of Potassium Fertilizer Application on Physiological Parameters and Yield of Cotton Grown on a Potassium Deficient Soil. *Zeitschrift Pflanzenernährung Bodenkunde* **1989,** *152*, 269–272.

Yazgi, A.; Degirmencioglu, A. Optimisation of the Seed Spacing Uniformity Performance of a Vacuum-Type Precision Seeder Using Response Surface Methodology. *Biosyst. Eng.* **2007,** *97*, 347–356.

Yilmaz, I.; Akcaoz, H.; Ozkan, B. An Analysis of Energy Use and Input Costs for Cotton Production in Turkey. *Renewable Energy* **2005,** *30*, 145–155.

Zhang, H.B.; Li, Y.; Wang, B.; Chee, P. W. Recent Advances in Cotton Genomics. *Int. J. Plant Genom.* **2008**.

Zwart, S. J.; Bastiaanssen, W. G. Review of Measured Crop Water Productivity Values for Irrigated Wheat, Rice, Cotton and Maize. *Agric. Water Manage.* **2004,** *69*, 115–133.

CHAPTER 10

Harvest and Postharvest Technology and Factors Affecting Fiber Quality

ABSTRACT

Fiber to yarn conversion process in cotton has been affected by several factors which include harvesting methods, postharvest technology, and other factors affecting fiber quality. The fiber quality of cotton can be improved by different breeding techniques. Research work on harvest methods and improvement of fiber quality is being done through various approaches and many advancements have been made that need to be assembled together. Therefore, in this chapter, recent approaches and their efficiency are being presented. A brief review of recent researches is given to introduce the current state of the research on harvesting methods and on improving fiber quality which can be reconsidered in novel ways.

10.1 HARVEST METHODS

10.1.1 HARVESTING

To maintain the good quality of cotton fiber it is very important to harvest the cotton at right stage, that is, when the cotton boll burst exposing cotton and before fiber quality get damage, reducing fiber yield due to unfavorable weather conditions. In developed country such as the United states, cotton is harvested in July, whereas in northern areas of the belt in October and the harvesting is done by mechanical method. In Texas and Oklahoma, stripper harvesters with rollers or mechanical brushes which remove the whole boll from the plant are used and in the other areas of the belt spindle pickers are mostly used. After harvesting, seed cotton is removed from the harvesters and in the form of moderately compact units placed in modules then stored before ginning.

10.1.2 COTTON FIBER MATURITY

Thibodeaux et al. (1986) used image analysis to determine and measure fiber maturity of cotton (*Gossypium* sp.). On microscope slides individual cotton fibers were arranged longitudinally and with image analyzer scanned optically. Then for measuring the projected fiber widths along its length correctly, fiber's video image is spontaneously separated into narrow bands. Finally it is found that in each fiber the degree of secondary wall development is associated with the average maximum to minimum ratio of these measurements.

10.1.3 METHODS OF HARVESTING

10.1.3.1 COTTON MECHANICAL HARVESTER

Peterson and Kislev (1986) specified that for getting higher wages labor has been attracted toward nonfarm occupations from agriculture. The two assumptions are tested using cotton mechanical harvester. For cotton pickers simultaneous-equation model of the labor market is estimated which revealed that the increased nonfarm wages causes 79% of decline in cotton hand picking—the pull effect and the rest 21% is accredited to the reduced cost push effect—of machine harvesting.

10.1.3.2 ROBOT-BASED COTTON HARVESTING

Ling and Chang (2006) stated that America adopted the massive scale single variety, mid-long, and long autumn fiber, etc., hence making it possible to pick and classify cotton entirely by machines, not like other developing countries such as China where this is not appropriate and because of the accessibility to plenty of labor hand harvesting of cotton is mostly assumed. It is identified that in China, the conflict between production efficiency and high fiber grade can be resolved by agricultural robot-based cotton harvesting which combines benefits of the handcraft based and the mechanism based. With the great progress of mathematics together with robot sectors and application of wise robot for cotton harvesting has the possibility to become customary operations so as to augment the yield of high quality cotton along with agricultural industrialization. Widespread cultivation of cotton production will be supplanted by intensivism progressively in the future.

10.1.3.3 BRUSH STRIPPER AND SPINDLE HARVESTER

Faircloth et al. (2004) assessed alternative harvesting methods of cotton in northeast Louisiana as the comparison of brush stripper and also a spindle harvester because to decrease input expenses producers find ways due to the increasing cost of production of cotton. In consequence of the fact that spindle harvesters are more costly in operation expensive to operate than brush strippers study was commenced for comparing the efficacies. Yield and high volume instrumentation (HVI) were analyzed to verify the economics related with each harvesting system. Furthermore, to scrutinize the properties of fiber related with cotton advanced lint they utilized the analysis information system (AFIS). HVI data from stripper-harvested plots showed reduction in micronaire values and improved color grade values and at two locations the overall dollar value per hectare was significantly enhanced from stripping owing to high yield data, HVI data, indicating a decline in the input costs.

10.2 POSTHARVEST TECHNOLOGY

After harvest with different technologies cotton bolls are processed to end with ginning and fabric production.

10.2.1 EFFECTS OF AMARANTH (AMARANTHUS PALMERI)

Smith et al. (2000) considered mechanically harvested and ginned dry land cotton (*Gossypium hirsutum*) in studying the Amaranth (*Amaranthus palmeri*) effects on yield attributes, harvesting, and ginning and fiber quality. It was noticed that harvesting time increased by 2- to 3.5-fold with all the weed densities but the decreased lint and seed yields were observed, when density in the field was high (3260 weeds/ha). Due to the large weeds the slower ground speeds and work stoppages owing to the time consumed in the hand removal of large weed stems that get stuck within the portions of the machine harvester. During ginning and lint-cleaning processes, the rest 2% was effectively removed. Thus, it brings about an increase in the harvesting time of stripper. Harvester allows in the rejection of more than 98% of the weedy plants material during ginning.

10.2.2 INTRASPECIFIC COMPETITIONS EFFECT ON PALMER AMARANTH

Morgan et al. (2001) conducted a field study near College Station, TX, to ascertain the competitive intervention of Palmer amaranth on cotton development, yield, and fiber properties and to quantify the Palmer amaranth growth which is affected by intraspecific competition. For each 9.1 m row of cotton plants palmer amaranth densities varies from 0 to 10 plants. At 10 weeks after cotton emergence (WAE), about 45% of cotton canopy volume gets reduced by palmer Amaranth and at the highest density cotton biomass by more than 50% at 8 WAE. However, for 1–10 Palmer Amaranth plants/9.1 m row, cotton yields declined linearly from 13% to 54% but Palmer Amaranth density did not have any effect on cotton lint properties. At any of the densities, intraspecific competitions could not have its effects on palmer Amaranth volume or on its biomass.

10.2.3 SPINNABILITY

Pynckels et al. (1995) determined very important parameter for a spinner, that is, spinnability of fibers which helps to predict the extent of spinnability of a given fiber quality using neural nets. Through neural network they explained in what way the spinnability or a given fiber quality on a rotor and also in a ring spinning machine can be expected to occur with a consistency of 95%. Here the simple method of neural net application is explained after the in-depth measurement of its structure and also its characteristics. By considering the machine settings and fiber quality parameters as input and the yarn tenacity and elongation as outputs to model the process neural network is applied. Later, for getting the best yarns the input parameters are optimized using genetic algorithm. The study indicates that as optimal input parameters functions concurrently optimization of yarn qualities can be simply attained and the outcomes are significantly better than present manual machine intervention. On the basis of genetic algorithm predictions available fiber qualities optimal mixture findings is given over to last part of this chapter.

10.2.4 YARN QUALITY

Viswanathan et al. (1989) critically evaluated a relationship between fiber quality parameters and hairiness of cotton. Owing to the practical restrictions of mechanical processing machines some amount of hairiness in yarn is usually considered as unavoidable. Thus, while choosing appropriate raw materials spinners are concerned to know the impact of yarn hairiness on fiber properties.

10.2.5 SHORT FIBER DETERMINATION

Zeidman et al. (1991) verified the short fiber content and explained its theoretical fundamentals in cotton. With specific emphasis in particular on the source of short fiber content (SFC), measured by number and weight and also other associated parameters of length, the mathematical fundamentals of fiber length distribution are studied in an integrated way. It is found that as a parameter of length SFC is linked to range and shape of the fiber length distributions also, further, by using the concept of similar distributions, the similarity in shape between different distributions is measured. Finally it is determined that if each component SFC and fiber characteristics are known to reveal the SFC of a binary mixture, presentation of equations is required.

10.2.6 WORKING PERFORMANCE OF AIR-BLOWING AND AIR-SUCTION COTTON-BOLL SEPARATION DEVICE

Chang Lin et al. (2017) conducted comparative study on working performance of air-blowing and air-suction cotton-boll separation device. Owing to the immature cotton bolls which can dye the cotton easily declining the cotton quality during cotton mechanization harvesting cotton-boll separation device is attached on stripper cotton harvester. Based on their main elementary structures, working principle, and the difference between suspending velocities of different materials two types of cotton-boll separation devices, that is, air-suction separation device and air-blowing separation device were built to increase separating efficiency, reduce immature cotton bolls dying the cotton and to solve other problems. With ANSYS Fluent software numerical stimulation of cotton-boll separation devices inner airflow velocities and pressures distributions was done and results were analyzed. During cotton conveying process stable zone of negative pressure was formed in which the airflow of air-suction cotton-boll separation device gathers and this device is better than air-blowing cotton-boll separation device. Furthermore, air-suction cotton-boll separation device can reduce 5% of cotton loss rate and increase more than 20% of cotton-boll separating rate compared with air-blowing cotton-boll separation device.

10.2.7 DUAL POWERED OPERATED TINY COTTON GINNING MACHINE

Cotton ginning is a major processing industry, whose main function is cleaning and ginning of the seed cotton, cleaning the lint and forming a bale with the help of electricity. Here, Sonewane et al. (2017) made a dual powered operated tiny cotton ginning machine which can be moved easily by workers with a purpose to develop a new cotton ginning concept or technique. It is designed to fabricate the single roller cotton ginning machine and it can be operated both electrically and manually. This machine can be used for the domestic purpose and cotton lint cleaning will be done by door to door.

10.2.8 COTTON FIBER MASS DENSITY DETECTIBLE MODEL

Hui (2012) assumed that cotton fiber mass density detectible model helps in detection of the cotton flow, for the pipeline of the online measurement through the pneumatic transportation, hence it is important to establish such detectible model. Here, cotton fibers attenuation for the different wavelengths of light is studied by managing the basic law of the light absorption. When the light source was 940 nm near infrared there was no significant relationship between the breed, moisture, and ambient temperature changes and test results suggested that this can be used to built cotton fiber mass density measurement system by selecting it as the emitter. The cotton fiber mass density detectible model was established by combining the cotton fiber decay rate of the light intensity measurements and equivalent region of the cotton fiber quality actual measurement and subjecting them to regression analysis. The measurement results average error reached 6.1% confirmed from the validation results of the model. This research helps to develop sensors of cotton flow rate and lays a foundation for online monitoring of cotton flow rate in cotton pipe.

10.2.9 MEASUREMENT OF COTTON COLOR USING LED

Vik et al. (2017) measured cotton color using LED. The Rd and +b are the color parameters of cotton fiber used worldwide. HVI is used generally to measure these parameters. Ceramic tiles and cotton samples are the cotton color standards given by USDA. In this study measurement of the color parameters of cotton standards using non-contact method with LEDs as light source and its comparison with color parameters of HVI by utilizing the different light sources spectral data are discussed.

10.2.10 ULTRAVIOLET/VISIBLE/NEAR-INFRARED REFLECTANCE SPECTROSCOPY TO ASSESS THE LINT COTTON WASTE

Liu et al. (2010) used ultraviolet/visible/near-infrared reflectance spectroscopy to assess the lint cotton waste recovered from cotton fiber and trash contents. To decrease the non-lint materials to the minimal level, with least fiber damage, lint cleaning is performed at cotton processing facilities. For cost and profit analysis, the resultant waste is of great concern because it contains some amount of cotton fiber that is of equal quality similar to that of fiber in the bale. In the cotton industry traditional methods, viz., Shirley Analyzer and HVI employed for quantifying the non-lint materials or trash are more labor intensive and time consuming. In the region of 220–2500 nm, cotton waste was scanned and using Shirley Analyzer reference value was measured and in different spectral ranges partial least squares regression models were established and then compared. Between trash and cotton fiber the model performance from a narrow NIR region of 900–1700 nm was approximately comparable to that from the full 226–2496 nm spectral region while in the visible and NIR regions obvious spectral differences were seen. In the meantime, the effectiveness of two NIR bands, at 900 nm and 1135 nm, in the valuation of fiber and trash components in cotton waste was indicated by developing simple two-band difference algorithms using two unique bands. Moreover, a 90% confidence interval was used for eliminating outlier samples from the calibration and validation sets. They took into consideration various sampling species between the reference and spectral measurements. The viability of this technique for the estimation of above contents was shown from recalibrated models.

10.2.11 PARAMETER OPTIMIZATION AND HEAVY IMPURITY SEPARATOR IN SEED COTTON CLEANING PROCESS

ZhaoGuo et al. (2016) considered parameter optimization and experiment of funnel-shaped heavy impurity separator in seed cotton cleaning process. The cotton loss rate and impurities separating rate are affected by four main parameters, viz., air velocity in seed cotton inlet, angle of adjustable deflector, air velocity in auxiliary air inlet, and angle of adjustable baffle. Under the premise of low cotton rates to make best use of the cleaning efficacy the effect of the four parameters to the four indicators should be studied when impurities are cleaned using funnel-shaped heavy impurity separator. Therefore, to find the best combination of these four parameters to fulfill the above four indicators, broad optimization method toward the above-mentioned four parameters was conducted. To standardize the target functions and create the wide-ranging target function using linear weighting method the "linear effect coefficient method" was applied then to reverse multi-objective optimization into a single objective nonlinear optimization problem, the evaluation function method was used and finally to conclude the best parameter combination of the funnel-shaped heavy impurity separator multi-objective nonlinear optimization method was applied in separating aspect. Under these conditions, seed cotton loss rate was 0.48%, nails loss rate was 96.59%, pebble loss rate was 85.96%, and cotton shell loss rate was 31.57%, while the optimal parameters were: seed cotton inlet air velocity 21 m/s, adjustable deflector angle 44°, auxiliary air inlet air velocity 2 m/s, adjustable baffle angle 45°.

10.2.12 BRIQUETTING COTTON STALKS AS A BIOFUEL

Saeidy et al. (2004) studied the technological fundamentals of briquetting cotton stalks as a biofuel. They stated that in Egypt, cotton residues support the growth of insects, for example, pink bollworm as an overwintering site which in turn damage the adjacent crops leading to valued losses in the quantity and the quality of the cotton crop. Thus, immediately after the harvest operation for killing these insects and to avoid growth of other disease carriers the farmers are forced by the Ministry of Agriculture to burn cotton residues on the field which causes both environmental pollution and harmful effects for human health due to the buildup of enormous amount of harmful gases and smoke clouds. Accordingly to prevent the farmers from burning these residues on the fields the Ministry of the Environment promulgated a law. In Egypt for the management of the cotton stalk residues the production of cheap, storable, pest controlled, and environmentally friendly domestic fuel for the Egyptian rural areas through briquetting technology seems to be an advantageous solution.

10.2.13 OPTIMIZED COMBINED METHOD FOR EXTRACTION OF NATURAL CELLULOSE FROM COTTON STALKS

Li et al. (2015) prepared textile fibers by combining alkali soaking, steam explosion, and lactase/mediator treatments to bark of cotton stalks. It is found that lignocellulosic byproducts of straws of rice and wheat had poor mechanical properties compared with natural cellulose fibers from bark of cotton stalks but these fibers have high fineness value and rigidity which make them difficult to be spun in to textile yarns and they are mostly used to reinforce thermoplastic composites. But the extraction of cellulose fibers from lignocellulosic byproducts through conventional method with strong alkaline solution (30–100 g/L NaOH) caused environmental pollution and high cost. So in this study by combining three steps, that is, alkali soaking, steam explosion, and lactase mediator system treatment, a new method is discussed. Here, first the cotton stalks bark is soaked in NaOH solution at 20°C for 24 h then steam exploded with alkali within 0.0875 s and to further degrade lignin steam exploded cotton stalk balk fibers were treated in lactase mediator system. The NaOH dosage, kind and dosage of the mediator, lactase dosage effects on the cotton stalk bark fibers lignin mass fraction were analyzed. With the help of chemical compositions analysis, scanning electron microscopy (SEM), tensile property test, moisture regain and water retention test, X-ray diffraction (XRD), and thermal stability analysis (TGA), the comparative investigation of thermal stability of before and after treated cotton stalk bark fibers by steam explosion and lactase mediator system, crystalline structure, chemical composition, morphology, mechanical property and hygroscopic property was made. NaOH 10 g/L, soaking at 20°C for 24 h, steam pressure of 0.8 MPa, holding time of 180 s, ratio of liquor to fiber of 15 mL/g, lactase dosage of 600 U/g, ABTS mediator amount accounting for 1% of dry weight of cotton stalk bark fibers, temperature of 55°C, pH value of 3.5, and treatment time of 10 h are the optimized processing parameters used. The results showed that the cotton stalk bark fibers obtained by this optimized combined method had 40% more yield with cleaner surface, lower moisture regain, and water retention and superior thermal stability which can be utilized in textile industry. Furthermore, this research can be used as a reference for extraction of natural cellulose from cotton stalks, wheat straws, and rice straws like lignocellulose byproducts.

10.2.14 RAW COTTON FOR THE PREPARING AND EVALUATING THE NANOCRYSTALLINE CELLULOSE AEROGELS

Shamskar et al. (2016) used raw cotton for the preparing and evaluating the nanocrystalline cellulose (NCC) aerogels. The cotton and cotton stalk bleached pulps were hydrolyzed using sulfuric acid to prepare NCC suspensions then with FT-IR spectroscopy, cotton as well as cotton stalk NCC presented rod like shapes of variable dimensions in TEM micrographs with Zeta potentials of −27.5 and −21.8 mV in the stable range and 91.47 and 93.89 m^2/g aerogels surface areas. There were no variations in the product characteristics of NCC and aerogels which were extracted from pulps of cotton and stalk. Therefore, for preparing NCC and aerogel the usage of cotton stalk is more ecofriendlier and cost-effective.

10.2.15 VALUE-ADDED PRODUCTS FROM COTTON GINNING FACTORIES WASTE

Nagarkar et al. (2016) used waste collected from cotton ginning factories to prepare value-added products. All through ginning operations 2–4%

of good quality cotton fibers accompanied by the trash are lost and this precleaning waste can be used to prepare cotton absorbent with good quality and from the post-cleaning waste high-grade pulp and good quality paper can be made.

10.2.16 COTTON AEROGELS AND COTTON–CELLULOSE AEROGELS FOR OIL ABSORPTION

Han Lin et al. (2017) used cotton aerogels and cotton–cellulose aerogels for oil absorption from environmental waste. Pure cotton and cotton–cellulose aerogels are acquired using mixed-blending method with polyamide–epichlorohydrin as strengthening additives which is a cost-effective method. Through a facile chemical vapor deposition these aerogels are silanized using methyltrimethoxysilane to provide aerogels with hydrophobic surface. The highest oil absorption capacity was found in cotton aerogel with an initial concentration of 0.25 wt% over 100 g g^{-1}. Furthermore, in various pollutant organics, the good absorption capacity of cotton/cellulose aerogels was demonstrated and using pseudo first-order model with different cotton concentrations the absorption kinetics of the aerogels are also examined. From equilibrium absorption and absorption kinetics, these were found to be effective in oil absorption.

10.2.17 COTTON FOR THE EXCLUSION OF OILS AND ORGANIC SOLVENTS

Wang et al. (2017) formed a simple and a multipurpose strategy for fabrication of robust and superhydrophobic/superoleophilic cotton for the exclusion of oils and organic solvents from polluted waters. On the cotton fiber surface in situ hydrolysis of tetraethoxysilane (TEOS) was done using a coupling agent polyvinylpyrrolidone (PVP), following hexadecyltrimethoxysilane modification to prepare super wettability cotton. To produce cotton fibers with excellent superhydrophobicity the SiO2-modified cotton fibers surface roughness could be well regulated by merely changing the molecular weight of PVP and the concentration of $NH_3.H_2O$ and for the separation of oil/water this prepared cotton fibers were utilized as super absorbents which absorbs up to 35 times and 50 times its own weight of n-hexane and chloroform repelling the water entirely. The cotton can be reused for at least five cycles.

10.3 FACTORS AFFECTING FIBER QUALITY

10.3.1 SAW AND ROLLER GINNED FIBER PROPERTIES

Funk and Gambel (2009) assessed saw and roller ginned naturally colored cotton fiber properties of two upland cotton cultivars. Naturally colored cottons do not require any extra dyeing, hence have economic and environmental demand but the fiber length and strength of naturally colored cottons are not similar to the white cotton cultivars. The roller ginning favored by the HVI and AFIS analysis brought 0.7 mm (1/32 in.) greater HVI fiber lengths and 33% less AFIS neps in fiber neps than saw ginning. Therefore, the higher cost of roller ginning may get justified with this increase in fiber value.

10.3.2 COTTON FIBER MEASUREMENT BY INFRARED SPECTROSCOPY

Xiao Feng et al. (2009) measured cotton fiber numbers by near-infrared spectroscopy and for precise measurements, they designed a near-infrared light test apparatus that used LEDs of 940 wavelength for its light source. From different harvest years and different color grades eight cotton fiber samples were selected for the relative attenuation of near-infrared light (NIR) and measuring the corresponding fiber numbers. From the results, the property of attenuation of cotton fiber by NIR was found to be associated with the light intensity input. As input increases, the relative attenuation decreases for the same cotton. For cotton fiber property of light, a positive exponent function of relative attenuation of light and the equation for the number of cotton fibers thus obtained gave better results than the classic Lambert's law from the data analysis. At a confidence level of 95% with R^2 of 0.99 equations the number of cotton fiber can be well assessed. Furthermore, on NIR light of cotton fiber property, the influences of harvest years and color grades were insignificant.

10.3.3 IMPACT OF CELLULOSE, MINERAL ELEMENTS AND PH OF FIBER CELLS ON THE FIBER QUALITY

Dutt et al. (2004) commenced colored cotton breeding for high yield and fiber quality and compared three types of cotton, viz., white, brown, and green for their yield and fiber quality. In all three types of cotton fiber the fiber cells at different growth stages were analyzed to know the impact of cellulose, mineral elements, viz., nitrogen (N), phosphorus (P), and potassium (K) and pH of fiber cells on the fiber quality. The quality of fiber is found to be closely related with cellulose content which might be the reason for higher quality of white fiber containing high cellulose and it is further compared with colored cotton fiber. In white and colored cottons the rapid and slow decline in pH is observed which may have some impact on elongation of fiber. Among the mineral contents, traits for fiber quality were found to be positively associated with potassium. Colored cotton fiber heterosis studies proposed that the yield and quality of colored fiber cotton can be increased by heterosis breeding.

10.3.4 SPINNING MACHINES

Strumillo et al. (2007) evaluated various spinning machines for their cotton yarns spun quality ring-, compact- and by use of BD 200S & R1 rotor spinning machines, cotton yarns, with linear densities of 15, 18, 20, 25, 30, and 40 were produced and their quality parameters were analyzed. From carded and combed middle-staple cotton slivers and rovings as primary half products were obtained. They determined the fundamental parameters selected for yarn quality, viz., tenacity, unevenness, linear densities, elongation at break, number of faults, etc. For any given linear mass the calculation of the cotton yarns quality parameters and the modeling of yarn parameters for various carded and combed yarn streams spun using various spinning machines was allowed by these dependencies measured.

10.3.5 COTTON FIBER PHYSICAL PROPERTIES

Özarslan (2002) examined the cotton fiber physical properties. The results unveil that with an increase in the moisture content from 8.33% to 13.78% (dry basis), there was a variation in average length, width, and thickness of seeds (9.02–9.19), (4.70–4.86), and (4.25–4.45 mm) correspondingly. From the analysis of rewetted cotton seed it is found that it increased (0.626– 0.635) the seed volume (95.4–109.6 mm^3), thousand seed weight (104.06–109.64 g), projected area (35.89–40.14 mm^2), and the terminal velocity from 8.46 to 8.67 m s^{-1} in the same moisture range. Furthermore, on four structural materials, viz., stainless steel (0.27–0.35), galvanized iron (0.38–0.41), plywood (0.39–0.42), and rubber (0.42–0.44), the static coefficient of friction of cotton seed raised linearly, whereas there was a decrease in bulk density (642–610 kg m^{-3}), true density (1091–1000 kg m^{-3}), porosity (41.16–39.00%), and the shelling resistance (65.0–50.2).

10.3.6 SPINDLE SPEED EFFECT ON COTTON QUALITY

Baker et al. (2010) studied effects of changing spindle speed on cotton quality. In southern New Mexico, under furrow-irrigated conditions three cotton varieties were grown and to remain the bolls intact hand harvesting is done. After conditioning, the cotton bolls in a controlled atmosphere they are subjected to a single cotton picker spindle working at a 1000–3000 rpm speed. Two spindle designs: a 12.7 mm (1/2 in.) round, tapered, barbed spindle, and a 4.8 mm (3/16 in.) straight and smooth square spindle were investigated. The portion of seed cotton expelled from the spindle ("fly-off") and the portion that is not picked was determined from mass measurements. The peak force required to pull out a silk cotton from spindle was determined using a force gauge, and the test bolls had a moisture content of 9–10%. The results unveiled that in removal of cotton from the boll, the smaller, straight spindle was more destructive and at any given speed compared with smaller straight spindle there was double fly-off from the barbed spindle. Furthermore, as the speed was increased there was also an exponential increase in fly-off, in each type of spindle fiber. With the increase of speed by 1000 rpm, the peak force requirement was approximately doubled for both spindles. The smaller straight required 50–100% more peak force for removal of seed cotton from the spindle than barbed spindle.

10.3.7 SPINDLE SPEEDS

In Mesilla Park, New Mexico an additional field experiments were performed for the 2005 and 2006 crop years. Under furrow-irrigated conditions, three cotton varieties were grown replicating four times. For the 2005 crop, a modified 1-row cotton picker with ground speed of 0.85 m/s and spindle speeds of 1500, 2000, and 2400 rpm was used for harvesting, whereas the spindle speeds of 2000, 3000, and 4000 rpm were used for the 2006 crop. For all varieties, where the spindle speed was 1500 rpm, there were more stalk losses. For speeds of 2000 rpm or more indicating that the picker needs at least 2000 rpm spindle speed to function effectively and for pima variety with speeds of 3000 and 4000 rpm stalk losses were more with high percentage of spindle twists in a 1000 g seed cotton sample than for a speed of 2000 rpm spindle speed. An increase in spindle speed from 2000 to 3000 rpm the number and percent spindle twists were nearly doubled. These twist numbers increased further in 4000 rpm of spindle speed. Among treatments significant differences were seen in HVI classing data except for upland lint samples, which were collected before lint cleaning. Furthermore, higher levels of trash were seen than with a 2000 rpm speed in these samples of 3000–4000 rpm spindle speeds. In the bale raw stock all the three varieties had prominent significant variations in both AFIS nep count and short fiber count. These were enhanced when spindle speed was increased from 2000 to 3000 rpm, however these disappeared at 4000 rpm, as the fiber has been processed further. In the raw stock, there were significant variations for APIS dust count and trash count, there were higher levels of dust and trash at 3000 rpm, but with additional processing these disappeared.

10.3.8 THE ASSOCIATION OF CELLULOSE, WAX AND PIGMENT CONTENT WITH FIBER QUALITY

Zong Ling et al. (2013) performed a field trial in China in 2009 using brown cotton lines (S9B12, S9B11, and Zongxu 1), light brown cotton variety (CCRI 51), dark brown cotton lines (Zong 263 and Zong 128), green cotton lines (G88 and G4560), and white line (Kuche T94-4) as a control. The different germplasms with dissimilar fiber colors were analyzed for their fiber quality, cellulose, wax and pigment content, and the fiber quality was found to be positively associated with cellulose content and negatively with wax content. Moreover, significantly different distribution of pigment between green and brown fiber cells and the less pigment content in the fiber cell wall was determined by electron microscopy study. In general, the fiber quality of some brown cotton was superior to white cotton while green cotton was with lowest fiber quality. Finally this study suggested that the cellulose, wax, and pigment content were associated with fiber quality of natural colored cotton.

10.3.9 FIBER AND YARN QUALITY OF MATERIAL PRODUCED WITH AND WITHOUT SEED COTTON CLEANER

Byler et al. (2013) evaluated the fiber and yarn quality of material produced with and without seed cotton cleaner in a commercial cotton gin. The bale of ginned lint is main product of cotton gins though valuable fiber occurs in various additional streams that come from a gin. To recover the fiber in the material that has been removed by the seed cotton cleaners, some gins have mounted equipment and from such a gin data and samples were collected using three cultivars. The data revealed that in a reclaimed material of about 8.6 kg (19.0 lb) of seed cotton cleaners, per bale, had more loose lint and motes, with a little quantity of seed cotton. The total reclaimed fibrous material comprised a huge amount of valuable fiber but its quality was lower than the lint gained from the seed cotton. When the seed cotton cleaning reclaimed material was sent back to the seed cotton before ginning, it provided less than 2% of the fiber indicating that it would be hard to identify possible differences because of the reclaimed material. However, in bale lint quality, in the quantity of card waste, or in the measurements of yarn quality that were made from the cotton, though some quality differences were identified because of variations in cultivars, these were not detected by HVI or AFIS, irrespective of whether there was a mixing or non-mixing of reclaimed material with seed cotton before and after ginning. Thus, their study has indicated that lint/yarn quality is affected to a large extent with the inclusion of reclaimed material.

10.3.10 LASER LINES SCAN IMAGING METHOD FOR WHITE FOREIGN COTTON FIBERS

Dong et al. (2015) made detection of white foreign fibers in cotton using laser line scan imaging method. From 12 types of white foreign fibers with lint cotton using optimized laser imaging system and the traditional imaging system of LED plus ultraviolet light separately, they acquired 730 frames of the images. The results unveiled that the 12 kinds of white foreign fibers could be easily distinguished from cotton with an effective detecting rate up to 93.7% or 92.9%, respectively, by applying a simple Prewitt edge detecting algorithm or fixed binary segmentation with the optimized imaging system.

10.3.11 DISTRIBUTION CHARACTERISTICS OF COTTON FIBER QUALITY

Shu Rong et al. (2016) used environmental materials from seven sites of early-maturing and 10 sites of medium-early maturing cotton of regional trials in the Northwest Inland in China to analyze the distribution characteristics of cotton fiber quality during 2005–2014. The association between spinning consistency index and fiber trait, cotton fiber quality performance, and environmental interaction patterns were determined with GGE model to draw biplots and this biplot was also used to zone potential ecological subregions. In the early maturing cotton cultivation area on the basis of cotton fiber quality the cultivation regions were separated into three ecological subregions and in the Northwest Inland for cotton spinning requirements regionalization of optimized cultivation area could offer multi-level raw cotton materials. In this study, the spinning consistency index exhibited significant and positive association with fiber length, strength and index uniformity, and other fiber traits were also found to be correlated with each other.

10.3.12 CELL ADHESION AND CELL DETACHMENT RELATIONS IN CULTIVARS

Hernandez-Gomez et al. (2017) investigated the cell adhesion and cell detachment relations in cultivars with dissimilar fiber qualities and developmental features of cotton fiber middle lamellae (CFML). During fiber development both cell adhesion and detachment processes occurs and the CFML is essential for this. In several cultivars belonging to four commercial species with different fiber properties, viz., *Gossypium hirsutum*, *G. barbadense*, *G. herbaceum*, and *G. arboreum* CFML structural and compositional analysis was undertaken to determine the association between fiber quality and the speed at which cotton fibers develop. In all these species, CFML is made up of de-esterified homogalacturonan, xyloglucan, and arabinan and through this CFML cotton fiber cell adhesion is a developmentally controlled process governed by genotype. Right from the beginning of fiber cell wall detachment to the commencement of secondary cell wall deposition the CFML of two *G. hirsutum* cultivars showed visible paired cell wall bulges with abundant Xyloglucan and in later stages is these regions were devoid of pectic arabinan. Therefore, it is concluded that compared with other cotton species in the *G. hirsutum* cultivars paired cell wall bulges, rich in xyloglucan, are significantly more obvious and during the transition phase the CFML of cotton fibers gets restructured.

10.3.13 TRANSGENIC BT COTTON EFFECT ON THE FIBER QUALITY

Rajput et al. (2017) evaluated the different transgenic Bt cotton technology for their effect on the fiber quality for 2 years in transgenic Bt cotton varieties CCR141, CCRI79, and Bollgard II. They evaluated different fiber quality traits and observed that in transgenic cotton varieties lint quality characters were not affected, though there was some effect in the fiber strength.

10.3.14 THE INFLUENCE OF HARVESTING, CONVEYING, AND CLEANING SYSTEMS ON FIBER CHARACTERISTICS

Porter et al. (2017) observed harvesting and conveying/cleaning systems on a brush-roll stripper harvester for the cotton fiber quality and foreign matter content. In 2011 and 2012 from four locations seed cotton samples were collected on a cotton stripper harvester together with handpicked seed cotton samples and foreign matter content was analyzed and ginned to get fiber for HVI and AFIS fiber analyses. Until the BFC location results display little difference between the initial entries in the machine confirming that for removing foreign matter the field cleaner was the most effective system on a cotton stripper. The results specified that the harvesting, conveying, and cleaning systems had negligible influence on fiber characteristics and not related with foreign matter. Until the seed cotton passed through the field cleaner most of the parameters remain unaffected, later the parameters gets reduced to hand-harvested seed cotton levels.

10.3.15 PICKING INTERVALS EFFECTS ON SEED–COTTON YIELD

Deho (2017) studied the picking intervals effects on seed–cotton yield and fiber quality. On the basis of boll-opening percentage picking intervals were followed, viz., at 30% boll opening stage 1st picking was done, 2nd at 50, 3rd at 70, and 4th picking was completed at 90% boll opening. For most of the traits, viz., boll weight, seed–cotton yield, ginning outturn, seed index, staple length, and micronaire, the variety Sadori showed higher values compared with other two varieties. Picking at 50% boll opening was found to be most appropriate stage for better yield and fiber quality.

10.3.16 DEVELOPMENTAL AND HORMONAL REGULATION OF FIBER QUALITY

Xiang et al. (2017) considered natural-colored cotton cultivars, viz., brown (Xiangcaimian 2), green (Wanmian 39), and white (Sumian 9) as control for the developmental and hormonal regulation of fiber quality. The white-fibered control fiber length was greater than both natural-colored cottons. The brown-fibered cotton was longer than green and the fiber strength, micronaire and maturation of natural-colored cotton were also higher in control. Slower growth during 10–30 days postanthesis (DPA) is the reason for the shorter fiber and lower fiber strength, micronaire, and maturation of natured-colored cotton. Compared with white-fibered cotton at 10 DPA, the natural-colored had lower indole-3-acetic acid (IAA) content and lower abscisic acid (ABA) content at 30–40 DPA. At boll opening stage, the 20 mg L^{-1} gibberellic acid (GA3) application improved the fiber strength, micronaire, maturation, and the IAA content at 20 DPA, indicating that the fiber quality gets affected with endogenous hormones levels and for improving the fiber quality in natural-colored cotton external hormones that increases the hormone content in fiber should be applied.

10.3.17 FIBER WAX AND CELLULOSE CONTENT EFFECTS ON COLORED COTTON FIBER QUALITY

Zhaoe et al. (2010) investigated the fiber wax and cellulose content effects on colored cotton fiber quality, in 48 lines and hybrids. The brown cotton lines with the dark brown fiber color had high fiber wax content and compared with white cotton the green cotton wax content was five to eight times greater and dark brown cotton had two times higher wax content, whereas the white cotton had the highest cellulose content next was brown cotton and green fiber cellulose content was the lowest. Among both different coloring types and different varieties with the similar color fiber the fiber length, fiber strength, and fiber uniformity differences were observed. In the color cotton lines the cellulose content showed a significant positive correlation with fiber length, fiber strength, fiber fineness, lint index, boll weight, and fiber uniformity, while it exhibited a significant negative correlation with wax content and fiber elongation.

10.3.18 FIBER MACROELEMENT CONTENT ASSOCIATION WITH COTTON FIBER QUALITY

ShuNa et al. (2010) carried out an experiment to provide an information on the poor fiber quality of colored fiber cotton and to study the association of colored cotton fiber quality with fiber macroelement content because in colored fiber cotton (*Gossypium hirsutum* L.) dying process can be ignored decreasing the harmful effluent liquor and it is ecofriendly for the textile industry. In this study, during the cotton fiber development, the macroelement, cellulose, and flavonoid content of three cotton varieties, brown fiber cotton (X008), green fiber cotton (S029), and white fiber cotton (Xuzhou 142) were analyzed. The results revealed that compared with white fiber cotton flavonoid content was 4–10 times higher, macroelement content was 1–2 times higher, and cellulose content was 15% lower in colored cotton fiber. A significant positive association was observed between fiber quality properties and fiber cellulose content and negative with respect to flavonoid contents. Furthermore, N, P, and Ca contents exhibited negative association with cellulose content and a positive correlation with flavonoid content indicating that the accumulation of higher flavonoid and less of cellulose content in fiber reduces the fiber quality of colored cotton and to form the flavonoid higher N, P, and Ca elements content might be essential.

10.3.19 EFFECTS OF POLYETHYLENE FILM WRAPPED COTTON BALES LONG-TERM STORAGE ON FIBER AND TEXTILE QUALITY

Hughs et al. (2011) investigated the effects of polyethylene film wrapped cotton bales long-term storage on fiber and textile quality. The objective of the test was to verify whether the storage location from raw fiber across textile processing have any significant effect on cotton quality factors. Here using specially formulated linear low-density polyethylene (LLDPE) film with UV inhibitors the bolls were covered and sampled at the time of ginning to determine HVI properties and before storage with a temperature and humidity recorder bolls were instrumented. No significant statistical differences were observed.

10.3.20 PHYSIOLOGICAL CHARACTERISTICS OF FIBER IN THE NATURAL COLORED-COTTON FIBER DEVELOPMENT SUBTENDING LEAF

Xiang et al. (2011) studied the physiological characteristics of fiber in the natural colored-cotton fiber development subtending leaf. The changes of metabolite and enzyme activities of brown cotton cultivar Xiangcaimian 2, green cotton cultivar Wanmian 39, and white fiber control Sumian 9 were measured. The results revealed that changing with an "S" curve fiber for Xiangcaimian 2 and Wanmian 39 cellulose accumulation was less. For colored cotton maximum accumulation, accumulative rate parameter and maximum accumulative rate for cellulose were also less. Further activity of sucrose invertase, sucrose synthase (SS), UGPase in fiber for colored cotton and SPAD, soluble sugar content, activity of sucrose invertase in leaf were reduced than those for white cotton. Therefore, in colored cotton enough nutrients were not available for fiber development resulting in poor quality fiber. The SS and UGpase activities and content of GAs were found to be involved in impairing fiber strength and its maturation in colored fiber cotton cultivars.

10.3.21 FRUITING-BRANCH POSITION, TEMPERATURE-LIGHT FACTORS, AND NITROGEN RATES EFFECTS ON COTTON

WenQing et al. (2011) carried out a field experiment in Nanjing to study the fruiting-branch position, temperature-light factors, and nitrogen rate effects on cotton (*Gossypium hirsutum* L.) fiber strength. Two cotton cultivars Kemian 1 with average fiber strength of 35 cN tex^{-1} and NuCOTN 33B with average fiber strength of 32 cN tex^{-1} were sown at two sowing dates applying three levels of nitrogen. Observations for the association between fruiting branch and temperature were recorded. The results showed that at optimum temperature–light factor stronger fiber was produced in the cotton bolls of middle branch than the in lower- and upper-branch, whereas the fruiting branch exhibited nonsignificant effects with decreased temperature–light. During rapid growth period, that is, cotton fiber secondary wall thickening period PTP (cumulative photo-thermal product) the temperature–light factor were linear with the rate (VRG) and duration (TRG) of fiber strength, whereas the observed cotton fiber strength (FSobs) were quadratic with PTP during steady growth period. At VRG 1.5 cN tex^{-1} d^{-1} and TRG 16 d, the strongest Strobs were produced for Kemian 1, while for NuCOTN 33B the strongest Strobs were obtained when VSG was 0.18 cN tex^{-1} d^{-1} and TSG was 24.0 d at PTP of 291 MJ m^{-2}. N fertilization had a compensatory effect on PTP and significantly influenced the cotton fiber strength. PTP for acquiring highest Strobs decreased with increased N. NA less than 480 kg N.hm-^{2} was more appropriate when PTP was less than 104 MJ m^{-2} and NA less than 240 kg N hm^{-2} when PTP was greater than 104 MJ m^{-2} was more appropriate for cotton fiber strength. Finally it is concluded that the cotton fiber strength formation was significantly affected with interaction between fruiting-branch and temperature–light factor.

10.3.22 SECONDARY METABOLISM AND CELL WALL STRUCTURE DURING FIBER DEVELOPMENT

Ghazi et al. (2009) studied two cotton species *Gossypium hirsutum* (L.) and *G. barbadense* (L.) cultivated for fiber. In both species for gene regulation, cell organization, and metabolism a major shift in great quantity of transcripts between fiber elongation and fiber thickening was found to be basically similar and a major percentage of the genome was found to be expressed during fiber elongation and consequent secondary cell wall thickening. Previously distinguished specific metabolic and regulatory genes signified that each stage had its own characteristic features. Throughout the diverse plant genera cell elongation conservation mechanisms and wall thickening was seen owing to similar type of genes and developmental expression in the fibers of these species and other fiber-producing plants, whereas during fiber elongation secondary metabolism and pectin synthesis and modification genes expression was significantly different. The gene profiles of the fiber thickening stage were indistinguishable, indicating that their dissimilar final fiber quality properties may be created at earlier stages of fiber development. In an inter-specific cotton recombinant inbred line (RIL) population-specific fiber properties exhibited high associations with representative phenyl propanoid and pectin modification genes expression levels, suggesting their role in determining fiber quality. Finally it is reported that the pathways in secondary metabolism and cell wall structure during fiber development can be identified by transcript profiling that may contribute to cotton fiber quality.

10.3.23 DRIP IRRIGATION REGIMES EFFECTS ON COTTON YIELD

Dagdelena et al. (2009) observed the different drip irrigation regimes effects on cotton yield, water use efficiency (WUE), and fiber quality. During 2004 and 2005 in the Aegean region of Turkey N-84 cotton variety was grown and on the same day treatment T_{100} received 100% of the soil water depletion and treatments T_{75}, T_{50}, and T_{25} received 75, 50, and 25% of the amount received by treatment T_{100}. The mean seasonal water use values varied from 265 to 753 mm and the mean seed cotton yield ranged from 2550 to 5760 kg ha^{-1}. In the experimental years, the T_{25} irrigation water use efficiency (IWUE) was highest, that is, 1.46 kg m^{-3} and the smallest IWUE, that is, 0.81 kg m^{-3} was in the T_{100} treatment; however, the high average cotton yield was provided by the full irrigation treatment (T_{100}). The results unveiled that under limited water supply conditions the use of drip irrigation method in cotton at 75% level (T_{75}) had many benefits in terms of saved irrigation water and the large WUE irrigation of cotton but from economic point of view, at T_{75} the saving of 25.0% irrigation water leads to decrease in the net income by 34.0% and areas having no water shortage and semi-arid climatic conditions well-irrigated treatments (T_{100}) could be applied with reasonable net income.

10.3.24 INFLUENCE OF POTASSIUM NUTRITION ON LINT YIELD AND FIBER QUALITY

Cassman et al. (2018) considered the Acala cotton for the potassium nutrition influence on lint yield and fiber quality. To assess this association on an irrigated, vermiculitic soil a single cultivar (1985) and two cultivars (1986 and 1987) were cultivated with 0, 120, 240, or 480 kg K ha^{-1} replicated design and to these treatments the response of seed-cotton yield was significant. The increased K supply results in higher lint percentage owing to increase in lint yield more than seed yield. Furthermore, this higher lint percentage stimulates improved fiber length and secondary wall thickness quantified as a micronaire index. Therefore, in regression analyses, the different dependent variables such as fiber length, micronaire index, fiber strength and percent elongation, and fiber length uniformity ratio were each positively associated to independent variables: (1) fiber K concentration at maturity, (2) leaf K concentration at early bloom, and (3) soil K availability index for both cultivars. It is resolved that the K requirement differs among genotypes and it is required to get high lint yield with standard quality and under field conditions it is an important determining factor of fiber quality.

10.3.25 FLOWERING DATE, BOLL POSITION, AND ENVIRONMENTAL FACTORS EFFECT ON FIBER QUALITY

Johnson et al. (2004) conducted a 2-year field study to evaluate the flowering date, boll position, and environmental factors effect on fiber quality and compared the effects of boll position on fiber properties through planting dates. In 1997, at all boll locations increased fiber length and micronaire values were found with panting earliness, whereas in 1999, the longest and most mature fiber was found to be correlated with boll locations number in the middle planting date. For both middle- and late-planted cotton boll distribution patterns were similar. Furthermore, increased fiber maturity was observed at high temperatures and adequate moisture level during boll development. Finally they reported that boll location and planting date are associated with fiber quality and in the Texas Coastal Bend early cotton (*Gossypium hirsutum* L.) planting has the potential to improve performance by avoiding drought.

10.3.26 IMPROVEMENT OF AUSTRALIAN COTTON FIBER QUALITY

Long et al. (2009) investigated Australian cotton genotypes for fiber quality and textile performance because the improvement of Australian cotton fiber quality is necessary to maintain the industry viability. The fiber quality and yarn performance of five *Gossypium hirsutum* L. and one *G. barbadense* L. genotypes were evaluated. The fiber maturity ratio, fiber linear density, and fiber diameter (ribbon width) were measured. Micronaire was found as an inferior indicator of yarn performance, for instance, the micronaire values of *G. hirsutum* L. breeding lines CHQX12B and CHQX377 was 4.4, but because of more mature and finer fiber of CHQX377 it spun stronger yarns and for *G. hirsutum* L. genotypes producing on average 104 neps g^{-1} per lint cleaner passage lint cleaning had the highest effect on nep (fiber knot) generation. Genotypes with the longest and finest fiber were used to produce strongest yarns and fiber quality was found to be negatively associated with yield. Cost benefit analysis revealed that for better fiber quality fiber yield was a dominant economic factor than price premiums and the use of alternate methods in determination of fiber fineness can improve Australian cotton value.

10.3.27 CLIMATE CHANGE AND ITS EFFECT ON COTTON PRODUCTION

Bange (2007) stated climate change effects on cotton growth and development. Cotton is an indeterminate and perennial plant. In arid regions having high temperatures the wild ancestors of cotton are naturally adjusted enduring longer dry weather periods. The currently cultivated cultivars are made adapted to rain-fed and irrigated production where intermittent water supply occurs by inheriting these traits from wild ancestors. The growth and development of cotton plant is more complex than other field crops. The understanding of the crop's response to climate and management is occasionally problematic because in cotton vegetative and reproductive growth takes place concurrently. The net increases in CO_2 concentration, reduced water availability, and increased atmospheric evaporative demand occurs due to the low rainfall and relative humidity which results in climate change effects on cotton growth and development that effect yield and fiber quality. These impacts are discussed in more detail below.

10.3.27.1 EFFECTS OF CLIMATE CHANGE ON COTTON PRODUCTION AND TRADE IN THE LEADING PRODUCING AREAS

Ton (2011) stated that cotton production and trade reliant on production site gets affected due to the climate change. The effects of climate change on cotton production and trade in the leading producing areas worldwide and the possibilities available to alleviate and adapt to these impacts are analyzed in this chapter. Cotton production contributes to climate change and also subjected to its impacts. The agricultural production, processing, trade, and consumption bring about 40% of the world's GHG emission of which 0.3% and 1% occurs from cotton production. The flowering and boll formation are the most sensitive stages to water availability in cotton, but because of its vertical tap root it has some resistance to high temperatures and drought. New cotton production areas should be established in places where it is not grown before. Increasing temperatures and atmospheric CO_2 supports plant development, except day temperatures exceed 32°C. However, the adaptation challenges are posed by increased pests, water stress, diseases, and weather extremes. On the whole in Xinjiang (China), Pakistan, Australia, and the western United States, the less availability of water for irrigation are the main negative effect of climate change on cotton production. In the Yellow River area (China), in India, the southeastern United States and southeastern Anatolia (Turkey) the climate change effects on rainfall will likely be positive while in in Brazil and West and Central Africa it is not known clearly.

10.3.27.2 CLIMATE CHANGE EFFECTS ON COTTON

Gerardeaux et al. (2013) discussed the positive effects of climate change by CO_2 enrichment and conservation agriculture on cotton in 2050 in Cameroon. This chapter foresees an unpredicted positive impact of climate change on cotton production in Cameroon. Owing to rising temperature and CO_2 and rainfall uncertainties in Africa global warming could threaten cotton production. As most of the African farmers grow cotton as their cash crop and have few or no possible substitutions this condition is worsened. Thus, the effect of climate change on cotton production needs to be evaluated. In this study, a process-based cotton growth and management stimulating crop model CROPGRO is used. From the ENSEMBLES project general climate models and regional climate models were combined in six regional climate projections sets and in North Cameroon this model is applied. From 2001 to 2005 and at an experimental station in 2010, the observations were made in farmer fields and with a data set the model was calibrated and validated. The results unveiled that surprisingly in North Cameroon from 2005 to

2050 if conservation agriculture systems are implemented the climate change will have a positive impact on cotton yields increasing 1.3 kg ha^{-1} $year^{-1}$ of yield. The crop cycles will get shortened by 0.1 day $year^{-1}$ if the projected increase of 0.05°C $year^{-1}$ in temperature occurs without having any adverse effect on yields. Furthermore, yields will improve by about 30 kg ha^{-1} owing to the fertilizing effect of CO_2 enrichment. These findings revealed that in North Cameroon climatic changes can be predicted by adapting different cropping systems and acclimatization techniques.

10.3.27.3 CLIMATE CHANGE EFFECTS ON COTTON PRODUCTION IN BURKINA FASO INSTITUTE

Diarra et al. (2016) considered the climate change effects on cotton production in Burkina Faso Institute. The key factors affecting the cotton yields and the possible effects of future climate change are evaluated by an econometric analysis. The results revealed that the yield gets reduced significantly with additional increase in global temperature and the production will also get affected with future rainfall changes. However, the rainfall effects are comparatively less than the temperature effects. Hence, new approaches should focus on development of heat-resistant cultivars instead of drought resistant ones so as to alleviate and acclimatize them to the climate change effects.

10.3.27.4 HIGHER TEMPERATURE EFFECTS ON COTTON LINT YIELD AND FIBER QUALITY

Pettigrew et al. (2007) studied the higher temperature effects on cotton lint yield production and fiber quality by conducting 3-year study and found the similar response in genotypes to the temperature regimes. Out of 3 years, in 2 years in warm regime 6% smaller boll mass, with 7% fewer seed per boll were produced resulting in 10% lower lint yield but the fiber strength was constantly 3% more than the control treatment. Further lower NAWB data was observed in warmer regime signifying a slightly advanced crop maturity. Ovule fertilization may get affected when temperatures become too hot, leading to reductions in seeds produced per boll, boll masses, and finally, lint yield.

10.3.27.5 GGE BIPLOT ANALYSIS FOR EVALUATING COTTON ENVIRONMENTS

NaiYin et al. (2013) based on cotton fiber micronaire selection evaluated regional cotton trial environments by using GGE biplot analysis. In the Yangtze River Valley cotton fiber micronaire was one of the key for fiber quality improvement and significant quality trait directly associated with yarn quality. In improving both micronaire selection efficacy and regional trial cost-savings the optimal arrangement of a regional trial scheme is critical which is based on identification of a representative trial location, and for representative location selection and trial location evaluation the most suitable statistical and visual tool is Biplot GGE analysis which is widely implemented in several crops in cultivar stability and trial location similarity analysis. But little has been done in terms of fiber micronaire selection-based regional trial location evaluation of cotton. Therefore, in this study using biplot GGE analysis ideal trial locations in cotton micronaire selection were selected. In the Yangtze River Valley in 2000–2010 in 15 trail locations to 27 independent sets of cotton variety regional trials the method was applied. The most ideal trail location was Jingzhou City of Hubei Province but the ideal locations were Huanggang City of Hubei Province, Nantong City of Jiangsu Province, and Jiujiang City of Jiangxi Province and it suggested that in the Yangtze River Valley for the cotton fiber micronaire-based identification of regional test points of optimal cultivar and for selection of eurytopic cotton cultivar these locations were most effective. However, for cotton fiber micronaire selection the cotton trial locations in Jiangsu and Zhejiang provincial coastal fields (Yancheng and Cixi) were not suitable in the Yangtze River Valley. This study set the theoretical basis for decision-making about fiber micronaire selection in regional/national cotton trials schemes in the Yangtze River Valley and fully demonstrated the GGE biplot analysis efficiency in regional trial environment.

10.3.27.6 SUITABLE TEMPERATURES FOR FIBER FINENESS

Zhen (2015) extracted fibers with lignin content of 4.5% and fineness of 28.3 dtex using 4% alkali from bark of cotton stalks for spinning application and the fibers are treated at temperatures of 150, 160, and 170°C, respectively. They found that with the increase in time the fineness of cotton stalk fibers decreased gradually and extremely longer time results in severe reduction in fiber length. It also results in breakdown of cotton bark stalks to a paste. Fibers with best fineness of 30.8, 28.3, and 28.1 dtex were extracted at 150, 160, and 170°C, these fibers had lowest lignin content of 5.1%, 4.5%, and 4.5%. This was attributed that increased time of exposure decrease the lignin content of cotton stalk fibers. With further increase in time and temperature the breaking strength and Young's modulus decreases severely. Therefore, the best quality fiber with fineness of 28.3 dtex, length of 39 mm, lignin content of 4.5%, and Young's modulus of 46 cN/dtex was extracted at using 4% alkali with time of 60 min, the breaking strength of fiber was 1.8 cN/dtex at 160°C, while at 170°C it decreased to only 1.4 cN/dtex. The high quality yarns that were extracted, by conventional method, viz., alkali, steam explosion, or H_2O_2, were found to have too coarse fineness (>45 dtex) or very high content of lignin (>11.8%). In this research, the cellulose structure in cotton stalk fibers was not changed due to alkali treatment so all the fibers had type I crystalline structure. All the fibers in this research had identical fineness, but owing to the removal of lignin at temperature of 160°C Young's modulus were low. Scanning electron microscope (SEM) images of blended yarns also exhibited that the fibers (stalks) with low content of lignin (4.5% and 5.5%) could effectively be spun to better quality yarns, having a fineness (22.4 tex) further, a decrease in the lignin content of (5.5–4.5%) increases the breaking strength and breaking elongation by 11.1% and 9.8%; however, there was a reduction of 75.1% and 29.6% in unevenness and hairiness index of blended fiber yarns. On the whole in this research the fibers with low lignin content, low Young's modulus, and 68 % lower fineness but with 93.5% higher breaking strength than that in references were extracted at 160°C which gives them a good potentiality for manufacture of yarns of high quality.

10.3.28 PART I. INTERPRETIVE SUMMARY

Cotton fiber quality is very important for each section of the cotton industry, and for dyeing performance, a physical characteristic, viz., specific fiber surface area acts as a predictor. The available area of solid surface per unit mass of material is specific surface area. The objective of this study was to find a method to measure the cotton fiber-specific surface area with methylene blue adsorption in liquid phase. As the surface area gets modified by the surrounding phase methods that were used so far to measure surface area had limitations. In N adsorption/desorption isotherm method before N adsorption the vacuum dried treatment modify whole surface and the surrounding phase interference was particularly problematic in this method and this method measures only the external surface area. The cotton which is not treated with any chemical before, analysis is referred as cotton fiber and for surface area determination of cotton and different natural solids: activated carbon, charcoal, graphite, and silica, the method of adsorption of methylene blue in liquid phase has been adopted extensively. In future, these results could be used for improving fiber quality measurements. Quality parameters of harvested cotton are improved by breeding process.

10.3.29 INDIVIDUAL COMPONENTS ASSOCIATED WITH THE LINT YIELD AND FIBER QUALITY

In upland cotton, in six genotypes of *Gossypium hirsutum* L., a sequence of individual components, for example, fiber length and the number of fibers formed on each seed are associated with the lint yield and fiber quality and this association is resolved by the Smith and Coyle (1997) in their study, the comparison is done in cotton grown at College Station, TX, between the F_1 population acquired by crossing parents with same general combining ability (GCA) with selected F_1 populations acquired by crossing parents with dissimilar direction of GCA for fiber quality and within boll lint yield components, in 1989 and 1992. Afterward, using direct measurement or through calculations within-boll lint yield components were measured and by high volume instrumentation fiber quality parameters ascertained. The results revealed a negative correlation between fiber strength, length, and also within-boll lint yield components which is found to be a result of repulsion phase linkage, however the influences of pleiotropy could not be ruled out.

10.3.30 COMBINING ABILITY AND HETEROSIS

Hüseyin et al. (2003) crossed six cotton genotypes in half diallel mating design and measured the combining ability and heterosis for yield components and fiber quality factors. Suitable parents for different characters, viz., DPL 5690 for number of bolls per plant, Acala SJ-5 for boll weight and fiber length; Nazilli-84 and Carmen for seed cotton yield and lint percentage; Tamcot CAMD-E for earliness and fiber fineness; and PD 6168 for fiber strength were selected. For further research, the Tamcot CAMD-E × Carmen, Nazilli84 × PD 6168, DPL 5690 × Tamcot CAMD-E, and Tamcot CAMD-E × PD 6168 cross-combinations are found to be promising.

10.3.31 GENETIC DIVERSITY OF THE COTTON CULTIVATED SPECIES

Tatineni et al. (1996) used morphological characteristics and RAPDs for studying the genetic diversity of the two primary cultivated species of cotton (*Gossypium hirsutum* L. and *G. barbadense* L.). From interspecific hybridization 16 near-homozygous elite cotton genotypes were obtained and their genetic diversity was studied with stable and highly heritable morphological characters at the phenotypic level and with the random amplified polymorphic DNA (RAPD) at the DNA level. Using polymerase chain reaction (PCR) DNA was amplified with 80 random decamer primers and 135 RAPDs were produced. In 1992 and 1993, total 19 morphological traits can easily differentiate typical *G. hirsutum* (upland cotton) from *G. barbadense* (Pima cotton) were selected and were measured. For the RAPDs genetic distance and morphological data average taxonomic distance dendrograms were constructed, the association among the genetic distance and taxonomic distance was 0.63 and grouping of all genotypes with these two methods provided identical results producing two clusters with one similar to *G. hirsutum* and one *G. barbadense.* Several genetically and phenotypically distant genotypes from typical *G. hirsutum* and *G. barbadense* were identified and this polymorphism shown by the genotypes can be used to identify economically important traits, for example, fiber quality in genetic mapping populations. In this experiment, genetic associations within a diverse array of *Gossypium* germplasm using RAPD analysis was determined for the first time showing the reliability of RAPD markers. Furthermore, the genotypes derived from *G. barbadense* could be used as a source of new alleles in *G. hirsutum.*

10.3.32 PROCESS PARAMETERS FOR OPEN-END SPUN YARNS

Hasani et al. (2012) based on the Taguchi method undertook a study for optimization of the process parameters for open-end spun yarns, having multiple performance characteristics by means of grey relational analysis. The process parameters comprising speed and diameter of rotor speed of opener, linear density of yarn and navel type are optimized using CVm%, hair number/meter, and yarn tenacity as quality characteristics. Cotton fibers (35%) and cotton waste from ginning machines (65%) were gathered and used as raw materials in this study. The results indicated that the most important parameter with significant on the multiple performance characteristics is rotor speed, finally the genotypes selected with good combining ability can be employed in three-way crosses, or modified backcross or recurrent selections to bring improvement in yield and fiber qualities.

10.3.33 FIBER QUALITY TRAITS ASSOCIATION MAPPING

Buriev et al. (2009) assumed fiber quality traits association mapping in *G. hirsutum* L. variety germplasm based on linkage disequilibrium. Here, 335 *G. hirsutum* germplasms were screened with 202 microsatellite marker primer pairs for genetic diversity, population characteristics, the extent of linkage disequilibrium (LD), and association mapping of fiber quality traits. In tested cotton variety accessions a genome-wide average LD stretched up to genetic distance of 25 cM at the significance threshold ($r^2 \geq 0.1$), while at $r^2 \geq 0.2$ genome-wide LD was decreased to ~6 cM suggesting the possibility of agronomically important traits association mapping in cotton. From the outcomes, linkage, selection, inbreeding, population stratification, and genetic drift were identified as potential factors for producing LD in cotton. ~20 SSR markers were found to be correlated with each trait of fiber quality using unified mixed liner model (MLM) incorporating population structure and kinship in two environments and these correlations were further established in general linear model and structured association test, accounting for population structure and permutation-based multiple testing. In both Uzbekistan and Mexican environments numerous common markers with significant correlations were ascertained. At "moderate to strong" and "strong to very strong" evidence levels the MLM-derived significant correlations between 7% and 43% were supported by a minimum Bayes factor signifying their effectiveness in marker-assisted breeding programs and association mapping general proficiency with cotton.

10.3.34 INTERSPECIFIC HYBRIDIZATION FOR GENETIC IMPROVEMENT OF FIBER STRENGTH IMPROVEMENT

Choudki et al. (2013) assumed genetic fiber strength improvement in diploid cotton (*G. herbaceum* L.) via interspecific hybridization using *G. anomalum* wild species. Here, for improving the fiber strength of *G. herbaceum* diploid cotton with wide adaptability and high degree of resistance to biotic and abiotic stresses using *G. anomalum* inter-specific cross-genetic introgression studies were started. Pedigree selection was followed from F_2 to F_{11} and during 2009–2010 these F_{11} lines were evaluated in augmented design-II at ARS, Dharwad Farm. Heritability, genetic advance, and variability parameters GCV of seed cotton yield, ginning outturn range, fiber strength, 2.5% span length, and micronaire value were compared. Seven selections PSCANOI-5, 42, 62, 160, 166, 170, and 173 had higher values in both fiber strength (18.05–21.85 g/tex) and seed cotton yield (1680.49–1896.69 kg/ha) with 36–53% increased yield over diploid cotton commercial variety DLSa-17, indicating that seed cotton yield and fiber quality traits of *G. herbaceum* cotton can be improved via interspecific hybridization using wild species *G. anomalum* as a donor parent.

10.3.35 SEQUENCING OF ALLOTETRAPLOID GOSSYPIUM HIRSUTUM L. ACC. TM-1 GENOME

Zhang et al. (2015) incorporated the whole-genome shotgun reads, bacterial artificial chromosome (BAC)-end sequences, and genotype-by-sequencing genetic maps to sequence the allotetraploid *Gossypium hirsutum* L. acc. TM-1 genome. In this study, 32,032 A-subgenome genes and 34,402 D-subgenome genes were assembled and annotated. Genome-wide expression dominance was not seen among the subgenomes but compared with D subgenome structural rearrangements, gene loss, disrupted genes, and sequence divergence were commonly seen in the A subgenome, indicating asymmetric evolution. The genomic signatures of selection and domestication showed a high correlation with the fiber improvement positively selected genes (PSGs) in the A subgenome and stress tolerance genes in the D subgenome. This study suggested that the allotetraploid cotton (*Gossypium hirsutum* L. acc. TM-1) sequencing aids in the fiber improvement of upland cotton and this draft genome sequence offers resources for developing the best quality cotton lines.

10.3.36 QUANTITATIVE TRAIT LOCI ANALYSIS OF GENOTYPE–ENVIRONMENT INTERACTIONS

Paterson et al. (2003) carried out quantitative trait loci (QTL) analysis of genotype–environment interactions having effect on cotton fiber quality. The influence of well-watered against water-limited growth conditions on the genetic control of fiber quality is defined. In one or more treatments 6, 7, 9, 21, 25, and 11 QTLs influencing fiber length, length uniformity, elongation, strength, fineness, and color (yellowness) were detected. It is found that the water management regimes specific variations and growing seasons ("years") general differences have significant effect on the genetic control of cotton fiber quality. In water-limited treatments 17 QTLs were identified, whereas in well-watered treatment only two QTLs were detected signifying that improvement of complex trait such as fiber quality might be more difficult under water-limited conditions than well-watered conditions. Furthermore, for conferring the adequate quality under both sets of these conditions as cotton is widespread in both the irrigated and also in rained production systems, there is a requirement of manipulation of more number of genes which will decrease the predictable rate of genetic gain.

10.3.37 CORRELATION AND HERITABILITY OF THE YIELD AND FIBER OR LINT QUALITY

Desalegn et al. (2009) made diallel crosses at Werer Agricultural Research Center, in Ethiopia to get 15 F_1 cotton hybrids, to find out the interrelationship and heritability of the yield and fiber or lint quality. Seed cotton yield exhibited a high genetic correlation with the boll weight (r = 0.99**), lint yield (r = 0.88**), and lint index (r = 0.96**), further the lint yield associated to a large extent to lint percentage and the number of seeds per boll (r = 0.96**) indicating that the high cotton lint yield was positively associated with more bolls per plant, high lint percentage, and a small seed size. All the fiber quality parameters were highly associated with fiber strength and its significant positive associations were observed with staple length 2.5% (r = 0.99**) and staple length 50% (r = 0.64**). Furthermore, the fineness indicator (micronaire) and the uniformity ratio ($r = 0.61^{**}$) exhibited a positive correlation. On the other hand, the fineness indicator (micronaire) ($r = -0.86^{**}$), short fiber index ($r = -0.85^{**}$), and uniformity ratio ($r = -0.99^{**}$) showed negative correlations with fiber length and with fiber strength the genetically negative associations of lint percentage and lint yield were quite high. However, they showed a positive association with the fiber-fineness indicator or micronaire ($r = 0.99^{**}$ and 0.79^{**}, respectively). For lint percentage (h2 = 97%), lint yield (h2 = 72%), lint index (h2 = 79%), and seed index (h2 = 86%) broad sense heritability estimates of the yield and yield components were also high along with strong correlation with other fiber quality parameters, suggesting that these characters can serve as indicators in breeding programs aimed at improving the cotton yield and fiber quality.

10.3.38 TRANSCRIPTION FACTORS FOR COTTON FIBER DEVELOPMENT

Xia et al. (2013) reviewed progress in studies related to cotton fiber development transcription factors. In different cotton fiber development stages, starting from boll initiation and enduring through secondary cell wall synthesis and maturity transcription factors performs significant regulatory roles. Many transcription factors comprising MYB, HD-ZIP, MADS, and TCP families associated to cotton fiber development have been stated and among them essential proteins involved in this process are MYB transcription factors. Trichome development is controlled by a MYB-bHLH-WD40 protein complex and has been well described at the molecular level. Cotton fiber development and *A. thaliana* leaf trichome differentiation are to be expected to share related regulatory mechanisms because several research studies with cotton fiber-related genes determined a close association between the seed fibers of cotton and leaf trichomes of *Arabidopsis thaliana* and owing to their different regulatory networks cotton fiber development is expected to be controlled by GL1-like MYB genes and MIXTA-like MYB genes further MIXTA-like MYB genes, for example, GhMYB25 and GhMYB25-like also have important regulatory roles, in bringing about the development of cotton fiber. An understanding of these molecular mechanisms of fiber cell differentiation and development has been provided by the wide-ranging research on transcription factors linked to cotton fiber cells.

10.3.39 MOLECULAR MAPPING AND CHARACTERIZATION OF TRAITS CONTROLLING FIBER QUALITY

Kohel et al. (2001) applied DNA-based molecular markers and a polymorphic mapping population for molecular mapping and characterization of traits controlling fiber quality in cotton. TM-1 (*G. hirsutum*) and 3-79 (*G. barbadense*) were crossed and from its interspecific cross-polymorphic mapping population was developed. In extra-long staple (ELS) cotton 3-79 for fiber quality properties 13 QTLs were detected. Among these 13 QTLs four QTLs have an effect on bundle fiber strength, three on fiber length, and six on fiber fineness. In the F_2 population for each fiber quality property about 30–60% of the total phenotypic variance was described collectively by these QTLs positioned on different chromosomes or linkage groups. In TM-1 genetic background 3–79 alleles were used to characterize the effects and modes of action of individual QTLs and the gene mode of these QTLs was found to be more recessive with less additive effect further for fiber fineness transgressive segregation was detected which can be helpful in improving the fiber fineness. In marker-assisted selection of these recessive alleles, the molecular markers linked to fiber quality QTLs would be most effectively used cotton breeding programs.

10.3.40 QTL MAPPING FOR COTTON FIBER QUALITY

Lin et al. (2005) carried out linkage map construction and QTL mapping with SRAP, SSR, and RAPD markers for cotton fiber quality. To work out the genetic basis of cotton fiber traits a genetic linkage map of tetraploid cotton, one of the most widely cultivated species, was generated by means of sequence-related amplified polymorphisms (SRAPs), simple sequence repeats (SSRs), and random amplified polymorphic DNAs (RAPDs). To screen the polymorphisms among *G. hirsutum* cv. Handan 208 and *G. barbadense* cv. Pima90, total of 238 SRAP primer combinations, 368 SSR primer pairs and 600 RAPD primers were used and total 205 SSRs, 107 RAPDs, and 437 SRAPs, viz., 749 polymorphic loci were revealed. "Handan208" × "Pima90" were crossed and using 749 polymorphic markers their 69 F_2 progeny were genotyped. Total 41 linkage groups each with at least three loci were observed from 566 loci. Using SSR markers with known chromosome locations 28 linkage groups were allocated to resultant chromosomes. The complete length of the map was 5141.8 cM with an average inter locus space of 9.08 cM and 135 loci (18.0%) with skewed segregation and with excess maternal parental alleles were detected from a test for significance of deviations from the expected ratio (1:2:1 or 3:1). From nine linkage groups with 16.18–28.92% of the trait variation two QTLs for fiber strength, four for fiber length, and seven for micronaire value constituting total 13 QTLs related with fiber traits were identified among which six QTLs were positioned in the A subgenome, six QTLs in the D subgenome, and one QTL in an unassigned linkage group. In molecular marker-assisted selection, three micronaire value QTLs clustered on LG1 would be very effective for improving this trait.

10.3.41 SALT TOLERANCE

Dinakaran et al. (2012) reviewed yield and fiber quality components in upland cotton *(Gossypium hirsutum L.)* under salinity and conducted an experiment in normal along with saline–alkaline condition to evaluate the 32 popular upland varieties for their salt tolerance. Using average electrical conductivity of 3.10 ds/m and bore well water irrigation saline conditions were created. Number of bolls per plant, boll weight, lint yield per plant, 2.5% span length, leaf area index, Na-K ratio and seed cotton yield exhibited high GCV and genetic gain indicating that simple selection could be used to improve these traits. In both normal and saline alkaline conditions, the high correlation between seed cotton yield and lint yield per plant was revealed from correlation and path analysis studies. Bartlett's rate index with uniformity ratio, 2.5% span was significantly and positively correlated with bundle strength, uniformity ratio with micronaire and elongation percent, specific leaf area with leaf area index. Furthermore, the characters boll weight (−0.347), ginning out turn (−0.528), 2.5% span length (−0.312), and uniformity ratio (−0.440) exhibited high correlation with seed cotton yield suggesting that selection for any one of these traits leads to simultaneous improvement of other traits along with seed cotton yield.

10.3.42 GENETICS OF FIBER QUALITY TRAITS

Ali et al. (2008) crossed five upland cotton varieties, viz., NIAB-78, CIM-499, LSS, RH-112, and NIAB Krishma, in a complete diallel mating system to study the genetics of fiber quality traits such as fiber length (FL), fiber strength (FS), fiber fineness (FF), fiber uniformity (FU), and fiber elongation (FE) using Mather and Jinks approach. For all the characters differences were significant ($P < 0.01$) and for genetic interpretation data of all the characters except FE was found to be partly adequate as revealed from adequacy tests. In all the traits dominant components (H_1 and H_2) of variation for FS and FU were nonsignificant. These were higher in magnitude than additive component (D), firmly supported by the value of $H1/D^{0.5}$. From $H_2/4H_1$ asymmetrical distribution of dominant and recessive genes for all the traits excluding FF in parents was confirmed and dominant genes were more than recessive genes in the parents. FF, FU, and FE exhibited moderately high narrow sense heritability ($h^2_{n.s}$), whereas SL and FS acquired low heritability and except FS for all the characters h^2 value was insignificant. Graphical representation confirmed that SL, FF, and FE were controlled by gene action which was additive in nature, while FS and FU by overdominance effects indicating that for improvement of FS and FU heterosis breeding would be rewarding and for SL, FF, and FE full sib or half sib family selection, pedigree and progeny test may be essential for achievement of genetic progress.

10.3.43 GENOME-SPECIFIC TRANSCRIPTS, TRANSCRIPTION FACTORS, AND PHYTOHORMONAL REGULATORS ACCUMULATION

Yang et al. (2006) considered allotetraploid cotton during early stages of fiber cell development and described the accumulation of genome-specific transcripts, transcription factors, and phytohormonal regulators. For 32,789 high-quality ESTs derived from Texas Marker-1 (TM-1) immature ovules (GH_TMO) of *Gossypium hirsutum* L. computational and expression analyses were reported. The ESTs represented about 15% of the unique sequences and accumulated in 4036 tentative consensus sequences (TCs) and 4504 singletons constituting total 8540 unique sequences in the cotton EST collection. The proportion of genes encoding putative transcription factors MYB and WRKY and genes encoding predicted proteins participating in auxin, brassinosteroid (BR), GA, abscisic acid (ABA), and ethylene signaling pathways was significantly higher in GH TMO ESTs than 178,000 existing ESTs obtained from elongating fibers and non-fiber tissues. During fiber cell initiation the cotton homologs linked to *MIXTA*,

MYB5, *GL2* and eight genes in the auxin, BR, GA, and ethylene pathways were stimulated, while in the naked seed mutant (*N1N1*) they were suppressed. They resulted in impairment of fiber formation. This data was in agreement with the well-recognized phytohormonal effects on fiber cell development in immature cotton ovules cultured in vitro. Furthermore, in cell fate determination an important role of phytohormones was suggested from the stimulation of phytohormonal pathway-related genes before the activation of *MYB*-like genes. In *G. hirsutum* L., an allotetraploid obtained from polyploidization between AA and DD genome species, an outcome reliable with the production of long lint fibers in AA genome species was selectively supplemented with AA subgenome ESTs of all functional classifications together with cell-cycle control and transcription factor activity. These consequences imply overall roles of genome-specific, phytohormonal, and transcriptional gene regulation throughout fiber cell development early stages in cotton allopolyploids.

10.4 CONCLUSION

In improving yield and fiber quality of cotton by the use of appropriate harvesting methods, postharvesting technology is required. In this chapter, an overview of the different cost-effective harvest methods that can be employed without losing yield or quality and the postharvest technique to improve yield and quality and to make the best use of cotton by products is given. Different factors affecting cotton fiber quality and recent advancements made in conventional and molecular breeding techniques that have been used to increase the favorable fiber quality traits in cotton are discussed.

KEYWORDS

- **cotton**
- **harvesting methods**
- **fiber quality**
- **breeding techniques**

REFERENCES

Ali, M. A.; Khan, I. A.; Awan, S. I.; Ali, S.; Niaz, S. Genetics of Fiber Quality Traits in Cotton (*Gossypium hirsutum* L.). *Austr. J. Crop Sci.* **2008,** *2*, 10–17.

Baker, K. D.; Hughs, E.; Foulk, J. Cotton Quality as Affected by Changes in Spindle Speed. *Appl. Eng. Agric.* **2010,** *26* (3), 363–369.

Bange, M. Effects of Climate Change on Cotton Growth and Development. *Austr. Cottongrower* **2007,** *28* (3), 41–45.

Buriev, Z. T.; Shermatov, S. E.; Scheffler, B. E.; Pepper, A. E.; Yu, J. Z.; Kohel, R. J.; Abdukarimo, A. Linkage Disequilibrium Based Association Mapping of Fiber Quality Traits in *G. hirsutum* L. Variety Germplasm. *Genetica* **2009,** *136*, 401–417.

Byler, R. K.; Delhom, C. D. Evaluation of Fiber and Yarn Quality with and Without Seed Cotton Cleaner Material Produced in a Commercial Cotton Gin. *Am. Soc. Agric. Biol. Eng.* **2013,** *29*, 621–625.

Cassman, K. G.; Kerby, T. A.; Roberts, B. A.; Bryant, D. C.; Higashi S. L. Potassium Nutrition Effects on Lint Yield and Fiber Quality of Acala Cotton. *Crop Sci.* **2018,** *30*, 672–677.

Chang Lin, C.; FanTing, K.; Lei, S.; Qing, X.; YuTong, Z.; YongFei, S.; MingSen, H. Comparative Experimental Study on Working Performance of Air-Blowing and Air-Suction Cotton-Boll Separation Device. *Int. Agric. Eng. J.* **2017,** *26*, 75–81.

Choudki, V. M.; Savita, S. G.; Sangannavar, P.; Vamadevaiah, H. M.; Khadi, B. M.; Patil, R. S.; Katageri, I. S. Genetic Improvement of Fiber Strength in Diploid Cotton (*G. herbaceum* L.) Through Interspecific Hybridization Using *G. anomalum* Wild Species. *Crop Res. (Hisar)* **2013,** *45*, 259–267.

Dagdelena, N.; Başalb, H.; Yılmaza, E.; Gürbüza, T.; Akçay, S. Different Drip Irrigation Regimes Affect Cotton Yield, Water Use Efficiency and Fiber Quality in Western Turkey. *Agric. Water Manage.* **2009,** *96*, 111–120.

Dai, J.; Dong, H. Intensive Cotton Farming Technologies in China: Achievements, Challenges and Countermeasures. *Field Crops Res.* **2014,** *155*, 99–110.

Deho, Z. A. Influence of Picking Intervals on Seed-Cotton Yield and Fiber Quality of Local Cotton Varieties. *Pak. J. Agri. Agric. Eng. Vet. Sci.* **2017,** *33*, 194–200.

Desalegn, Z,; Ratanadilok, N.; Kaveeta, R. Correlation and Heritability for Yield and Fiber Quality Parameters of Ethiopian Cotton (*Gossypium hirsutum* L.) Estimated from 15 (diallel) Crosses. *Kasetsart J. Nat. Sci.* **2009,** *43*, 1–11.

Dinakaran, E.; Thirumeni, S.; Paramasivam, K. Yield and Fiber Quality Components Analysis in Upland Cotton (Gossypium hirsutum L.) Under Salinity. *Ann. Biol. Res.* **2012,** *3*, 3910–3915.

Dutt, Y.; Wang, X. D.; Zhu, Y. G.; Li, Y. Breeding for High Yield and Fiber Quality in Coloured Cotton. *Plant Breed.* **2004,** *123*, 145–151.

Diarra, A.; Barbier, B.; Zongo, B.; Yacouba H. Impact of Climate Change on Cotton Production in Burkina Faso. *Afr. J. Agric. Res.* **2017,** *12* (7), 494–501.

Faircloth, J. C.; Hutchinson, R.; Barnett, J.; Paxson, K. An Evaluation of Alternative Cotton Harvesting Methods in Northeast Louisiana—A Comparison of the Brush Stripper and Spindle Harvester. *J. Cotton Sci.* **2004,** *8*, 55–61.

Funk, P. A.; Gamble, G. R. Fiber Properties of Saw and Roller Ginned Naturally Colored Cottons. *J. Cotton Sci.* **2009,** *13* (2), 166–173.

Gérardeaux, E.; Benjamin, N. S.; Palaï, U.; Guiziou, C.; Oettli, P.; Naudin, K. Positive Effect of Climate Change on Cotton in 2050 by CO_2 Enrichment and Conservation Agriculture in Cameroon. *Agron. Sust. Dev.* **2013,** *33*, 485–495.
Ghazi, Y.; Bourot, S.; Arioli, T.; Dennis, E. S.; Llewellyn, D. J. Transcript Profiling During Fiber Development Identifies Pathways in Secondary Metabolism and Cell Wall Structure That May Contribute to Cotton Fiber Quality. *Plant Cell Physiol.* **2009,** *50*, 1364–1381.
Han Lin, C.; Bo Wen, G.; Pennefather, M. P.; Nguyen, T. X.; Nhan Phan Thien; Duong, H. M. Cotton Aerogels and Cotton-Cellulose Aerogels from Environmental Waste for Oil Spillage Cleanup. *Mater. Des.* **2017,** *130*, 452–458.
Hasani, H.; Tabatabaei, S. A.; Grey, G. A. Relational Analysis to Determine the Optimum Process Parameters for Open-End Spinning Yarns. J. *Eng. Fibers Fibrics* **2012,** *7*, 81–86.
Hernandez-Gomez, M. C.; Runavot, J. L.; Meulewaeter, F.; Knox, J. P. Developmental Features of Cotton Fiber Middle Lamellae in Relation to Cell Adhesion and Cell Detachment in Cultivars with Distinct Fiber Qualities. *BMC Plant Biol.* **2017,** *17*, 102–115.
Hughs, S. E; Gray Gamble; Carlos, B.A; Dennis, C.T. Long Term Storage of Polythene Wrapped Cotton Bales and Effects on Fiber and Textile Quality. *J. Cotton Sci.* **2011,** *15*, 127–136.
Hui, G.; Rong, M.; ShuAi, L.; Yang, L.; XinJie, L. Optimization of Optical Measurement Model for Mass Density of Cotton Fiber. *Trans. Chinese Soc. Agric. Eng.* **2012,** *28*, 253–258.
Hüseyin, B.; Ismail, T. Heterosis and Combining Ability for Yield Components and Fiber Quality Parameters in a Half Diallel Cotton (*G. hirsutum* L.) Population. *Turkish J. Agric. Forestry* 2003, *27*, 207–212.
Johnson, A. S.; Landover, J. A.; Fernandez, C. J.; Davidonis, G. H. Cotton Fiber Quality Is Related to Boll Location and Planting Date. *Agron. J.* **2004,** *96*, 42–47.
Kaewprasit, C.; Hequet, E.; Abidi, N.; Gourlot, J. P. Quality Measurements Application of Methylene Blue Adsorption to Cotton Fiber Specific Surface Area Measurement. *J. Cotton Sci.* **1998,** *2*, 164–173.
Kohel, R. J.; Yu, J.; Park, Y. H.; Lazo, G. R. Molecular Mapping and Characterization of Traits Controlling Fiber Quality in Cotton. *Euphytica* **2001,** *121*, 163–172.
Ling, W.; Chang, J. I. Technical Analysis and Expectation for Cotton Harvesting Based on Agricultural Robot. *Cotton Sci.* **2006,** *18*, 124–128.
Long, R. L.; Bange, M. P.; Gordon, S. G.; Sluijs, M. H. J.; Naylor, G. R.S.; Constable, G. A. Fiber Quality and Textile Performance of Some Australian Cotton Genotypes. *Crop Sci.* **2009,** *50*, 1509–1518.
Lin, Z.; He, D.; Zhang, X.; Nie, Y.; Guo, X.; Feng, C. Linkage Map Construction and Mapping QTL for Cotton Fiber Quality Using SRAP, SSR and RAPD. *Plant Breed.* **2005,** *124*, 180–187.
Liu, Y. L.; Gamble, G. R.; Thibodeaux, D. Assessment of Recovered Cotton Fiber and Trash Contents in Lint Cotton Waste by Ultraviolet/Visible/Near-Infrared Reflectance Spectroscopy. *J. Near Infrared Spectrosc.* **2010,** *18*, 239–246.
Morgan, G. D.; Baumann, P. A.; Chandler, J. M. Competitive Impact of Palmer Amaranth (*Amaranthus palmeri*) on Cotton (*Gossypium hirsutum*). *Dev. Yield* **2001,** *15*, 408–412.

Nagarkar, R. D.; Saxena, S.; Ambare, M. G.; Shaikh, A. J. Preparation of Value Added Products from Waste Collected from Cotton Ginneries. *Agric. Mech. Asia. Africa Latin Am.* **2016,** *47* (1), 24–27.

Özarslan, C. PH—Postharvest Technology: Physical Properties of Cotton Seed. *Biosyst. Eng.* **2002,** *83*, 169–174.

Peterson, W.; Kislev, Y. The Cotton Harvester in Retrospect: Labor Displacement or Replacement. *J. Econ. Hist.* **1986,** *46*, 199–216.

Paterson, A. H.; Saranga, Y.; Menz, M.; Jiang, C. X.; Wright, R. QTL Analysis of Genotype × Environment Interactions Affecting Cotton Fiber Quality. *Theoret. Appl. Genet.* **2003,** *106*, 384–396.

Pettigrew, W. T. The Effect of Higher Temperatures on Cotton Lint Yield Production and Fiber Quality. *Crop Sci.* **2007,** *48*, 278–285.

Ton, P. In *Cotton & Climate change: Impacts and Options to Mitigate and Adapt*. Ton Consultancy, Ceramplein 58, 1095 BX Amsterdam, The Netherlands, 2011.

Porter, W. M.; Wanjura, J. D.; Taylor, R. K.; Boman, R. K.; Buser, M. D. Tracking Cotton Fiber Quality and Foreign Matter Through a Stripper Harvester. *J. Cotton Sci.* **2017,** *21*, 29–39.

Pynckels, F.; Kiekens, P.; Sette, S.; Langenhove, L.; Impe, K. Use of Neural Nets for Determining the Spinnability of Fibers. *J. Textile Inst.* **1995,** *86*, 425–437.

Rajput, L. B.; Cui Jin Jie; Zhang Shuai; Luo Jun Yu; Wang ChunVi; Lv Li Min. The Effect of the Different Transgenic Bt Cotton Technology on the Fiber Quality. *J. Basic Appl. Sci.* **2017,** *13*, 166–170.

Saeidy, El Technological Fundamentals Of Briquetting Cotton Stalks as a Biofuel, Gärtnerische Fakultät, 2004, pp. 231–245.

Smith, D. T.; Baker, R. V.; Steele, G. L. Palmer Amaranth (*Amaranthus palmeri*) Impacts on Yield, Harvesting, and Ginning in Dryland Cotton (*Gossypium hirsutum*). *Weed Technol.* **2000,** *14*, 122–126.

Smith, C. W.; Coyle, G. G. Association of Fiber Quality Parameters and Within-Boll Yield Components in Upland Cotton. *Crop Sci.* **1997,** *37*, 1775–1779.

Shamskar, K. R.; Heidari, H.; Rashidi, A. Preparation and Evaluation of Nanocrystalline Cellulose Aerogels from Raw Cotton and Cotton. *Industrial Crops and Products*, **2016,** *93*, 203–211.

Sonewane, L. K.; Sheikh, S. M.; Shahare, A. Dual Powered Operated Tiny Cotton Ginning Machine: A Review. *Int. J. Res. Appl. Sci. Eng. Technol.* **2017,** *5*, 116–119.

Strumillo, J. L.; Cyniak, D.; Czekalski, J.; Jackowski, T. Quality of Cotton Yarns Spun Using Ring-, Compact-, and Rotor-Spinning Machines as a Function of Selected Spinning Process Parameters. *Fibers Textiles Eastern Europe,* **2007,** *60*, 24–30.

Tang, S. R.; Xu, N. Y.; Yang, W. H.; Wei. S. J.; Zhou. Z. G. *Ecological Regionalization of Cotton Fiber Quality in the Northwest Inland Region Using Gge Analysis*; Science Press, Beijing, China, 2016.

Tatineni, V.; Cantrell, R. G.; Davis, D. D. Genetic Diversity in Elite Cotton Germplasm Determined by Morphological Characteristics and RAPDs. *Crop Sci.* **1996,** *36*, 186–192.

Thibodeaux, D. P.; Evans, J. P.; Thibodeaux D. P. Cotton Fiber Maturity by Image Analysis. *Textile Res. J.* **1986,** *51*, 102–111.

Vik, M.; Khan, N.; Vikova, M. LED Utilization in Cotton Color Measurement. *J. Nat. Fibers* **2017,** *14*, 574–585.
Viswanathan, G.; Munshi, V. G.; Ukidve, A. V. A Critical Evaluation of the Relationship Between Fiber Quality Parameters and Hairiness of Cotton Yarns. *Textile Res.* **1989,** *65*, 570–579.
Wang, D.; Yin, B. B.; Liu, X.; He, X. C.; Su, Z. W. Laser Line Scan Imaging Method for Detection of White Foreign Fibers in Cotton. *Trans. Chinese Soc. Agric. Eng.* **2015,** *31*, 9–18.
Wang, Q.; Yu, M; Chen, G.; Chen, Q.; Tai, J. Facile Fabrication Of Superhydrophobic/Superoleophilic Cotton for Highly Efficient Oil/Water Separation. *Bioresoures* **2017,** *12*, 643–654.
Wang, Z. G.; Feng, X. Y.; Wang. H. P. Parameter Optimization and Experiment of Funnel-Shaped Heavy Impurity Separator in Seed Cotton Cleaning Process. *Trans. Chinese Soc. Agric. Eng.* **2016,** *32*, 30–36.
Wen Qing, Z.; YaLi, M.; BingLin, C.; YouHua, W.; WenFeng, L.; ZhiGuo, Z. Effects of Fruiting-Branch Position, Temperature-Light Factors and Nitrogen Rates on Cotton (*Gossypium hirsutum* L.) Fiber Strength Formation. *Scientia Agric. Sinica* **2011,** *44*, 3721–3732.
Wu, X.; Li, Y. E.; Shang, G.; Xiao. X. Progress in Studies on Transcription Factors Related to Cotton Fiber Development. *Cotton Sci.* **2013,** *25*, 269–277.
Xiang, Z.; Da Peng, H.; Yuan, L.; Yuan, C.; Abidallha, E. H. M. A.; ZhaoDi, D.; De Hua, C.; Lei, Z. Developmental and Hormonal Regulation of Fiber Quality in Two Natural-Colored Cotton Cultivars. *J. Integr. Agric.* **2017,** *16*, 1720–1729.
Xiao Feng, H.; GuoXin, W.; Shou Dong, X. Measurement of Cotton Fiber Numbers By Near Infrared Spectroscopy. *Trans. Chinese Soc. Agric. Eng.* **2009,** *25*, 119–123.
Xu, N. Y.; Li. J.; Zhang. G. W.; Zhou. Z. G. Evaluation of Regional Cotton Trial Environments Based on Cotton Fiber Micronaire Selection by Using Gge Biplot Analysis. *Chinese J. Eco-Agric.* **2013,** *21* (10), 1241–1248.
Yang, S. S.; Cheung, F.; Lee, J. J.; Ha, M.; Wei, N. E. Accumulation of Genome-Specific Transcripts, Transcription Factors and Phytohormonal Regulators During Early Stages of Fiber Cell Development in Allotetraploid Cotton. *Plant J.* **2006,** *47*, 761–775.
Yuan, S. N.; Hua, S. J.; Ni. M.; Li. Y. Y.; Wen. G. J.; Shao. M. Y.; Zhang, H. P.; Zhu. S. J.; Wang. X. D. Relationship Between Fiber Macroelement Content and Fiber Quality in Colored Cotton. *Scientia Agric. Sinica* **2010,** *43*, 4169–4175.
Zeidman, M. I.; Batra, S. K.; Sasser, P. E. Determining Short Fiber Content in Cotton. Part I: Some Theoretical Fundamentals. *Textile Res. J.* **1991,** *61*, 106–113.
Zhang, T.; Hu, Y.; Chen, Z. J. Sequencing of Allotetraploid Cotton (*Gossypium hirsutum* L.) acc. TM-1 Provides a Resource for Fiber Improvement. *Nat. Biotechnol.* **2015,** *33*, 531–537.
Zhang, X.; Liu, X. F.; Lü, C. H.; Dong. Z. D.; Chen. Y.; Chen. D. H. Physiological Characteristics Associated with Fiber Development in Two Different Types of Natural Colored-Cotton Cultivars. *Acta Agron. Sinica* **2011,** *37*, 489–495.
Zhang, L.; Xia, S. J.; Ma, B. M.; Liao, X. R.; Hou. X. L. Textile Fibers Prepared by Combined Alkali Soaking, Steam Explosion and Laccase/Mediator Treatments to Bark of Cotton Stalks. *Trans. Chinese Soc. Agric. Eng.* **2015,** *31*, 292–299.

Zhaoe, P.; DongLei, S.; JunLing, S.; ZhongLi, Z.; YinHua, J.; BaoYin, P.; ZhiYing, M.; XiongMing, D. Effects of Fiber Wax and Cellulose Content on Colored Cotton Fiber Quality. *Euphytica* **2010,** *173*, 141–149.

Zhen, D.; Xiu Liang, H. Extraction and Characterization of Fibers with Low Lignin Content From Bark of Cotton Stalks for Spinning. *Trans. Chinese Soc. Agric. Eng.* **2015,** *31*, 309–314.

Zong Ling, R.; Guo Xi, W.; Shou Pu, H.; Xiong Ming, D. Differences in Fiber Quality and Fiber Ultrastructure Between Different Natural Coloured Cotton Lines and Varieties. *Cotton Sci.* **2013,** *25*, 184–188.

CHAPTER 11

Research Advances in Breeding and Biotechology

ABSTRACT

Cotton being an important cash crop, significant advances have been achieved both in breeding and biotechnology which are discussed in this chapter.

11.1 BREEDING

11.1.1 EFFECT OF BT COTTON ON GROWTH PARAMETERS

Dinakaran et al. (2010) reported that Bt. photosynthetic pigments of Bt cotton secrete insecticidal proteins which result in higher basic growth parameters of Bt cotton than non-Bt cotton. Their study aimed at determining the effectiveness of Bt and conventional non-Bt cotton under three changed temperatures with respect to growth parameters, biochemical parameters, enumerating rhizosphere soil microflora, and germination of seed. Similarly, due to the effect of Cry 1 Ac (endotoxin) protein in the rhizosphere soil of Bt cotton microbial population was found to be lower than non-Bt cotton. For biosafety assessment to prove this reduction further study at molecular level needs to be carried out. The Bt and non-Bt cotton both require ambient temperature of 21–30°C for seed germination.

11.1.2 BT COTTON TECHNOLOGY INTRODUCTION IN AFRICA

In May 2000, a teamwork between Monsanto and Burkina Faso's national cotton companies made possible the commercial release of second-generation insect-protected biotech cotton. In 2009, the major biotechnology introduction on the African continent was attained by planting about 125,000 ha in local varieties with Bollgard II from Monsanto Co. by Burkina Faso producers and Burkina Faso has become one of the more advanced and pre-emptive sub-Saharan African countries with respect to biotechnology. In this chapter, Vitale et al. analyzed the impact of Bt cotton commercial release on rural households in Burkina Faso. The data collected from 160 cotton producers indicated that Bt cotton technology is very successful and farmers have realized significant benefits through pesticides reductions, higher effective average yield of 18.2% over conventional cotton, and the profit gains are in a magnitude of $39.00 per ha. Finally, it can be concluded that Bt cotton has created large and sustainable benefits without any significant difference in production costs as the reduction in pesticide costs offsets the increased costs of Bt seeds which contributes to positive economic and social development in Africa.

11.1.3 ORGANIZATION AND EVOLUTION OF COTTON FIBER GENES

Xu et al. (2010) made a comprehensive analysis of configuration and advancement of the fiber improvement genes by constructing a combined genetic and physical map to validate the functions of fiber development genes. For this study, 103 transcription factors of fiber, 259 fiber development genes, and 173 SSR-contained fiber ESTs constituting 535 cotton fiber development genes were examined at the subgenome level. Overall, 499 fiber-related contigs of which 397 contigs attached onto individual chromosomes covering around 151 Mb in physical length or nearly 6.7% of the tetraploid cotton genome were selected and assembled. Results from this study reported that most of the genes defining the cotton fiber development are dispersed through At and Dt subgenomes of tetraploid AD cottons. The outcomes indicated that Dt subgenome have more transcription factors than At, while At subgenome have more fiber development genes compared with Dt. This study proposed a novel functional assumption for tetraploid cotton by mapping additional fiber quantitative traits loci (QTLs) in Dt subgenome than At subgenome. After crossing the two diploid *Gossypium* genomes, the At genome remains alike to its fiber yielding diploid A genome ancestor in function and contributes maximum genes for fiber development; however, Dt subgenome provides more transcription factors due to the non-fiber producing D genome ancestor, and regulate the At subgenome fiber genes expression. Previously published mapping results would explain this assumption. The integrated map of fiber development genes will enable ultimate replication and detection of individual full-length fiber genes and these candidate genes may possibly aid in detailed analysis of the physiological mechanisms involved in fiber differentiation, elongation, and maturation.

11.1.4 TRAINING OF PLANT BREEDING STUDENTS AT TECOMAN WINTER NURSERY

Kothari et al. (2011) stated to acquaint plant breeding students with cotton germplasm possessions diversity contained by National Plant Germplasm System (NPGS), the project was coordinated in which winter nursery cotton germplasm at Tecoman, Mexico, was utilized for plant breeding training and research. To augment the graduate school experience and students' professional development, from Texas A&M University and Texas Tech University, students group in 2009 spent 5 days at the winter nursery in Tecoman, Mexico, where students participated in national cotton gemplasm collection characterization and rouging operation which helped in increasing their proficiency in understanding new phenotypic traits, the usefulness and resource worth of the national germplasm collection, and to get exposed to international travel and unaware cultures.

11.1.5 COMBINING ABILITY AND HETEROSIS STUDIES

Dhamayanthi et al. (2011) crossed two tetraploid cotton species *Gossypium hirsutum* and *G. barbadense* to obtain 49 interspecific F_1 combinations. Evaluation of these crosses along with the parental lines in terms of seed cotton yield, its component traits, and ginning percent aided in the identification of best heterotic combinations. Male lines ICB-260 and Pima S4 and female line BRS-53-53 are identified as best general combiners for seed cotton yield and its per se performance but for majority of the yield components B-4 line was found to be the best. Out of 49 crosses, five crosses having more than 40% heterosis with superior grain yield and yield constituents along with significant specific combining ability (SCA) effect were identified and require further research for profitable utilization of heterosis.

11.1.6 COMPARATIVE STUDY OF BT AND CPTI COTTON

Cui et al. (2011) carried out a comparative study between the transgenic Bt cotton, *Bacillus thuringiensis* (Bt), cowpea trypsin inhibitor gene (CpTI) pyramided cotton, and a conventional nontransgenic variety to determine their efficiency in conferring resistance to cotton bollworm, *Helicoverpa armigera* (Hübner) (Lepidoptera: Noctuidae). The plants were screened under laboratory and field conditions. The results revealed that the Bt + CpTI showed a significant upregulation of defense-related genes compared with Bt cotton and conventional nontransgenic variety. Simultaneous coexpression of both the genes in Bt + CpTI cotton was found to be more efficient in reducing the survival rate of first, second, and third-stage larvae than on Bt cotton. *H. armigera* fed on Bt or Bt + CpTI cotton from the sixth stage onward do not produce offspring's as both types of transgenic restrict the development of sixth-stage larvae to adults. However, both Bt cotton and Bt + CpTI cotton do not differ in seasonal trends in level of resistance in different plant structures and food conversion efficiency. The bollworm densities was found to be greatly inhibited on both Bt and Bt + CpTI cotton compared with conventional nontransgenic variety from the field experiments and the results from laboratory work suggest that pyramided transgenic plants were more competent at restricting the *H. armigera* but superior control under field conditions need to be confirmed.

11.1.7 GENETIC DIVERSITY STUDY USING METROGLYPH ANALYSIS

Haidar et al. (2012) carried out an investigation with an objective to study genetic diversity available in 13 local elite cotton genotypes and two exotic lines for the identification of superior genetically diverse and cotton leaf curl virus (CLCV) tolerant lines using metroglyph analysis. To recognize significance of the genotype with respect to each character index scores were assigned. Metroglyph analysis distributed the cotton genotypes into six clusters, these distributions of genotypes into different clusters suggested the presence of genetic divergence in the germplasm and the crosses between diverse parents belonging to different clusters gives superior recombinants for yield and CLCV tolerance. On the basis of this grouping, six genotypes, that is, FH-1000, CIM-443, NIBGE-3, NIBGE-115, NIBGE-160, and NIBGE-253 tolerant to CLCV disease were identified and the effective hybridization can be initiated with genotypes of diverse groups to produce better segregants which could be used for the production of cotton varieties with high yielding capacity in future.

Drought is one among the main abiotic stresses affecting the plants and has been given a major importance in crop improvement programs. Several approaches are carried out by breeders to obtain drought-tolerant varieties of agronomically important crops but such approaches are limited in cotton. Among various traits having relevance to drought tolerance the traits like roots, water use efficiency, and cellular-level tolerance are important. Since drought is controlled by quantitative genes many cotton species, viz., *G. arboreum* and *G. herbaceum* can be searched for drought-related QTLs and other stress adaptive traits which can be pyramided in background of cultivated cotton. To increase cotton yield and fiber quality under water stress conditions, a combination of traditional breeding, breeding based on molecular markers and genetic engineering are essential in the near future. In this chapter, the prospects of these strategies to achieve drought tolerance without losing yield and fiber quality and other traits which can increase tolerance are reviewed. It accentuates the need of knowing the ecophysiology of cotton genome structure and quantitative and association genetics study for crop improvement.

11.1.8 POTENTIAL BENEFITS AND PROBLEMS OF BT COTTON

Despite rapid and widespread adoption of Bt cotton by most of the farmers, the benefits of genetically modified (GM) cotton continue to be disputed. In the present study, Mukherjee and Neeta (2012) focused on the yield and economic performance of Bt cotton and non-Bt cotton so as to recognize the potential benefits of Bt cotton and problems related to its cultivation in Gujarat and the nation as a whole. As health effects are major issue in case of genetically modified crops, an investigation was carried out with nearly 100 Agro Products dealers, Bt cotton cultivating 200 farmers, and conventional cotton cultivating 200 farmers as sample to know the influence of GM crops on environment, health, and the soil. The results showed the primary benefits of Bt cotton over conventional cotton has been decreased due to usage of insecticides, more actual yield, and considerably greater profits increasing the living standard of farmers. However, the negative effects of Bt cotton health, environment, and soil, that is, Bt cotton may cause some allergy to farmers and it causes reduction in soil fertility which cannot be refused.

11.1.9 CRY 1AC AND CRY 2AB GENES AGAINST SPODOPTERA LITTORALIS

Cry 1Ac and Cry 2Ab genes from Bt are transferred to the American cotton *Gossypium hirsutum* using gene particle gun method during the coordinate project between Monsanto company and Egyptian Ministry of Agriculture, Agricultural Research Center (ARC) including Cotton Research Institute (CRI), Agricultural Genetic Engineering Research Institute (AGERI), and Plant Protection Research Institute (PPRI). Afterward by crossing American cotton and the Egyptian cotton varieties, the two genes were transferred to the three Egyptian cotton varieties. Dahi (2012) performed a study on these three GM Egyptian cotton varieties *Gossypium barbadense* L. (Giza 80, Giza 90, and Giza 89) to assess their effect against *Spodoptera littoralis* (Boisd.). The results revealed that Cry 1Ac and Cry 2Ab genes of *Bacillus thuringeinsis* affects the various biological features like larval duration, pupal weight, pupal duration, emergence percentage, malformed adult percentage, male and female longevity, and sex ratio of *S. littoralis*, increasing the mortality percent to 97.7, 97.7, and 99.0% for Giza 80, Giza 90, and Giza 89, respectively. Also the egg laying capacity and egg fertility percent for female moths ensued from larvae nurtured on Bt cotton was reduced.

11.1.10 CORRELATION AND PATH ANALYSIS

Alkuddsi et al. (2013) performed correlation and path coefficient analysis studies to elucidate the nature of interrelationship among different characters affecting cotton yield in intra-*hirsutum* cotton hybrids at the Agricultural Research Station, Bavikere, UAS, and Bangalore with two replications in a randomized complete block design (RCBD). Six *hirsutum* non-Bt lines (RAH 318, RAH 243, RAH 128, RAH 146, RAH 97, and RAH 124) and eight *hirsutum* non-Bt testers (SC 14, SC 18, SC 7, SC 68, RGR 32, RGR 24, RGR 58, and RGR 37) were used in the experiment. From the correlation studies, seed cotton yield per plant was significantly and positively correlated with days to 50% flowering (0.359), plant height (0.443), bolls per plant (0.590), mean boll weight (0.422), and ginning percent (0.379), while seed index exhibited significantly negative correlation with yield per plant at genotypic level. Similarly significant and positive phenotypic correlations were located among seed cotton yield per plant and mean boll weight (0.327). The path coefficient analysis revealed that highest direct effect on seed cotton yield was exhibited by monopodia per plant (0.619) and mean boll weight (0.321) on seed cotton yield, whereas high positive indirect effect on seed cotton yield was showed by sympodia per plant via seed index (0.374). The results of this study revealed that the high and low positive correlation with seed cotton yield was exhibited by mean boll weight and sympodia per plant, respectively.

11.1.11 THEORETICAL PARAMETERS FOR BREEDING HYBRID LINE WITH HIGH PHOTOSYNTHETIC EFFICIENCY

Feng et al. (2013) identified theoretical parameters for breeding hybrid line with high photosynthetic efficiency and to improve utilization of F_2 in hybrid cotton. By measuring the leaf area index (LAI), leaf inclination angle (MTA), and light interception rate (LIR) at different growing periods in two cotton hybrids in a field experiment, the canopy structure and matter production features of F_1, F_2, and their parents were estimated. The outcomes unveiled that the F_1 in two cotton hybrids exhibited over-parent heterosis for LAI and mid-parent heterosis for LIR while MTA was influenced by paternal inheritance. The photosynthate accumulation of F_1 in two cotton hybrids was mainly affected by the numerical value of their parents and over-parent heterosis, whereas in F_2 of two cotton hybrids the canopy structure and photosynthate accumulation was mainly influenced by the related parameters of F_1. Therefore, to produce cotton hybrids with improved photosynthetic performance and increased light use efficiency for augmenting the yield potential further along with choosing the parents based on optimized canopy structure, we should make a well combination between father plant with larger MTA and mother plant with larger LAI.

Two introduced cotton species, Pima cotton (*G. barbadense*) and upland cotton (*G. hirsutum*), were acclimatized on the main Hawaiian Islands. These species were able to hybridize and produce fertile hybrids with native Hawaiian cotton *G. tomentosum* which is a major global risk to native floras. Lehman et al. (2014) evaluated the continuance of hybrids among alien Pima cotton, *Gossypium barbadense* (Malvaceae) and endemic Hawaiian cotton, *G. tomentosum*, in Hawaii using morphological and molecular (microsatellite markers) techniques and verified a herbarium which designated that hybrids occurred at an adjoining sites until as late as 1980. However, they could not relocate original hybrid populations indicating that hybrid plants have not continued and no recent gene flow has taken place.

11.1.12 GENETIC MALE STERILITY STUDY

On the basis of data from literature, genetic male sterility (GMS) found to be a best, economical, and alternative method for hybrid seed production technique in cotton and especially in diploid. However, the diverse features of GMS, viz., its development and utilization in producing hybrid seeds, constraints involved in its utilization, physiological and biochemical indices related to GMS, its inheritance, genetic effects of heterosis, cytological aspects of microsporogenesis breakdown, and environmental effects on expression of GMS need to be studied in future line of work. In this study, Mehetre (2015) reviewed the problems associated with hybrid seed production in upland and cultivated diploid cottons using various male sterility systems. They confirmed that the cost of seed production can be reduced with increased purity of seeds with the use of GMS, photoperiod sensitive genic male sterility (PGMS), thermosensitive genic male sterility (TGMS), and environmental male sterilities (EGMS) lines in *G. arboreum* as this method avoids laborious emasculation process adding to the production of hybrid seed.

11.1.13 EFFECT OF IRRADIATION TREATMENT

Haidar et al. (2016) crossed a local variety NIAB-78 with REBA-288 exotic line with pollen irradiated with gamma rays at 10 Gray (Gy) before cross-pollination to generate novel genetic variability and pick out the required new cotton mutants. For evaluating the effect of irradiation treatment and induction of mutations different generations were raised and significant variations from control/parents were detected. From M_1 and M_2 populations, mutants with desirable traits were selected and evaluated in M_3–M_6 generations. Finally, an elite mutant M-7/09 with early maturity, high yield, short stature, good boll bearing, and better tolerance to Burewala strain (CLCuV-B) CLCV virus disease was selected and named as NIAB-2008.

Consequently, it was concluded that low dose pollen irradiation technique is economical and efficiently induces mutations to improve the yield and its component traits, fiber quality, and disease tolerance in cotton.

11.1.14 BROWN AND WHITE LINT CULTIVARS

Bunpet et al. (2009) crossed two white lint cultivars, Takfa 2 and Kaset 2, as female parent and three brown lints cultivars B624, B651, and New Hairy Brown and 1 green lint cultivar as male parent to get F_1 hybrid seeds and to get F_2 seeds the F_1 seeds are selfed. The ratio among white lint cottons with brown lint cottons and white lint cottons with green lint cottons were compared using F_1 hybrid cross-seeds and F_2 hybrid seeds. Statistical analysis by Chi-square test was carried out and the results indicated that the white lint and green lint cotton plants F_2 hybrid phenotype ratio was not 1:3, whereas in white lint and brown lint F_2 hybrids the dominant gene exhibited complete dominance over the recessive gene and the phenotype ratio was 1:3 which is in agreement with Mandel's law of segregation.

11.1.15 COMBINING ABILITY AND NATURE OF GENE ACTION

Vekariya et al. (2017) made an effort to identify good parents and hybrids based on combining ability and nature of gene action governing various traits. Based on SCA, the hybrids DGMS 2 × HD 528, DGMS 1 × HD 432, DGMS 34 × HD 517, and DGMS 34 × HD 523 were superior for seed cotton yield and yield related traits, while the hybrids DGMS 34 × HD 517, DGMS 34 × HD 523, and DGMS 2 × HD 528 exhibited superior performance in terms of number of bolls, boll weight, and seed cotton yield and other characters. Based on the GCA effect for seed cotton yield, number of bolls, boll weight, number of monopods, and days to first flower the best general combiner was male parent HD 517, whereas for ginning out turn, boll weight, number of seeds per boll, and days to first flower for earliness HD 534 was identified as good combiner and for seed cotton yield, number of bolls and number of seeds per boll the female line DGMS 34 was good general combiner. Predominance of nonadditive gene action for all the traits was studied based on the extent of GCA and SCA variances, suggesting that the hybrid breeding need to be followed for utilization of yield advantage.

11.1.16 OPTIMUM PLANT GEOMETRY AND FERTILITY LEVELS

Meena and Kumhar (2017) performed a study at Agricultural Research Station, Borwat Farm, Banswara to find out the optimum plant geometry and fertility levels for interspecific cotton hybrids during Kharif 2010. Three cotton hybrids, viz., JKCHB-214, RAHB-170, and DCH-32 were planted using 90 × 60 and 90 × 45 cm plant geometries and three 75, 100, and 125% levels of RDF were applied. The results showed the increase in seed cotton yield under closer spacing, whereas other yield attributing parameters for instance bolls per plant and boll weight were better under wider spacing. Comparable seed cotton yield was obtained with 100% RDF (1555 kg ha^{-1}) and 125% RDF (1602 kg ha^{-1}) application, however under the definite agroclimatic zone IV b of Rajasthan to get higher yield from interspecific hybrid cotton the combination of RDF 75% and 90 × 60 cm plant geometry appeared to be best.

11.1.17 CORRELATION AND PATH COEFFICIENT ANALYSIS

Reddy et al. (2017) conducted a study with 55 genotypes of upland cotton to know the interrelationship between yield and yield contributing characters by means of correlation and path coefficient analysis. At both phenotypic and genotypic levels, seed cotton yield/plant exhibited significant and positive correlation with plant height, bolls/plant, boll weight, 2.5% span length, bundle strength, and lint yield/plant. Path analysis indicated the bolls/plant; boll weight and lint yield/plant exhibits high and direct effect on seed cotton yield/plant. In breeding to improve the seed cotton yield/plant, the plants with more bolls per plant and high boll weight should be selected.

11.1.18 COMBINING ABILITY AND HETEROSIS STUDY

Pundir et al. (2017) crossed 4 lines and 10 testers following line × tester mating design to produce 40 hybrids of desi cotton (*G. arboreum* L.) and studied combining ability and heterosis for seed cotton yield and its contributing traits. The genotypes GMS 1, GAK 20, HD 450, HD 432, and HD 324 were found with high mean performance and GCA effect for seed cotton yield and other traits and categorized as good general combiners. The hybrids GAK 20 × HD 432, GMS 1 × HD 432, and GMS 1 × GCD 22 contains good general combiners as parents and recorded high significant heterobeltosis and SCA effects with high seed cotton yield.

11.1.19 GENETIC DIVERSITY STUDY

The wild cotton accessions contain vast genetic diversity for important characteristics like fiber quality or disease resistance to microorganisms compared with elite cotton germplasm, therefore secondary or tertiary gene pools were being searched by the cotton breeders for novel alleles and wild Australian *Gossypium* species (tertiary gene pool) is a complementary basis for new alleles. To trace the introgression of exotic genes into cultivated cotton, chromosome-specific molecular markers can be used. Lopez-Lavalle et al. (2011) used 114 genotypes with 291 AFLP loci from F_2 population to generate Australian wild C-genome species *Gossypium sturtianum* genetic linkage map. The map length was 1697 cM with 5.8 cM average distance between markers. Using information from cultivated mapped marker 29 SSR and RFLP-STS markers were allocated to chromosomes and 51 AFLP primer pairs and 38 RFLP-STS and 115 SSR mapped markers of cotton showed the polymorphism at the whole structural level representing the linkage group similarities between *G. sturtianum* and A and D subgenomes of cotton. The study indicated that in future to associate the *G. sturtianum* genetic map with the nurtured species comparative approach can be used which is a useful source of markers.

11.2 BIOTECHNOLOGY

11.2.1 COTTON GENOME STRUCTURE

Yang et al. (2017) made a study on the research progress of cotton genome structure. They reported that a complicated polyploidization process has occurred in cotton during its evolution which made the genomic structure of cotton complicated and in recent years the evolution of cotton at genomic level and its subgenomic groups have been distinguished with the development of sequencing technology and through other research advancements including cotton whole genome doubling, comparative genomics, and evolution of cotton genome.

11.2.2 MICRO (MI) RNAS AGAINST CLCUBV

Shweta et al. (2018) found in cotton the resistance to destructive cotton leaf curl disease caused by cotton leaf curl Burewala virus (CLCuBV, genus Begomovirus) and can be developed using micro (mi) RNAs and using translational inhibition or cleavage of viral mRNA genes, the efficiency of these miRNA targets against CLCuBV have been verified. In this chapter, the ghr-miR168 was identified as the most effective miRNAs among number of potential miRNAs capable of targeting C1, C3, C4, V1, and V2 vital genes of CLCuBV genome detected on the basis of threshold free energy and highest complementarity scores. Furthermore, it was found that C1 and C4 genes overlapping transcripts of virus were targeted with ghr-miR395a and ghr-miR395 miRNAs.

11.2.3 GENOMIC VARIATION OF TETRAPLOID COTTON

For improving the diversity of cotton and for utilization of exceptional characteristics of wild cotton, identifying the tetraploid cotton genomic variation is essential. Shen et al. (2017) using specific length amplified fragment sequencing (SLAF-seq) detected 139,176 high-quality DNA polymorphisms, 111,795,823 reads, and 467,735 specific length amplified fragment (SLAF) tags. Among the reference genome (TM-1) and the five species of tetraploid cotton 132,880 SNPs and 6296 InDels were identified and phylogenetic trees revealing high concurrence to the phylogeny of diploid and polyploid cottons was reconstructed using these new data sets. With the help of single-strand conformation polymorphism (SSCP) method some SNPs and InDels can be useful in introgression genetics and breeding with *G. hirsutum* cv. Emian22 as the recipient parent and the other species as donor parent have been validated.

11.2.4 TRANSGENIC COTTON WITH HIGH RESISTANCE TO GLYPHOSATE

Zhang et al. (2017) developed transgenic cotton with high glyphosate resistance. A new gene G2-*aroA* encoding 5-enolpyruvylshikimate-3-phosphate synthase (EPSPS) using *Agrobacterium* was incorporated into cotton variety K312. The transgenic cotton plants regenerated from callus tissue culture using kanamycin selection were confirmed by polymerase chain reaction (PCR) and Southern and Western blot analyses and results showed the target gene gets incorporated into the chromosome of cotton and express successfully at the protein level making the plants tolerant to glyphosate.

11.2.5 SWEET GENE FAMILY OF COTTON

Li et al. (2018) evaluated SWEET gene family of cotton and detected 55 putative *G. hirsutum* SWEET genes based on gene structural features and phylogenetic analysis, the GhSWEET genes were categorized into four clades. Extension of the cotton SWEET gene family in *Gossypium arboreum*, *Gossypium raimondii,* and *G. hirsutum* particularly in Clade III and IV is mainly due to the tandem duplications, polyploidation, and whole genome duplication as indicated by chromosomal localization and homologous genes analysis. Artificial selection, expression profiles, and promoter regions cis-acting regulatory elements analysis confirmed that the SWEET genes contains MtN3_saliva domain and are sugar efflux transporters affecting plant improvement and responses to biotic and abiotic stresses.

11.2.6 GENETIC DIVERSITY STUDY USING SSR PRIMERS

Zhao et al. (2013) screened 17 accessions of low gossypol cotton and three accessions of regular cotton germplasm resources to study genetic diversity using 73 core primer pairs from 400 pairs of cotton SSR primers. Two to nine alleles, with the average of 3.24, were amplified by primers and the polymorphic information content (PIC) was about 0.0905–0.7564, with the average of 0.399. The kernel gossypol content is a quantitative trait controlled by many genes and both presence of gland and kernel gossypol content should be considered simultaneously during breeding. The low gossypol cotton contains 10 times less gossypol than regular cotton. The 20 accessions were divided into three groups, with low gossypol germplasm in each group during cluster analysis and results showed the accession Suyan 606 (1.97 mg/kg) contains lowest amount of gossypol and the highest gossypol content was found in Jimian 27 (380.00 mg/kg). Furthermore, association analysis revealed 34 marker-trait associations on 15 chromosomes of cotton and finally concluded that by using low gossypol germplasm resources in China with abundant genetic variations cotton cultivars with characteristics like bollworm resistance, low gossypol, and excellent agronomical characters can be bred in future breeding programs.

11.2.7 POTASSIUM TRANSPORTER GENES KUP/HAK/KT IDENTIFICATION

Identification and cloning of potassium transporter genes KUP/HAK/KT perform a key role in the absorption of K+ aids in breeding varieties with high potassium efficiency in cotton. Chao et al. (2018) amplified CDS sequence of GhHAK5 transporter gene from upland cotton variety Baimian 1 by homologous cloning. The molecular weight of the protein was 91.23 kD with 816 amino acids residues, its isoelectric point was 8.15. The sequence was 2451 bp long containing KUP/HAK/KT gene family symbolic amino acid sequence GXXXGDXXXSPLY and a conserved "K-trans" (Pfam02705) domain phylogenetic tree analysis revealed that GhHAK5 along with AtHAK5 and OsHAK5 belongs to cluster I with close association. GhHAK5 is a consistent potassium transporter involved in K+ uptake, its expression is high in roots and low in leaf, stem, petal, fiber, and sepal and it is found in plasma membrane as confirmed by subcellular localization experiment.

11.2.8 CLONING AND EXPRESSION ANALYSIS OF NEW LEAF-SPECIFIC PROMOTER

Si et al. (2018) carried out a cloning and expression analysis of leaf-specific promoter (LSP), a new LSP which helps in the localization and expression of other exogenous genes in cotton leaves. The use of such promoters in improving traits in cotton using genetic engineering reduces the physiological side effects. By using cDNA microarray and RT-PCR screening leaf-specific high activity expression gene was detected and using genome walking a 2 kb DNA fragment with upstream promoter sequence was acquired and named as LSP. Further to study its function the promoter-driven GUS pGh10424pro: GUS plant expression vector was transformed by *Agrobacterium tumefaciens* into Arabidopsis and analyzed by GUS staining which showed that this gene was mainly expressed in leaves.

11.2.9 EFFICIENCY OF CRISPR/CAS9

Janga et al. (2017) studied efficiency of clustered, regularly interspaced, short palindromic repeats (CRISPR)/CRISPR associated (Cas) 9 protein system, the most simple and effective tool for genome editing in eukaryotic cells to target a gene within the genome of cotton. A transgenic cotton line with an integrated green fluorescent protein (GFP) into its genome was generated, the cells that had experienced targeted mutations with CRISPR/Cas9 activity were identified by the loss of GFP fluorescence. Furthermore, three independent sg RNAs-guided Cas9-mediated cleavage provided the examples of different types of indels of the GFP gene. In this study, CRISPR/Cas9 was explained for the first time in cotton genome and gave useful information of important native genes which can be potential targets to improve cotton plant traits in further breeding programs.

11.2.10 NF-YB TRANSCRIPTION FACTOR

Chen et al. (2018) reported constituents of the NF-YB transcription factor in *Gossypium hirsutum* L. They have significant roles in varied processes, for instance, seed development, drought tolerance, and flowering time correlated with plant growth and development and many events of duplication which happened over the course of evolution were the major force for upland cotton NF-YB gene extension. In *G. arboretum*, *G. raimondii*, *G. hirsutum*, *Arabidopsis thaliana,* cacao, rice, and sorghum, a systematic phylogenetic analysis was carried out and 150 NF-YB genes were separated into five clades (α-ε). Of which α was the biggest clade, and γ comprises the LEC1 type NF-YB proteins. The extended group of NF-YB genes in *G. hirsutum* was due to the segmental duplication within the A subgenome (At) and D subgenome (Dt) and good collinearity was found between paralogues of NF-YB genes via Syntenic analyses. Through quantitative real-time PCR (qRT-PCR) analysis it was established that the majority of NF-YB genes contained one exon and the same motif patterns were shown by the genes from same clade and most of the NF-YB genes were expressed universally and merely a small number of genes have a tissue-specific expression like GhDNF-YB22 gene which is found to be predominantly expressed in embryonic tissues and have an effect on embryogenesis in cotton. The primary broad description of the GhNF-YB gene family in cotton was made in this study.

11.2.11 CHROMOSOME SEGMENT SUBSTITUTION LINES (CSSLS) AND TRANSCRIPTOME ANALYSIS

Lu et al. (2018) through their chromosome segment substitution lines (CSSLs) and transcriptome analysis reported that the XLOC_036333 [mannosyl-oligosaccharide-a-mannosidase (MNS1)], XLOC_029945 (FLA8), and XLOC_075372 (snakin-1) are the key genes involved in the cotton fiber strength regulation and can be used as candidate genes to study the fiber strength development molecular mechanisms and to improve the cotton fiber quality via molecular breeding in future. High-yielding upland cotton cultivar CCRI45 was crossed with a Sea Island cotton cultivar Hai1 with better fiber quality to form CSSL population. For further analysis, one CSSL with lesser fiber strength than CCRI45 (MBI7285) and two CSSLs having greater fiber strength compared with CCRI45 (MBI7747 and MBI7561) were selected and after anthesis they were sequenced at four different points for all four transcriptomes and based on their function 44,678 genes were identified and clustered. Finally, from MBI7747 and MBI7561 high quality CSSLs and low quality CSSL (MBI7285), 2200 common differentially expressed genes (DEGs) related with a number of fiber strength metabolic pathways, that is, upregulated DEGs involved in polysaccharide metabolic regulation, localization of single organism, organization of cell wall, and biogenesis and downregulated DEGs involved in regulation of microtubule, the cellular response to stress, and the cell cycle were recognized.

11.2.12 GENOMICS ANALYSIS OF MYB FAMILY TRANSCRIPTION FACTORS

Salih et al. (2016) through their comparative genomics analysis reported that MYB family transcription factors plays varied roles in cotton growth and evolution and provided ability to infer the MYB regulatory networks which may help in developing new breeding methods to improve cotton fiber development. During cotton fiber development a complete genome-wide characterization and expression analysis of the MYB transcription factor in *Gossypium hirsutum* was conducted. Out of 1986 MYB and MYB-associated putative proteins 524 non-redundant cotton MYB genes were distinguished and categorized into four subfamilies including 1R-MYB, 2R-MYB, 3R-MYB, and 4R-MYB and divided into 16 subgroups using phylogenetic tree analysis, indicating that most of the (69.1%) GhMYBs genes are included in the 2R-MYB subfamily in upland cotton.

11.2.13 GENOMIC INFORMATION OF COTTON APHID-BORNE VIRUS

Zhang et al. (2014) undertook a study to augment the genomic information of cotton aphid-borne virus which would help in further studying the mechanism of cotton aphid transmission of plant viruses using high-throughput sequencing. From seedling and mid-summer growth stages of cotton aphid virus gene sequences were extracted, identified, and classified through high-throughput sequencing technology. From two transcriptomes of cotton aphid, total 18 kinds of viral genome sequences were acquired of which 17 virus genomic information had never been stated and 13 kinds were positive strand RNA virus types. The mid-summer aphids were found to contain lower FPKM value with only eight kinds of viruses, whereas the seedling aphids carried all the 18 kinds of viruses with much higher types and number of viruses.

11.2.14 PLANT AQUAPORINS

The large major intrinsic protein (MIP) family contains plant aquaporins and includes five subfamilies, namely, small basic intrinsic proteins (SIP), plasma membrane intrinsic proteins (PIP), NOD26-like intrinsic proteins (NIP), tonoplast intrinsic proteins (TIP), and the newly determined X intrinsic proteins (XIP). They help to transport water and other small molecules across cell membranes of cotton and other plants. In the present study, Park et al. (2010) made a thorough detection of 71 aquaporin genes in upland cotton (*G. hirsutum*) using molecular cloning and bioinformatic homology search results showed that the cotton aquaporins contains highly similar 28 PIP and 23 TIP members and 12 NIP and 7 SIP diverse members and from a distinct 5th subfamily one XIP member was found. With semi-quantitative reverse transcription (RT)-PCR expression analyses was conducted to study the aquaporin genes physiological roles in cotton which indicated the high sequence resemblance and varied functions of many cotton aquaporin genes. Therefore, in cotton, to change the water use properties, the aquaporin genes recognized in this study can be used as potential targets.

11.2.15 SSR MARKERS IN COTTON GENOME STUDY

More than 9000 SSR markers have been mapped on the cotton chromosomes in the past 15 years which are important resource for cotton genetics and breeding. The analysis of these SSR markers helps in studying the architecture of the cotton genome. In this chapter, Lin et al. (2010) studied cotton chromosomes to know the arrangement of SSRs mapped on them. The findings revealed that the different predominant SSR types were provided from different sources which were allocated unevenly with C5, C11, and C19 having more SSRs and C02, C04 chromosomes with less SSRs in cotton which helps to supplement the genetic map of cotton with motif-specific SSRs. For AT/TA of di-, tetra-, penta-, and hexanucleotide, novel SSR motifs could be created from EST-SSRs compared with enriched libraries and BAC SSRs.

11.2.16 MOLECULAR MECHANISM IN THE FORMATION OF BROWN COTTON

In this chapter, brown fiber Asian cotton (*Gossypium arboreum*) Cixizimian and white fiber Yuyaozhongmian were selected to study the brown pigment formation and accumulation molecular mechanism in brown cotton considered as ecologically environmental protection cotton. With sequencing and bioinformatics analysis, GaTT12a (Gen Bank accession number: JX013908) gene of one cDNA 500 bp-expressed fragment of brown cotton fiber was cloned with the RT-PCR method. The analysis results revealed that the gene belongs to MATE super-gene family, its molecular weight is 52.7 kD and encodes 490 amino acid residues. The gene was found to be 75% homologous in amino acid sequence with brown pigment synthesizing testa gene of *Arabidopsis thaliana* and exhibited superdominant expression both in the brown cotton fiber, testa of brown and white cotton, but not detected in white cotton fiber, indicating that the GaTT12a gene has same synthesis metabolic pathway in both testa and fiber and included in the synthesis of brown pigment in cotton fiber.

11.2.17 BIBAC LIBRARY IN COTTON

Lee et al. (2013) verified the first BIBAC library in cotton and its related species in terms of its utility and quality by studying the genes for fatty acid metabolism in seed, development of fiber, biosynthesis of fiber cellulose, cotton–nematode interaction, and bacterial blight resistance which is an efficient tool for integrative physical mapping, detailed genome sequencing, and extensive functional analysis of the upland cotton genome. Randomly, 10,000 BIBAC ends (BESs) from the library, that is, one BESs for each 250 kb length of upland cotton genome was selected and sequenced. The results revealed that the genome contains over 77% transposable elements with predominating retro element Gypsy/DIRS1 family. About 1006 new SSRs out of total 1269 were identified supplying surplus markers for cotton genome research. Furthermore, comparative analysis exhibited that *G. raimondii* comprises a D genome (D5) but D and A subgenomes of upland cotton do not have any significant differences between them and *G. raimondii* genome correlation and at the genome sequence level the upland cotton is more divergent compared with *G. raimondii.*

11.2.18 SSR MARKERS TO IDENTIFY THE GENETIC PURITY

Liu et al. (2013) verified the use of SSR markers to identify the genetic purity and stability of cotton varieties and screened 12 conventional *Gossypium hirsutum* cotton genotypes using 78 pairs of core SSR primers. At the molecular level, three setups of non-homozygous SSR alleles were formed by comparing SSR loci of all tested cotton varieties. The genetic purity of three varieties, viz., C5, C9, and C11 was above 98% and two varieties C3 and C6 showed genetic purity of 67.31% and 31.79%, out of 12 genotypes screened indicating the effect of SSR markers on genetic purity of cotton.

11.2.19 RAV GENE FAMILY IN COTTON

Lu et al. (2014) carried out an RAV gene family genome-wide analysis in cotton. They conducted RT-PCR and phylogenetic analysis, the results of RT-PCR under á *Verticillium dahliaeá* stress showed that RAV genes of *G.*

hirsutum are related to stress response. The phylogenetic analysis divided the plantáRAVáproteins into four groups indicating the involvement of cottoná RAVá gene in one specific whole genome duplication of cotton, suggesting that the 10 RAV genes may be present in the cotton diploid ancestor and highly conserved with 4, 5, 8, and 9 chromosomes of D5 subgenome of *Gossypium raimondii* and additional 10áRAVágenes were observed in the A2 subgenome of *Gossypium arboreum*. The differential expression of áRAVá genes in various tissues of *Gossypium hirsutum* was indicated by Blast search of the NCBI EST and unigene database.

Bajwa et al. (2015) conducted a 3-year field experiment to identify the gene influencing cotton fiber growth and quality. With an *Agrobacterium*-mediated gene transformation, *Gossypium hirsutum* GhEXPA8 fiber expansion gene is transferred into a local variety NIAB 846 and a neomycin phosphor transferase (NPTII) was used as a selection marker to screen these transgenic plants. Southern blot analyses and real-time PCR verified its integration. Transgenic cotton fiber cellulose contents was measured by using cellulose assay and the transgenic plants with GhEXPA8 gene exhibited higher fiber lengths and micronaire values than control *G. hirsutum* NIAB 846 variety, indicating that the genetic modifications can be used to improve cotton fiber length and quality.

11.2.20 GENETIC DIVERSITY USING SSR MARKERS

Abbas et al. (2015) conducted disease-screening field experiments to estimate genetic diversity and to detect DNA markers particularly simple sequence repeats (SSRs) related to cotton leaf curl disease resistance. From National Institute for Biotechnology and Genetic Engineering, Faisalabad, Pakistan, five highly tolerant, four highly susceptible, and one immune cotton genotypes of varied origin were chosen and screened with 322 SSRs acquired from *Gossypium raimondii* bacterial artificial chromosome end sequences. The genetic similarity was 81.7–98.7% and only 65 polymorphic primer pairs were identified. Using unweighted pair-group method with arithmetic means (UPGMA) analysis, the dendrogram was made which divided the genotypes into two clusters with tolerant and susceptible genotypes separately. Only PR-91 and CM-43 SSR markers were amplified in tolerant genotypes indicating their correlation with disease resistance.

11.2.21 RNA INTERFERENCE (RNAI) IN BOLL WORM CONTROL

Tian et al. (2015) investigated the use RNA interference (RNAi) in boll worm control. In the juvenile hormone (JH) synthesis mevalonate pathway of cotton bollworm, a rate-limiting enzymatic reaction catalyzing reductase (HMGR) gene coenzyme 3-hydroxy-3-methylglutaryl was targeted by transferring double-stranded RNAs (dsRNA) using *Agrobacterium tumefaciens*-mediated transformation. RT-PCR and qRT-PCR analysis showed the transgenic cotton lines have very high dsHMGR transcriptional level and larvae feeding on these transgenic lines exhibited 80.68% less transcription of HMGR gene compared with larvae fed on wild type. Furthermore, the embryo development nourishment source gene vitellogenin expression was reduced by 76.86%, indicating the dsHMGR transgenic plant reduces the net weight gain besides interrupting the cotton bollworm larvae growth.

11.2.22 BT AND CONVENTIONAL COTTON LEAF ANATOMY

Sundaramurthy (2015) studied the effect of the bollworm-resistant alien gene cry1Ac construct from *Bacillus thuringiensis* Berliner variety kurstaki on the leaf anatomy of a cotton hybrid. The leaf anatomy of both Bt and conventional cotton was compared using light microscope method. The results revealed that the Bt-cotton hybrid contains accessory vascular bundles with more number of xylem vessels around main vascular strands and two-fold higher density of stomata with enlarged guard cells on the adaxial surface of leaf. Compared with the conventional cotton, the size of the cells in transgenic cotton was reduced in size, indicating that these changes improves the water and nutrients uptake from the soil which augments the photosynthesis to support growth, seed cotton yield, and fiber development.

11.2.23 TRANSGENIC RNAI COTTON PLANTS

Ni et al. (2017) by focusing on the suppression of juvenile hormone synthesizing JH acid methyl transferase (JHAMT) and JH transporting JH-binding protein (JHBP) to organs designed two forms of transgenic RNAi cotton plants generating double-stranded RNA (dsRNA) against

Helicoverpa armigera, the universal lepidopteran pest and on two controls, Bt cotton, the Bt-resistant strain larvae and associated susceptible strain larvae were verified, two types of RNAi cotton (targeting JHAMT or JHBP) and two pyramids (Bt cotton plus each type of RNAi) showed that both Bt cotton and RNAi acted separately against susceptible strains and both RNAi cotton and Bt + RNAi pyramid effective against Bt-resistant insects indicating pyramided cotton substantially delays resistance against *Helicoverpa armigera* compared with Bt cotton alone.

11.2.24 VIP3ACAA AND CRY1AC COMBINATION AGAINST H. ARMIGERA

Chen et al. (2017) used one susceptible and two Cry1Ac-resistant strains of *H. armigera* to study the cross-resistance and Vip3AcAa and Cry1Ac interactions. The observed mortality for each of Vip3AcAa and Cry1Ac combination verified was equal to the assumed mortality in the three strains revealing independent activity of these two toxins. High mortality rate was observed in larvae fed on vip3AcAa chimeric gene and the cry1Ac gene cotton plant compared with non-Bt and Cry1AC containing cotton plants fed larvae. These outcomes stated that the development of insect resistance to Cry1Ac toxin did not affect the Cry1Ac-resistant strains sensitivity to Vip3AcAa and in favor of pest resistance management in China, Vip3AcAa protein is a tremendous choice for pyramid strategy.

11.2.25 MICROCLONING AND MICRODISSECTION OF THE AH01 CHROMOSOME

For the improvement of cotton genome research and breeding through gene cloning, marker development, and to acquire DNA libraries containing RGAs, chromosome microdissection technique is one of the most important techniques. Cao et al. (2017) carried out a resistance gene analogs microcloning and microdissection of the Ah01 chromosome of upland cotton. From this Ah01 chromosome, three nucleotide sequences PS016 (KU051681), PS054 (KU051682), and PS157 (KU051680) were identified with rice disease-resistance homologues primers and the PCR (LA-PCR) linker adaptor. Blast findings specified that three sequences are the nucleotide binding site-leucine rich repeat (NBS-LRR) type RGAs and results of clustering confirmed three RGAs as NBS-LRR class of RGAs in upland cotton.

11.2.26 SAD GENE FAMILY

Shang et al. (2017) ascertained stearoyl-acyl carrier protein desaturase (SAD) is the only one identified enzyme from SAD gene family which converts the plants saturated fatty acids into unsaturated fatty acids and performs main role in verifying the composition of cottonseed oil fatty acid. In this chapter, cotton SADs were found to be present in two classes from bioinformatic and phylogenetic analyses and the developmental and spatial regulation of SADs were displayed by expression patterns. After 20–35 days of anthesis in developing ovules, GhSAD2 and GhSAD4 favorably expressed which is found to be significantly different among cotton cultivars with high- and low-oil content, suggesting these two genes are related with cottonseed oil. In addition to this association, analysis showed that the amount of oleic acid (O), linoleic acid (L), and O/L value in cottonseed was closely correlated with GhSAD4-At expression confirming its part in composition of cottonseed oil. From this study 9, 9, 18, and 19 SAD gene in the genomes of four sequenced cotton species: diploid *Gossypium raimondii* (D5), *G. arboreum* (A2), tetraploid *G. hirsutum* acc. TM-1 (AD1), and *G. barbadense cv. Xinhai21* (AD2) were identified.

11.2.27 COTTON SNP80K ARRAY UTILIZATION IN GENOTYPING COTTON ACCESSIONS

Cai et al. (2017) conducted a study to describe a high-throughput cotton SNP80K array and its application in various cotton cultivars genotyping. The array on the Illumina Ifinium platform was produced using 100 cotton cultivars resequencing data and selected 82,259 SNP markers of them 77,774 SNP loci (94.55%) were effectively created on the array, in 352 cotton accessions 77,252 (99.33%) had call rates of >95%, and 59,502 (76.51%) were polymorphic loci. The high level of genotyping precision, good repeatability, and wide-ranging applicability of cotton SNP80K array was proved from application tests using parent/F_1 combinations with similar genetic backgrounds of 22 cotton accessions. Furthermore, for genome-wide association studies (GWAS), 54,588 SNPs (MAFs >0.05) related to 10 attributes of salt stress were incorporated into 288 cultivars of *G. hirsutum* and for three salt stress traits eight major SNPs were identified. The phylogenetic analysis of 312 cotton cultivars and landraces classified the different landraces in different clusters suggesting that these landraces were significant donors in developing breeding populations of modern *G. hirsutum* cultivars in China.

11.2.28 LIGNOCELLULOSIC SCW

The present description of a usual primary or secondary cell wall is inappropriate to all plant species cell types. The information about the secondary cell wall (SCW) control and deposition is mostly based on the lignocellulosic SCW *Arabidopsis* model. MacMillan et al. (2017) used cotton as a model and performed cotton stem and seed fiber model chemical analysis and RNA deep sequencing to study the refinements of gene regulation involved in formation of different SCW. Some NAC transcription factors tissue specific and developmentally controlled expression which are main top tier regulators of SCW development in xylem and/or seed fiber was shown by comparison of cotton xylem and pith transcriptomes in addition to developmental series of seed fibers indicating the lignocellulosic SCW of cotton xylem and pith cell walls are very distinct in composition from that specified for other plant species comprising *Arabidopsis*.

11.2.29 GHPPO1 GENE IN COTTON

Cheng et al. (2017) reported GhPPO1 gene with polyphenol oxidase (PPO), a class of copper-containing oxidoreductase, in cotton (*Gossypium arboreum*) is involved in direct defense response against herbivorous insects. In this chapter, a recombinant vector pTRV2-GhPPO1 was formed and transformed by virus-mediated gene silencing (VIGS) into *Agrobacterium tumefaciens* GV3101. After gene silencing by using qRT-PCR, the expression of GhPPO1 gene in cotton plants fed to 2nd instar larvae of cotton bollworm (*Helicoverpa armigera*) was evaluated. The results indicated that the cotton bollworm weight fed with gene silencing treated cotton leaf increased by 86% after 72 h compared with bollworm fed with common leaf showing that the direct defenses ability of treated cotton was weakened and bollworm weight increased. Furthermore, the GhPPO1 gene had a significant silence effect in cotton and its expression was inhibited by tobacco rattle virus (TRV) system so compared with the control group, the GhPPO1 gene expression was only 20% indicating that in cotton GhPPO1 gene was a form of defense mechanism.

11.2.30 H+-PYROPHOSPHATASE GENE (TSVP)-OVEREXPRESSING TRANSGENIC COTTON

Zhang et al. (2017) carried out a study using *Thellungiella halophila* vacuolar H+-pyrophosphatase gene (TsVP)-overexpressing transgenic cotton and wild-type cotton plants to ascertain the emergence time, rate of emergence, rate of survival, carbon assimilation ability during bud stage, seed cotton yield, and fiber quality of cotton in field trials with saline condition. Transgenic cotton with TsVP-overexpressing emerged 2 days earlier and exhibited better emergence and survival rate, superior fiber quality, and average of 14.81% more yield than wild type plants in saline field. In greenhouse environment when more than 100 mM of NaCl concentration was used, TsVP-overexpressing plants showed 50% shorter time of emergence compared with wild-type plants. This study reveals that *Thellungiella halophila* vacuolar H+-pyrophosphatase gene (TsVP) expression enhances tolerance to salinity and augments yield of seed cotton in saline fields.

11.2.31 GHNAC63 EXPRESSION PATTERNS IN RESPONSE TO ABIOTIC STRESSES

JUNGBRUNNEN1 homologous gene GhNAC63 contains stresses-related Cis-elements such as drought-related, hot-related, phytohormone-related, and light responsive-related elements in its promoter. Its highest expression level is observed in the middle cotyledon development stage with a dominant expression in fiber, flower, stem, leaf, and cotyledon. Gh6NAC3 contains 289 amino acid residues with a molecular weight 33.12 kD and 870 bp length. The N-terminal of the gene was found to be conserved with NAC transcription factors NAM domain, while its C-terminal is variant with most of the NAC transcription factors. GuoYaNing et al. (2017) conducted a study in various culture conditions to examine GhNAC63 expression patterns in response to abiotic stresses and in different cotton tissues, different cotyledon developmental stages by using qRT-PCR analysis. For this study, 35S::GhNAC63 expression vector was constructed and transferred through Agrobacterium LBA4404 in to Arabidopsis by flower dipping method and its phenotype was observed in homozygous T4 stage. To knockdown GhNAC63 expression by VIGS, another vector pYL156::GhNAC63 was constructed with pYL156-pYL192 system, and transferred into cotton. The results revealed that under closed conditions, methyl jasmonate (MeJA), SA, drought, and abscisic acid (ABA) in leaf treatments upregulated GhNAC63 and other treatments downregulated its expression in both leaf and root. Under open culture condition, GhNAC63 expression was highly induced by ethylene treatment and showed different expression patterns under drought, salt, ABA, and MeJA. The transgenic Arabidopsis leaves became yellow and wilted after ethylene and drought treatment, indicating the overexpression of GhNAC63 made these plants susceptible to drought and ethylene treatments, whereas VIGS induced GhNAC63 silencer led to its low expression level in cotton. Finally, it was concluded that GhNAC63 negatively regulates the ethylene or drought treatments in *Arabidopsis* and plant development in cotton, thus low expression level of GhNAC63 could make cotton seedlings stronger and the information from this study can be used to supplement NAC transcription factor studies and to generate new stresses induced senescence related material in upland cotton.

11.2.32 GENES AND QTL IN FIBER DEVELOPMENT

Diouf et al. (2018) identified five candidate genes, that is, Gh_D03G0889 connected to qFM-D03_cb, Gh_D12G0093, Gh_D12G0410, Gh_D12G0435 related with qFS-D12_cb, and Gh_D12G0969 connected to qFY-D12_cb from two most important QTL cluster regions, that is, cluster 1 (chromosome17-D03) and cluster 2 (chromosome26-D12) which were highly associated with protein kinase and phosphorylation functions of fiber development using GO functional annotation. In this study, an intra-specific cross was produced by crossing high fiber quality upland cotton accession CCRI35 as female parent and Nan Dan Ba Di Da Hua (NH), upland cotton accession having good yield characteristics as male parent. From $F_{2:3}$ population, 277 genotypes were used for constructing a genetic map with the help of 5178 single nucleotide polymorphism (SNP) markers for 11 different traits, total 110 QTLs were characterized on a map of 4768.098 cm length with a mean distance of 0.92 cm, out of these QTL only 30 were found to be stable under two environments. These 30 QTLs were observed in two major clusters from which the genes for fiber yield are identified. The results obtained in the present study provide resources to identify genes for fiber quality and yield related pathway confirming their reliability and authenticity by structural annotation and fine mapping.

11.2.33 GENETIC MAP CONSTRUCTION

Kirungu et al. (2018) crossed *Gossypium davidsonii* and *Gossypium klotzschianum* wild diploid cottons to produce an interspecific hybrid and 188 $F_{2:3}$ population genotypes were utilized to form a genetic map. This map had increased recombination length of 1480.23 cm, with a mean length of 2.182 cm linking adjoining markers compared with other D genome cotton species maps. High significant collinearity was found between the markers on the developed map and the physical map of *G. raimondii* with two types of duplications. The map was screened with nearly 12,560 SWU simple sequence repeat (SSR) primers and 8%, that is, 1000 polymorphic markers were obtained out of which 928 polymorphic primers were scored effectively and with an asymmetrical distribution 728 markers were found to be linked across the 13 chromosomes. From the results, 27 key genes with varied functions in development, plant hormone signaling, and protection responses were recognized. The accomplishments of this study in obtaining the $F_{2:3}$ populations and formation of its genetic map can be utilized in breeding superior cultivars from wild cotton plants.

11.3 CONCLUSION

The application of biotechnology to cotton has evolved traditional breeding methods to get plants with desired traits such as increase yield, disease resistance, and enhanced fiber quality. Modern biotechnology vastly increases the precision and reduces the time with which desirable changes in plant characteristics can be made and greatly increases the potential source from which desirable traits can be obtained. In this chapter, the recent scientific developments that underpin modern biotechnology to improve cotton are summarized.

KEYWORDS

- **cotton**
- **research advances**
- **breeding**
- **biotechnology**
- **microflora**
- **Genetic diversity**

REFERENCES

Abbas, A.; Iqbal, M. A.; Mehboob-ur-Rahman; Paterson, A. H.; Budak, H.; Cattivelli, L.; Spangenberg, G. Estimating Genetic Diversity Among Selected Cotton Genotypes and the Identification of DNA Markers Associated with Resistance to Cotton Leaf Curl Disease. *Turkish J. Bot.* **2015,** *39*, 1033–1041.

Alkuddsi, Y.; Rao, M. R. G.; Patil, S. S.; Joshi, M.; Gowda, T. H. Correlation and Path Coefficient Analysis Between Seed Cotton Yield and its Attributing Characters in Intra *hirsutum* Cotton Hybrids. *Mol. Plant Breed.* **2013,** *13*, 214–219.

Bajwa, K. S.; Shahid, A. A.; Rao, A. Q.; Bashir, A.; Aftab, A.; Husnain, T. Stable Transformation and Expression of Ghexpa8 Fiber Expansin Gene to Improve Fiber Length and Micronaire Value in Cotton. *Front. Plant Sci.* **2015,** *6*, 838.

Birch A. N.; Asacuberta j. C.; Schrijver, A.; Gralak, M.; Guerche. P. Scientific Opinion on an Application by Monsanto (EFSAGMO-NL-2013-114) for the Placing on the Market of a Herbicide-Tolerant Genetically Modified Cotton MON 88701 for Food and Feed Uses, Import and Processing Under Regulation (EC) No 1829/2003. *EFSA J.* 2017, *15* (3), 4746.

Bunpet, U.; Klaipongpan, N.; Boonrumpun, P. Inheritance of Color Lint Genes in Cotton. Thai National AGRIS Centre, 2009.

Cao, X. C.; Liu, Y. L.; Liu, Z.; Liu, F.; Wu, Y. L.; Zhou. Z. L.; Cai, X. Y.; Wang, X. X.; Zhang, Z. M.; Wang, Y. H.; Luo, Z. M.; Peng, R. H.; Wang, K. B. Microdissection of the Ah01 Chromosome in Upland Cotton and Microcloning of Resistance Gene Anologs from the Single Chromosome. *Hereditas* **2017,** *154*, 13.

Cai, C. P.; Zhu, G. Z.; Zhang, T. Z.; Guo, W. Z. High-density 80 K SNP Array is a Powerful Tool for Genotyping *G. hirsutum*. *BMC Genomics* **2017,** *18,* 1–14.

Chao, M. N.; Wen, Q. Y.; Zhang, Z. Y.; Hu, G. H.; Zhang, J. B.; Wang, G.; Wang, Q. L. Sequence Characteristics and Expression Analysis of Potassium Transporter Gene GhHAK5 in Upland Cotton (*Gossypium hirsutum* L.). *Acta Agron. Sinica* **2018,** *44*, 236–244.

Chen, W. B.; Lu, G. Q.; Cheng, H. M.; Liu, C. X.; Xiao, Y. T.; Xu, C.; Shen, Z. C.; Soberón, M.; Bravo, A.; Wu, K. M. Screening of Exotic Cotton Germplasms Against Transgenic Cotton Co-Expressing Chimeric Vip3AcAa and Cry1Ac Confers Effective Protection Against Cry1Ac-Resistant Cotton Bollworm. *Transgenic Res.* **2017,** *26*, 763–774.

Chen, Y. L.; Yang, Z.; Xiao, Y. Q.; Wang, P.; Wang, Y.; Ge, X. Y.; Zhang, C. J.; Zhang, X. L.; Li, F. G. Genome-Wide Analysis of the NF-YB Gene Family in *Gossypium hirsutum* L. and Characterization of the Role of GhDNF-YB22 in Embryogenesis. *Int. J. Mol. Sci.* **2018,** *19*, 483.

Cheng, H.; Zhang, S.; Luo, J. Y.; Rong, W.; Cui, J. J.; Wang, D. Y. Study on Insect Resistance of GhPPO1 Gene in Cotton (*Gossypium arboretum*) by VIGS Technique. *J. Agric. Biotechnol.* **2017,** *25*, 722–728.

Cui, J.; Luo, J.; Werf, W. V. D.; Yan, M.; Xia, J.. Effect of Pyramided Bt and CpTI Genes on Resistance of Cotton to *Helicoverpa armigera* (Lepidoptera;Noctuidae) Under Laboratory and Field Conditions. *J. Econ. Entomol.* **2011,** *104*, 673–684.

Dahi, H. F. Field Performance of Genetically Modified Egyptian Cotton varieties (Bt Cotton) Expressing an Insecticidal Proteins Cry 1Ac and Cry 2Ab Against Cotton Bollworms. *Nat. Sci.* **2012,** *10*, 78–85.

Dhamayanthi, K. P. M.; Kranthi, K. R.; Venugopalan, M. V.; Balasubramanya, R. H.; Kranthi, S.; Singh, S.; Blaise. In Study of Interspecific Hybrids (*Gossypium hirsutum* × *G. barbadense*) For Heterosis and Combining Ability. World Cotton Research Conference-5, Mumbai, India, Nov. 2011, pp. 51–55.

Dinakaran, J; Kumar, N. S.; Nagadesi, P. K.; Shanna, V. K.; Pati, R. Comparative Analysis of Bt and Non-Bt Hybrid Cotton with Special Reference to Growth, Biochemical Characters and Rhizospheric Microbes. *Plant Arch.* **2010,** *10*, 101–106.

Diouf, L.; Magwanga, R. O.; Gong, W.; He, S.; Pan, Z.; Jia, Y.; Kirungu, J. N.; Du, X. QTL Mapping of Fiber Quality and Yield-Related Traits In an Intra-Specific Upland Cotton Using Genotype by Sequencing (GBS). *Int. J. Mol. Sci.* **2018,** *19*, 441.

Feng, G.; Gan, X.; Yang, M.; Yao, Y.; Luo, H.; Zhang, Y.; Zhang, W. Canopy Structure and Matter Production Characteristics of F_1, F_2, and Their Parents in Two Cotton Hybrids. *Acta Agron. Sinica* **2013,** *39*, 1635–1643.

Guo, Y.; Dou, L.; Ma, Q.; Zhao, F.; Pang, C.; Wei, H.; Wang, H.; Fan, S.; Yu, S. Cloning and Functional Analysis of GhNAC63 Gene in Upland Cotton (*Gossypium hirsutum*). *J. Agric. Biotechnol.* **2017,** *25*,173–185.

Haidar, S.; Aslam, M. NIAB-2008: A New High Yielding and Long Staple Cotton Mutant Developed Through Pollen Irradiation Technique. *Int. J. Agric. Biol.* **2016,** *18*, 865–872.

Haidar, S.; Aslam, M.; Mahmood-ul-Hassan; Hassan, H. M. Genetic Diversity Among Upland Cotton Genoypes for Different Economic Traits and Response to Cotton Leaf Curl Virus (CLCV) Disease. *Pak. J. Bot.* **2012,** *44*, 1779–1784.

Janga, M. R.; Campbell, L. M.; Rathore, K. S. CRISPR/Cas9-Mediated Targeted Mutagenesis in Upland Cotton (*Gossypium hirsutum* L). *Plant Mol. Biol.* **2017,** *94*, 349–360.

Kirungu, J. N.; Deng, Y.; Cai, X.; Magwanga, R. O.; Zhou, Z.; Wang, X.; Wang, Y.; Zhang, Z.; Wang, K.; Liu, F.. Simple Sequence Repeat (SSR) Genetic Linkage Map of D Genome Diploid Cotton Derived from an Interspecific Cross Between *Gossypium davidsonii* and *Gossypium klotzschianum*. *Int. J. Mol. Sci.* **2018,** *19*, 204.

Kothari, N.; Hague, S. S.; Frelichowski, J.; Nichols, R. L.; Jones, D. C. Utilization of Cotton Germplasm in the Winter Nursery at Tecoman, Mexico for Plant Breeding Training and Research. *J. Cotton Sci.* **2011,** *15*, 271–291.

Lee, M. K.; Zhang, Y.; Zhang, M.; Goebel, M.; Kim, H. J.; Triplett, B. A.; Stelly, D. M.; Zhang, H. B. Construction of a Plant Transformation–Competent BIBAC Library and

Genome Sequence Analysis of Polyploidy Upland Cotton (*Gossypium hirsutum* L.). *BMC Genomics* **2013,** *14*, 208.

Lehman, A.; Pender, R.; Morden, C.; Wieczorek, A. M. Assessment of Persistence of Hybrids Between Alien Pima Cotton, *Gossypium barbadense* (Malvaceae), and Endemic Hawaiian Cotton, *G. tomentosum,* in Hawaii. *Pacific Sci.* **2014,** *68*, 85–96.

Lin, Z.; Yuan, D.; Zhang, X. Mapped SSR Markers Unevenly Distributed on the Cotton Chromosomes. *Front. Agric. China* **2010,** *4*, 257–264.

Li, W.; Ren, Z.; Wang, Z.; Sun, K.; Pei, X.; Liu, Y.; He, K.; Zhang, F.; Song, C.; Zhou, X.; Zhang, W.; Ma, X.; Yang, D. Evolution and Stress Responses of *Gossypium hirsutum* SWEET Genes. *Int. J. Mol. Sci.* **2018,** *19*, 769.

Liu, G.; Wang, F.; Gong, Y.; Ma, H.; Zhang, J. A New Method for Identification of Genetic Purity of Cotton Varieties by SSR Marker. *Cotton Sci.* **2013,** *25*, 382–387.

Lopez-Lavalle, L. A.; Matheson, B.; Brubaker, C. L. A Genetic Map of an Australian Wild Gossypium C Genome and Assignment of Homoelogies with Tetraploid Cultivated Cotton. *Genome* **2011,** *54*, 779–794.

Lu, Q.; Shi, Y.; Xiao, X.; Li, P.; Gong, J.; Gong, W.; Liu, A.; Shang, H.; Li, J.; Ge, Q.; Song, W.; Li, S.; Zhang, Z.; Harun-or-Rashid, M.; Peng. R.; Yuan, Y.; Huang, J. Transcriptome Analysis Suggests that Chromosome Introgression Fragments from Sea Island Cotton (*Gossypium barbadense*) Increase Fiber Strength in Upland Cotton (*Gossypium hirsutum*). *Genetics* **2017,** *7*, 3469–3479.

Lu, H.; Zhou, Z.; Chen, H.; Ling, J.; Liu, F.; Cai, X.; Wang, X.; Wang, C.; Wang, Y.; Wang, K. Genome-Wide Analysis of RAV Gene Family in Cotton. *Cotton Sci.* **2014**, *26*, 471–482.

MacMillan, C. P.; Birke, H.; Chuah, A.; Brill, E.; Tsuji, Y.; Ralph, J.; Dennis, E. S.; Llewellyn, D.; Pettolino, F. A. Tissue and Cell-Specific Transcriptomes in Cotton Reveal the Subtleties of Gene Regulation Underlying the Diversity of Plant Secondary Cell Walls. *BMC Genomics* **2017,** *18,* 1–18.

Meena, H.; Kumhar, B. L. Performance of Inter Specific Cotton Hybrids Under Various Plant Geometries and Nutrient. *Int. J. Forestry Crop Improve.* **2017,** *8*, 49–52.

Mehetre, S. Constraints of Hybrid Seed Production in Upland and Cultivated Diploid Cottons: Will Different Male Sterility Systems Rescue: A Review. *J. Res. Dev.* **2015,** *29*, 181–211.

Mukherjee, R.D.; Sinha, N. GM Crops in India with Reference to Bt Cotton: Opportunities and Challenges. *J. Environ. Res. Dev.* **2012,** *7*, 188–193.

Ni, M.; Ma, W.; Wang, X. F.; Gao, M. J.; Dai, Y.; Wei, X.; Zhang, L.; Peng, Y.; Chen, S.; Ding, L.; Tian, Y.; Li. J.; Wang. H.; Wang. X.; Xu, G.; Guo, W.; Yang, Y.; Wu, Y.; Heuberger, S.; Tabashnik, B. E.; Zhang, T.; Zhu, Z. Next-Generation Transgenic Cotton: Pyramiding RNAi and Bt Counters Insect Resistance. *Plant Biotechnol. J.* **2017,** *15*, 1204–1213.

Park, W.; Scheffler, B. E.; Campbell, B. T. Identification of the Family of Aquaporin Genes and Their Expression in Upland Cotton (*Gossypium hirsutum*). *BMC Plant Biol.* **2010,** *10*, 142.

Pundir, S. R.; Sangwan, O.; Nimbal, S.; Sangwan, R. S.; Siwach, S. S.; Mandhania, S.; Jain, S. Heterosis and Combining Ability for Seed Cotton Yield and Its Component Traits of Desi Cotton (*Gossypium arboreum* L.). *J. Cotton Res. Dev.* **2017,** *31*, 24–28.

Reddy, K. B.; Reddy, V. C.; Ahamed, M. L.; Naidu, T. C. M; Rao, V. S. Character Association and Path Coefficient Analysis for Yield and Component Traits in Upland Cotton (*Gossypium hirsutum* L.). *J. Cotton Res. Dev.* **2017,** *31*, 29–33.

Salih, H.; Gong, W.; He, S.; Sun, G.; Sun, J.; Du, X. Genome-wide characterization and expression analysis of MYB transcription factors in *Gossypium hirsutum*. *BMC Genet.* **2016,** *17*, 129.

Shang, X.; Cheng, C.; Ding, J.; Guo, W. Identification of Candidate Genes from the Sad Gene Family in Cotton for Determination of Cottonseed Oil Composition. *Mol. Genet. Genomics* **2017,** *292*, 173–186.

Shen, C.; Jin, X.; Zhu. D.; Lin, Z. Uncovering SNP and Indel Variations of Tetraploid Cottons by SLAF-seq. *BMC Genomics* **2017,** *18*, 247.

Shweta; Yusuf Akhter; Khan, J. A. Genome Wide Identification of Cotton (*Gossypium hirsutum*)-Encoded microRNA Targets Against Cotton Leaf Curl Burewala Virus. *Gene* **2018,** *638*, 60–65.

Si, A.; Yang, W.; Xie, Z.; Tian, Q.; Dong, Y.; Li, Y. Ma, P. Cloning and Expression Analysis of Cotton Leaf-Specific Promoter. *Southwest China J. Agric. Sci.* **2018,** *31*, 646–652.

Sundaramurthy, V. T. Expression of *Bacillus thuringiensis* Toxin Affects the Leaf Anatomy of a Cotton Hybrid. *Int. J. Biores. Stress Manage.* **2015,** *6*, 87–92.

Tian, G.; Cheng, L. L.; Qi, X. W.; Ge, Z. H.; Niu, C. Y.; Zhang, X. L.; Jin, S. X. Transgenic Cotton Plants Expressing Double-Stranded RNAs Target HMG-CoA Reductase (HMGR) Gene Inhibits the Growth, Development and Survival of Cotton Bollworms. *Int. J. Biol. Sci.* **2015,** *11*, 1296–1305.

Vekariya, R. D.; Nimbal, S.; Batheja, A.; Sangwan, R. S.; Mandhania, S. Combining Ability and Gene Action Studies on Seed Cotton Yield and its Related Traits in Diploid Cotton (*Gossypium arboreum* L.). *Electr. J. Plant Breed.* **2017,** *8*, 1159–1168.

Vitale, J. D. The Commercial Application of GMO Crops in Africa: Burkina Faso's Decade of Experience with Bt cotton. *J. AgroBiotechnol. Manage. Econ.* **2010,** *13*, 115–123.

Xu, Z. Y.; John, Z. Y.; Jaimin Cho; Yu, J.; Russell, J. K.; Percy, G. R. Polyploidization Altered Gene Functions in Cotton (*Gossypium* spp). *PLOS One* **2010,** *5*, 230–239.

Yang, N. S.; Wang, J. P,; Wang, X. Y. Research Progress on Evolution of Cotton Genome Structure. *Genomics Appl. Biol.* **2017,** *36*, 1090–1095.

Zhao, J.; Xiao, S. H.; Wu, Q. J.; Liu, J. G. Genetic Diversity of Cotton Germplasm with Low Gossypol, Jiangsu. *J. Agric. Sci.* **2013,** *29*, 1211–1220.

Zhang, K. W.; Song, J. L.; Chen, X. G.; Yin, T. T.; Liu, C. B.; Li, K. P.; Zhang, J. R. Expression of the *Thellungiella halophila* Vacuolar H+-Pyrophosphatase Gene (TsVP) in Cotton Improves Salinity Tolerance and Increases Seed Cotton Yield in a Saline Field. *Euphytica* **2016,** *211*, 231–244.

Zhang, S.; Lu, L. M.; Wang, C. Y.; Luo, J. Y.; Li, C. H.; Cui, J. J. Study on Cotton Aphid-Borne Viruses via High-Throughput Sequencing. *Cotton Sci.* **2014,** *26*, 539–545.

Zhang, X. B.; Tang, Q. L.; Wang, X. J.; Wang, Z. X. Development of Glyphosate-Tolerant Transgenic Cotton Plants Harboring the G2-aroA Gene. *J. Integr. Agric.* **2017,** *16*, 551–558.

Index